Heat Shock Proteins

Volume 10

Series editors
Alexzander A.A. Asea, Ph.D.
Professor and VD for Research Innovations
Deanship for Scientific Research, University of Dammam
Dammam, Saudi Arabia

Stuart K. Calderwood, Ph.D.
Professor and Director
Division of Molecular and Cellular Radiation Oncology
Department of Radiation Oncology, Harvard Medical School of Beth Israel
Deaconess Medical Center
Boston, MA, USA

Heat Shock Proteins: key mediators of Health and Disease. Heat shock proteins (HSP) are essential molecules conserved through cellular evolution required for cells to survive the stresses encountered in the environment and in the tissues of the developing and aging organism. These proteins play the essential roles in stress of preventing the initiation of programmed cell death and repairing damage to the proteome permitting resumption of normal metabolism. Loss of the HSP is lethal either in the short-term in cases of acute stress or in the long-term when exposure to stress is chronic. Cells appear to walk a fine line in terms of HSP expression. If expression falls below a certain level, cells become sensitive to oxidative damage that influences aging and protein aggregation disease. If HSP levels rise above the normal range, inflammatory and oncogenic changes occur. It is becoming clear that HSP are emerging as remarkably versatile mediators of health and disease. The aim of this series of volumes is to examine how HSP regulation and expression become altered in pathological states and how this may be remedied by pharmacological and other interventions.

More information about this series at http://www.springer.com/series/7515

Alexzander A.A. Asea • Stuart K. Calderwood
Punit Kaur
Editors

Heat Shock Proteins and Plants

Springer

Editors
Alexzander A.A. Asea
Department of Neuroscience
University of Dammam
Damman, Saudi Arabia

Punit Kaur
Department of Radiation Oncology
University of Texas MD Anderson
Cancer Center
Houston, TX, USA

Department of Microbiology, Biochemistry
& Immunology
Morehouse School of Medicine
Atlanta, GA, USA

Stuart K. Calderwood
Division of Molecular and Cellular Biology
Department of Radiation Oncology
Beth Israel Deaconess Medical Center
and Harvard Medical School
Boston, MA, USA

ISSN 1877-1246 ISSN 1877-1254 (electronic)
Heat Shock Proteins
ISBN 978-3-319-46339-1 ISBN 978-3-319-46340-7 (eBook)
DOI 10.1007/978-3-319-46340-7

Library of Congress Control Number: 2016959191

Printed on acid-free paper

This Springer imprint is published by Springer Nature
The registered company is Springer International Publishing AG
The registered company address is: Gewerbestrasse 11, 6330 Cham, Switzerland

Preface

The sedentary nature of plants constantly exposes them to extreme climatic changes in various geographic regions. Their ability to overcome these adverse conditions, grow, and thrive is a result of evolutionary changes in various morphological and physiological mechanisms that enable plants to survive extremely stressful conditions.

The importance of plants in human survival cannot be overstated, not only in terms of an important source of food, but also for its critical therapeutic value. Since the dawn of time, the use of plants for therapeutic purpose has been recorded in all major cultures.

Heat-shock proteins (HSP) are stress proteins known to provide cytoprotection and play important roles in protein folding/unfolding and modulate cellular immune responses. HSP are found in all plant species and are associated to plant biotic stresses and are often referred to as stress defense proteins. In addition, HSP serve a critical role in the plant's response against key crop phytopathogens around the world.

Heat-Shock Proteins and Plants provides the most up-to-date and concise reviews and progress on the role of heat-shock proteins in plant biology, structure, and function and is subdivided into chapters focused on Small Plant HSP (Part I), Larger Plant HSP (Part II), and HSP for Therapeutic Gain (Part III). This book is written by eminent leaders and experts from around the world and is an important reference book and a must-read for undergraduate, postgraduate students, and researchers in the fields of agriculture, botany, crop research, plant genetics and biochemistry, biotechnology, drug development and pharmaceutical sciences.

Dammam, Saudi Arabia
Boston, MA, USA
Houston, TX, USA
Atlanta, GA, USA

Alexzander A.A. Asea
Stuart K. Calderwood
Punit Kaur

Part I
Small Plant Heat Shock Proteins (HSP)

Chapter 1
Small Heat Shock Proteins: Roles in Development, Desiccation Tolerance and Seed Longevity

Harmeet Kaur, Bhanu Prakash Petla, and Manoj Majee

Abstract Small heat shock proteins are one of the five classes of heat shock proteins, a family named after their expression in response to heat shock. Despite their name some members of this family have been shown to express during a gamut of non-stress conditions in a variety of plant species. Small HSPs have been known to accumulate during plant developmental stages like pollen development, seed maturation stages, early seed germination and also in storage organs. Interestingly, aging induced accumulation of small HSPs has also been observed in a few species. The spatial and temporal accumulation pattern of small HSPs also correlates well with other seed abundant proteins like late embryogenesis abundant (LEA) proteins. Regulation of these developmental stages responsive and non-stress induced small HSPs is also distinct from the heat stress regulated transcript induction in terms of involvement of some novel and exclusive transcription activators like ABI3 and HsfA9. Small HSPs are known to function as molecular chaperone and thus their role in plant development especially during seed development has been discussed in the light of their functional implication during these stages.

Keywords Pollen • Seed development • Chaperone • LEA protein • Glassy matrix

Abbreviations

ACD	alpha crystalline domain
ATP	adenosine triphosphate
CDT	controlled deterioration treatment
DAP	days after pollination
DPI	days post imbibition

H. Kaur • B.P. Petla • M. Majee (✉)
National Institute of Plant Genome Research,
Aruna Asaf Ali Marg, JNU Campus, New Delhi 110067, India
e-mail: manojmajee@nipgr.ac.in

A.A.A. Asea et al. (eds.), *Heat Shock Proteins and Plants*,
Heat Shock Proteins 10, DOI 10.1007/978-3-319-46340-7_1

GUS β-glucuronidase
HSE heat shock element
HSF heat shock transcription factor
HSP heat shock proteins
LEA late embryogenesis abundant
ROS reactive oxygen species

1.1 Introduction

Small heat shock proteins (HSPs) are the major class of heat shock protein repertoire during heat stress in plants. This class typically contains a highly conserved ~90 amino acid long alpha crystalline domain (ACD) at the c-terminal end (Waters et al. 1996). This class comprises of heat shock proteins ranging between 15 and 42 kDa in monomeric molecular weight, although they exist as large oligomeric assemblies of 2–48 subunits (Vierling 1997; Waters et al. 1996; Basha et al. 2012). These proteins are encoded by six nuclear gene families in plants making them the most complex class of HSPs in plants. These families vary on the basis of sequence similarity among member genes and their respective intracellular localization. Two cytosol localized (CI and CII) and rest organelle localized (e.g. Endoplasmic reticulum, chloroplast, mitochondrial) small heat shock protein families have been identified in plants (Waters et al. 1996). Small HSPs act by binding to cellular proteins during stress conditions and prevent their aggregation or misfolding. Although they are not themselves involved in folding they assist by maintaining the target proteins in a conformation which can be readily refolded once the constraining factors cease to exist (Lee et al. 1995, 1997). Earlier, small HSPs were thought to express almost exclusively in vegetative tissues only under heat stress, but studies have demonstrated the role of these proteins in diverse stresses like cold, dehydration, salinity and oxidative stress (Waters et al. 1996; Wang et al. 2004, 2005). A plethora of studies both in animal and plant systems provide evidence for the physiological role of Heat shock proteins in general and small HSPs in particular during developmental processes (Waters et al. 1996).

Small HSPs are functional molecular chaperones and protect the substrate proteins against thermal aggregation or denaturation (Lee et al. 1995, 1997). These proteins are highly capable of binding substrate proteins with non-native conformations. This binding possibly involves hydrophobic interactions and does not require ATP thus making them ATP independent chaperones. Although they are not directly involved in refolding reactions and only facilitate refolding by ATP dependent chaperones (Ehrnsperger et al. 1997; Lee et al. 1997; Lee and Vierling 2000).

1.2 Small HSPs and Their Role in Development

Small heat shock proteins (sHSPs) are ubiquitously produced by all organisms including plants in response to increased temperature and certain other stresses to protect organisms from stress induced damage or to repair damage caused by stress

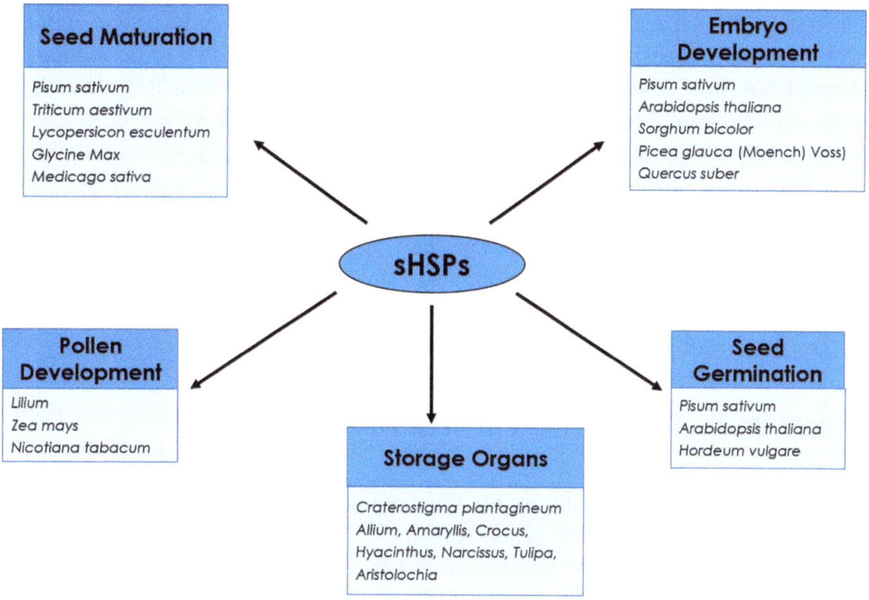

Fig. 1.1 Role of small HSPs during plant development processes

(Lindquist and Craig 1988; Vierling 1997). Recent studies have demonstrated the function of small HSPs during development in plants which range from pollen development, seed development and germination to their accumulation in storage organs as well (Fig. 1.1). These developmental processes related roles of small HSP family are discussed here in greater detail.

1.2.1 Role of Small HSPs in Pollen Development

Small HSPs have been shown to play important role during various developmental stages in the plant. Class II small HSPs have also been shown to be expressed in meiotic microsporocytes in lily (Bouchard 1990) and in developing maize anthers (Bouchard and Walden 1990; Dietrich et al. 1991). The transcript of one such gene *HSP18* was found to accumulate during meiosis phase of pollen development in lily and maize (Bouchard 1990; Bouchard and Walden 1990). This mRNA accumulation was observed in absence of high temperature and was independent of the heat stress mediated induction of small HSP mRNAs. The role of sHSPs in developmental stages was also observed as the higher expression of HSP18 in yeast (*Saccharomyces cerevisiae*) during late prophase of meiosis (Kurtz et al. 1986). Studies on plants, yeast and Drosophila meiotic stages linked accumulation of HSPs sheds light on the development related roles in meiotic cells (Winter and Sinibaldi 1991). Later studies done in tobacco by Zarsky et al. (1995) clearly demonstrated the transcriptional activation of a small HSP gene, *NtHSP18P* in the course of initiation of embryogenesis in pollen. *NtHSP18P* mRNA was shown to be induced at normal non heat stress temperatures just before anthesis, which is marked by a dehydration phase during pollen development. A decade later Volkov et al. (2005) analysed the differential accumulation of small HSPs during pollen development and heat stress in tobacco. They identified eleven cytosolic small HSP coding cDNAs which were not expressed under normal temperatures but exclusively during pollen development. The protein products of these genes were also shown to be expressed throughout pollen developmental stages in a coordinated manner during early and late stages of pollen development.

1.2.2 Role of Small HSPs in Seed Development

Small HSP family of proteins are not only synthesised in response to stress conditions but also have a developmental response in organs like seeds. First reports of the presence of small HSP transcripts in seeds came from the work in *Pisum sativum* (DeRocher and Vierling 1994, 1995) where these transcripts were present in the embryonic axes and cotyledons of pea seeds and declined rapidly within a few hours of germination. Helm and Abernethy (1990) studied the HSP levels in wheat embryos and dry seed and described the presence of transcripts as well as proteins from several classes of HSPs in addition to the small HSPs. Vierling and Sun (1989) also identified few class I cytosolic mRNAs in pea seeds and similar observation was made in tomato fruits during ripening (Fray et al. 1990).

Cytoplasmic classes of small heat shock protein have been shown to accumulate at slightly elevated temperatures in pea and alfalfa indicating their role in heat stress adaptation (DeRocher et al. 1991; Hernandez and Vierling 1993). However, Hernandez and Vierling (1993) studied the accumulation of class I cytoplasmic small HSPs in various legumes viz. *P. sativum, G. max, V. unguiculata, P. acutifolius, M. sativa* and *A. constricta*, both under stress and in different plant organs. Their results showed the expression of small HSPs even in plants grown in field or green house under optimal growth conditions. They also showed that in addition to temperature stress, small HSPs normally express during certain developmental stages like in flower and seed development. Interestingly, these HSPs were absent in young pods but started accumulating at 29DAP and were also found in dry seeds and this expression is deemed important for their reproductive success.

1.2.3 Role of Small HSPs During Seed Germination

Studies demonstrate that developmentally induced small HSP (HSP17.6 and HSP17.9) mRNAs in sunflower seeds disappeared soon during imbibition but the proteins were interestingly present at elevated levels even 3 days post imbibition

(Coca et al. 1994). Although the major seed storage proteins degrade uptil 3DPI, the HSP17.6 and HSP17.9 protein products disappear afterwards. The persistence and organellar localization of these proteins hints at their involvement in seed reserve mobilization as discussed by Coca et al. (1994). In pea, small HSPs continue to exist in the cotyledons and axes for a number of days during imbibition and early germination before coming to a decline (DeRocher and Vierling 1994). Interestingly, barley HSP26 protein which is localized to plastid is found to accumulate in early germination stage which might have been translated from the HSP26 mRNA present in the early embryo (Kruse et al. 1993). Wheat chloroplastic HSP26 has also been shown to have high transcript during seed germination which was also confirmed by promoter: GUS construct expression. GUS expression driven by *TaHSP26* promoter was high during first 24 h after seed germination and declined later highlighting its role in early seed germination phase (Chauhan et al. 2012).

1.2.4 Role of Small HSPs in Storage Organs

Small HSPs have not only been shown to be involved in meiosis, microsporogensis, seed development, seed germination and somatic embryogenesis, they have also been demonstrated to accumulate in vegetative organs not experiencing heat stress e.g. In desiccation tolerant plant *Craterostigma plantagineum*, polypeptides crossreactive to small HSP antibodies were detected in roots and lower shoot parts and similar observations were also made for mulberry where small HSP accumulates in response to seasonal cold acclimation in the endoplasmic reticulum enriched fractions of cortical parenchyma cells (Alamillo et al. 1995; Ukaji et al. 1999). Interestingly, small HSPs have been detected in bulbs and tubers of a variety of species like *Allium, Amaryllis, Crocus, Hyacinthus, Narcissus, Tulipa* and potatoes and also in twigs of *Acer* and *Sambucus* and tendrils of liana, *Aristolochia*. The expression in twigs and tendrils coincides with the dormant period in winter and fades out in spring. Potato tubers and *Aristolochia* tendrils were shown to have high proteins content which was located in the central vacuoles of storage parenchyma cells. A possible functional correlation was investigated between the accumulation of storage proteins and small HSPs and studies on transgenic tobacco plants demonstrated that in embryos of these plants which lack protein bodies and accumulation of storage proteins, small HSPs were also absent (Lubaretz and Niedin 2002). Interestingly, this association among the storage proteins and accumulation of small HSPs at the same stage also reflects in mature seeds where both these protein families are known to coexist and their expression has also found to be in sync (Almoguera and Jordano 1992; Coca et al. 1994) which is discussed in more detail later in this chapter. Thus, the existence of small HSPs in storage organs of perennial plants appears to have a correlation with the presence of other storage proteins (Lubaretz and Niedin 2002).

1.3 Molecular Regulation of Small HSP Accumulation During Development

Heat stress leads to a rapid increase in HSP gene transcription most of them being the small HSPs and the mRNA levels reach 20,000 copies in a cell (Schöffl and Key 1982). Heat shock elements (HSE) are the cis acting elements present in the 5′ upstream regions of the heat shock responsive genes and are known to confer heat shock induced expression of these genes. These HSE provide binding sites for heat shock transcription factors thus regulating the heat stress induced transcription activation (Goldenberg et al. 1988; Parker and Topol 1984; Wu 1984a, b). This heat stress responsive role of small HSP family proteins was complemented by the developmental stage specific expression of certain members of this family as transcripts for some small HSPs have been detected in wheat, sunflower and pea grown in the absence of heat stress (Helm and Abernethy 1990; Almoguera and Jordano 1992; Coca et al. 1994; DeRocher and Vierling 1994; Carranco et al. 1997).

Developmental and heat stress regulated expression of small HSPs might be overlapping but should be distinctly controlled as not all small HSP classes/members are developmentally expressed. In order to decipher these complex regulatory mechanisms, promoters and 5′ upstream regions were studied from small HSP genes expressed in response to both heat stress and development. Promoter studies with promoter: GUS reporter constructs made with two such genes from soybean and sunflower (*GmHSP17.3B* and *HaHSP17.7G4*) demonstrated their expression during zygotic embryogenesis (Prändl et al. 1995; Prändl and Schöffl 1996; Coca et al. 1996; Carranco et al. 1999). Subsequent studies on *HaHSP17.4G4* promoter transgenic lines revealed the involvement of two discrete regulatory mechanisms, an HSE/HSF dependent mechanism and another HSE independent mechanism suggested to be regulated by developmental stage specific novel transcription activators (Almoguera et al. 1998). Experiments with different mutations within the HSE core element did not affect the tissue specific expression pattern of the GUS reporter gene during early seed maturation however the binding between this mutated HSE and the HSF transcription factor was highly compromised. Thus the seed maturation stage responsive transcription by *HaHSP17.4G4* promoter was determined to be HSE independent (Almoguera et al. 1998). One such example for the developmental regulation of small HSP family genes in animals comes from larval development in Drosophila where the promoters of HSP23 and HSP27 genes are regulated by ecdysterone receptors. These ecdysterone receptors have DNA binding domain and are activated by addition of ecdysterone hormone. This type of activation is independent of HS induced HSE mediated activation of HSPs (Luo et al. 1991). Similar evidence comes from the study of Arabidopsis seed mutants, *aba1* and five *abi* loci mutants, (Wehmeyer et al. 1996) in which *abi3* mutants had lower accumulation of small HSPs and the *abi3* null mutants were found to be desiccation intolerant. Although studies with other *abi* mutants showed the presence of wild type levels of small HSPs; but other desiccation intolerant Arabidopsis mutants *abi3-6, fus3-3, lec1-2* and line 24 did have severely reduced levels of HSP17.4 (major small HSP present in Arabidopsis seeds). Therefore it can be concluded that small HSPs may not be solely responsible for providing desiccation tolerance but are definitely necessary for seed desiccation tolerance (Wehmeyer et al. 1996; Wehmeyer and Vieling 2000). Apart from HSF, some other transcriptional activators are also speculated to be involved in seed developmental regulation of small HSPs.

In Arabidopsis, ABI3 has been implicated as one such activator of *HSP17.4* during seed development. Wehmeyer and Vierling (2000) showed the levels of *HSP17.4* to be untraceable in the *abi3-6* knockout mutant. Promoter: GUS constructs transformed in *abi3-6* mutant background also demonstrated a very faint GUS staining in seeds which reveals very low transcriptional activation in the absence of ABI3 activator. However, in other seed transcription factor mutants *lec1-2*, and *fus3-3*, noticeable levels of HSP17.4 was present and when the same construct was transformed in these backgrounds, the seeds had better GUS staining in comparison to *abi3-6* mutant. Interestingly, the heat stressed embryos from these *HSP17.4*:GUS transformed in mutant backgrounds stained evenly and throughout for GUS thus clearly indicating the independent regulation of small HSPs during stress and development (Wehmeyer and Vieling 2000). In addition to this ABI3 has been shown to regulate the developmental expression of HsfA9 which is a heat shock transcription factor with unique expression in later seed development stages (Kotak et al. 2007). Experiments reveal the negligible accumulation of HsfA9 mRNA and protein alongwith other seed specific small HSPs (HSP17.4-CI, HSP17.7-CII) in ABI3 knockout mutant lines which was overcome by ABI3 overexpression in transgenic plants. ABI3 was also shown to activate the HsfA9 promoter while, HSfA9 could activate the promoters of few small HSP genes thus clearly establishing HsfA9 as a specialized member of the heat shock transcription factor family regulating developmental expression of small HSPs during seed maturation (Kotak et al. 2007) (Fig. 1.2).

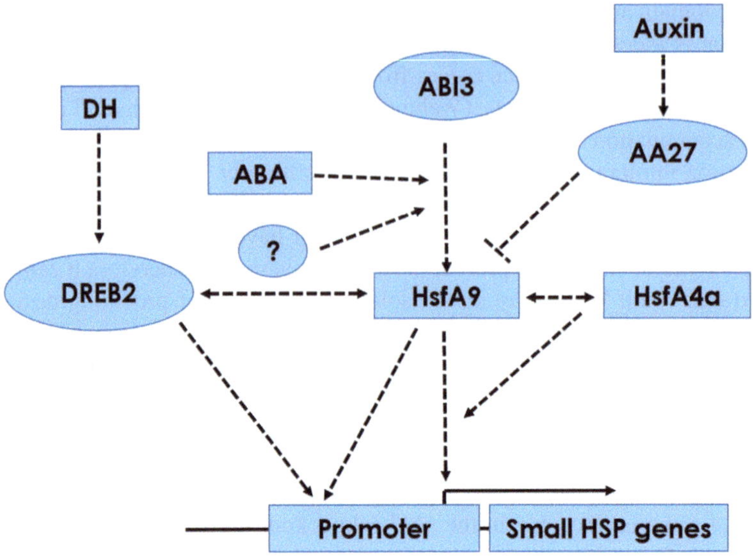

Fig. 1.2 Proposed model for developmental regulation of small HSPs in plants. HsfA9 is centrally positioned as it has been shown to be developmentally regulated by ABI3, a seed specific transcription factor and also is co-regulated by other transcription factors HsfA4a and DREB2. HsfA9, DREB2 and HsfA4a play important role both individually as well as in co-ordination for the regulation of downstream small HSP genes. Phytohormones, ABA and Auxin are also implicated for this developmental regulation of HSfA9 positively and negatively respectively

1.4 Concerted Expression of Small HSPs with Other Seed Storage Proteins

Seed development is a complex process which involves the accumulation of various macromolecules like carbohydrates, lipids, mRNAs and proteins in the seeds. All these components play important roles in preparing the seed for surviving desiccation and finally to complete germination successfully. Small HSP mRNAs are often expressed with some other seed specific protein mRNAs like LEA (Late Embryogenesis Abundant) proteins. Studies identified a small HSP (*HaHSP17.6*) and two LEA protein (D-113 and Emb-1) like mRNAs from sunflower seeds. These mRNAs were found to have similar expression profiles and the highest expression was noticed in mid maturation stage of embryos and dry seeds which was independent of the stress. Coordinated transcript accrual of small HSP with LEA proteins indicates the presence of shared transcription regulatory elements in their upstream regions (Almoguera and Jordano 1992). Interestingly enough in yeast a previously known small HSP (HSP12) was reclassified as a LEA protein. This study demonstrated that HSP12 accumulates in yeast cells during nutrient limited stationary phase when yeast cells are preparing for sporulation which is akin to seed maturation stage in plants when LEA protein synthesis takes place. Detailed amino acid sequence homologies and the hydropathy plots also put forth another evidence for the reclassification (Mtwisha et al. 1998). Pea mitochondrial proteins, LEAM and HSP22 are also an example of the concerted expression of these two groups of proteins during seed development, desiccation tolerance and germination (Bardel et al. 2002; Macherel et al. 2007). Beech seeds stored for 1–8 years at −10 °C revealed the presence of dehydrin and dehydrin like proteins in addition to the LEA proteins and small HSP of 22 kDa. Dehydrins were found to express in both cotyledons and embryonic axes of dry stored seeds. In addition to dehydrins, small HSPs which might be induced by the oxidative stress and ROS generation during long term storage of beech seeds might function by decreasing the intracellular ROS and are positively correlated with germinability of stored beech seeds (Arrigo 1998; Kalemba and Pukacka 2008).

1.5 Small HSPs: Key Players in Acquisition of Seed Desiccation Tolerance

Small HSPs form a remarkable group of proteins and function in varied cellular processes. In addition to their obvious role in heat shock (although not all small HSPs have a heat induced expression) they are also synthesized during somatic embryogenesis, microsporogenesis, pollen formation and seed maturation (Bouchard 1990; Bouchard and Walden 1990; Zarsky et al. 1995; Vierling 1997; Coca et al. 1994; DeRocher and Vierling 1994). Small HSPs are known to function as molecular chaperones and their expression across abiotic stresses and various physiological stages makes them ideal candidates to counteract cellular damage caused by these conditions. Certain small HSPs have been shown to be induced in water stress and the expression of small HSPs during final stages of seed development and maturation highlights their importance in acquisition of desiccation tolerance (Almoguera et al. 1993; Coca et al. 1994; DeRocher and Vierling et al. 1994). Small HSP expression in pollen, in seed desiccation stages and in storage organs argue for their underlying unified role during quiescent stages of plant development marked with a decrease in water content and suppression of metabolism. All these stages are characterised by development of desiccation tolerance and are associated with accumulation of oligosaccharides, osmolytes and unique proteins like the late embryogenesis proteins (LEAs) and heat shock proteins (HSPs) (Hoekstra et al. 2001). Studies have suggested that during such desiccation events the cells form a glassy matrix consisting of soluble sugars which immobilize macromolecules offering protection to membranes and proteins (Leopold et al. 1994; Crowe et al. 1997; Bernal-Lugo and Leopold 1998; Buitink et al. 1998). Small HSPs are also hypothesized to play a role in glassy matrix formation coincident with their presence in late embryo maturation (Wehmeyer and Vieling 2000; Kalemba and Pukacka 2008). As a mechanism of action small HSPs are known to hold proteins under denaturing conditions and prevent their irreversible aggregation in addition to assisting in protein folding and helping with intracellular transport (Waters et al. 1996; Lee et al. 1997; Hendrick and Hartl 1995; Haslbeck et al. 2005).

1.6 Small HSPs Maintain Seed Viability During Aging

Recent studies have suggested the role of small HSPs in providing seed vigor and germinability under stress conditions. Transgenic plants overexpressing small HSPs have been reported to display improved stress tolerance such as thermotolerance (Perez et al. 2009; Sanmiya et al. 2004; Zhou et al. 2012), osmotic stress tolerance (Sun et al. 2001), chilling tolerance (Guo et al. 2007), and seed longevity (Prieto-Dapena et al. 2006; Kaur et al. 2015). In contrast, transgenic plants with reduced small HSP levels were less tolerant to thermal shock (Charng et al. 2006) and are susceptible to pathogens (Maimbo et al. 2007).

Studies on *Nelumbo nucifera* have revealed astonishing seed viability of more than 1300 years (Shen-Miller et al. 1995) which was attributed to the presence of several thermostable proteins including heat shock proteins (Shen-Miller et al. 2013). Ectopic expression of *NnHSP17.5* in Arabidopsis resulted in increased seed vigor under artificial aging conditions, where transgenic seeds showed remarkable germination. Interestingly, ectopic expression of *NnHSP17.5* in Arabidopsis resulted in increased SOD (superoxide dismutase) activity, a ROS scavenging enzyme, under artificial aging conditions, suggesting that small HSPs might play important roles in germination vigor by limiting free radical induced cellular damage (Zhou et al. 2012).

The presence of increased small HSP protein levels during initial stages of germination (3–4 DPI), indicates the role of small HSPs in rehydration and early seed establishment contributing to seed vigor (Wehmeyer et al. 1996; Sun et al. 2001; Zhou et al. 2012; Kaur et al. 2015; Koo et al. 2015). Abundance of small HSPs in dry seeds, during storage and germination was reported in pea (DeRocher and Vierling 1995), sunflower (Coca et al. 1994), Arabidopsis (Wehmeyer et al. 1996; Sun et al. 2001, 2002), *Nelumbo* (Zhou et al. 2012) and rice (Kaur et al. 2015) supporting the suggested function of small HSPs in seed vigor and viability.

Artificial aging reduces seed vigour by damaging proteins via carbonylation and reducing translation efficiency by damaging several components of protein translation machinery (Rajjou et al. 2008). Interestingly, HSPs were proven to play a protective role during protein carbonylation (Cabiscol et al. 2002) and in translation (Basha et al. 2004) thereby providing seed vigor and better germinability after aging. Beech (*Fagus sylvatica* L.) seeds which were aged at −10°C for 8 years showed increase in the levels of 22 kDa small HSP as a factor of time (Kalemba and Pukacka 2008). This accumulation of small HSPs during natural or artificial seed aging suggests protective roles of small HSPs during long term seed storage. A proteomic analysis of 6 day CDT (Controlled deterioration treatment) rice seeds identified small HSP18.2 to be induced after treatment. Arabidopsis seeds with seed specific overexpression of rice small HSP, *OsHSP18.2* displayed improved seed viability under CDT. The remarkable germination in OsHSP18.2 overexpressing lines correlated well with the reduced ROS levels in transgenic seeds as compared to the control (Kaur et al. 2015).

1.7 Small HSPs Impart Vigor and Better Germinability to Seeds Under Stress

Recent studies have demonstrated that constitutive overexpression of small HSPs not only improves germination under aging but also improves germination under stress conditions. Zou et al. (2012) have demonstrated that overexpression of *OsHSP17.0* and *OsHSP23.7* in rice resulted in improved seed vigor and seeds were able to germinate better under stress conditions. Transgenic rice seeds showed remarkable difference in the germination percentage under osmotic stress and salt stress showing reduced accumulation of MDA and electrolyte leakage, increased root/shoot length and increased proline content compared to wildtype seedlings. Increase in the germination vigor in rice seeds expressing *OsHSP17.0* and *OsHSP23.7* was attributed to reduction in membrane damage and increase in the protective molecules such as proline. Wheat chloroplastic small HSP26 when ectopically expressed in Arabidopsis resulted in increased seed vigor. Transgenic Arabidopsis seeds expressing *TaHSP26* were able to germinate and reach maturity even under continuous heat stress of 35 °C. Antisense plants showed reduced heat tolerance even to non-lethal heat stress providing a very strong evidence for the role of small HSPs in providing vigor to the seeds under stress conditions (Chauhan et al. 2012).

Ectopic expression of *LimHSP16.45* from David Lily (*Lilium davidii*) in Arabidopsis resulted in increased seed germination vigor under stress conditions. Over expression lines were able to germinate under heat stress showing germination upto 90 %, whereas wild type showed germination rates less than 60 %. *LimHSP16.45* overexpression in Arabidopsis also resulted in better germination in salt stress conditions showing 90–100 % germination, while wildtype showed only 70–80 % germination rate. Under oxidative stress overexpression lines showed better root growth in terms of root length whereas wildtype seedlings showed reduced root length. Improved seed vigor in the overexpression lines was associated with the increase in the levels of SOD and Catalase activities in transgenic seeds which might be supported by the chaperone activity of small HSPs (Mu et al. 2013). This shows that small HSP play a significant role in maintaining cell viability under stress conditions thereby improving seed vigor and viability. Increase in seed vigor and their ability to germinate under various stress conditions was demonstrated with the heterologous expression of *OsHSP18.2* in Arabidopsis. Transgenic Arabidopsis seeds showed improved germination under heat, dehydration and salt stress compared to controls. Strong evidence suggest that OsHSP18.2 is an aging responsive protein and plays a very important role in maintaining seed vigor and longevity. It also plays a crucial role in seedling emergence by protecting proteins from structural damage and restricting ROS accumulation (Kaur et al. 2015).

In tobacco light is one of the important factors that help in breaking seed dormancy. Light dependency of tobacco seeds was alleviated when *NtHSP18.2*,

NtHSP18.3 and *NtHSP17.6* small HSPs, were overexpressed. A similar light independent germination was also observed when seeds were subjected to heat stress prior to germination (Koo et al. 2015). These evidences suggest the small HSPs might also be involved in breaking dormancy through mechanisms still not known.

1.8 Conclusion

The small HSP class of heat shock proteins has been very well characterized for their heat shock related roles; however, their interesting accumulation patterns during various stages of development in plants e.g. pollen and seed development and seed germination has led to intriguing investigations into their diverse roles and functional capabilities. These developmental stages share a unifying underlying physiological state of desiccation or loss of cellular water (during pollen and seed development) or rehydration (during seed germination). Hence, the associated accumulation of small HSPs during these stages may possibly prepare these plant organs to face the imminent loss of water. Moreover, the chaperone function of small HSPs might also contribute by preventing the irreversible denaturation of important cellular proteins during water deficit conditions as encountered in these stages of plant life cycle. Small HSPs have also been implicated in the formation of glassy matrix in the cells during late seed maturation stages along with other important seed specific proteins like LEA proteins. Thus, the small HSPs are a unique class of heat shock proteins which are equally important for their roles in protection towards heat stress or other environmental stresses and in plant development and various key stages of plant life cycle.

Acknowledgements B.P.P. and H.K. thank the Council of Scientific and Industrial Research and National Institute of Plant Genome Research, Government of India, for research fellowships.

References

Alamillo J, Almogura C, Bartels D, Jordano J (1995) Constitutive expression of small heat shock proteins in vegetative tissues of the resurrection plant *Craterostigma plantagineum*. Plant Mol Biol 29:1093–1099

Almoguera C, Jordano J (1992) Developmental and environmental concurrent expression of sunflower dry-seed-stored low-molecular-weight heat-shock protein and Lea mRNAs. Plant Mol Biol 19:781–792

Almoguera C, Coca MA, Jordano J (1993) Tissue-specific expression of sunflower heat-shock proteins in response to water-stress. Plant J 4:947–958

Almoguera C, Prieto-Dapena P, Jordano J (1998) Dual regulation of a heat shock promoter during embryogenesis: stage-dependent role of heat shock elements. Plant J 13(4):437–446

Arrigo AP (1998) Small heat shock proteins: chaperones that act as regulators of intracellular redox state and programmed cell death. J Biol Chem 379:19–26

Bardel J, Louwagie M, Jaquinod M, Jourdain A, Luche S, Rabilloud T, Macherel D, Garin J, Bourguignon J (2002) A survey of plant mitochondria proteome in relation with development. Proteomics 2:880–898

Basha E, Lee GJ, Demeler B, Vierling E (2004) Chaperone activity of cytosolic small heat shock proteins from wheat. Eur J Biochem 271(8):1426–1436
Basha E, O'Neill H, Vierling E (2012) Small heat shock proteins and alpha-crystallins: dynamic proteins with flexible functions. Trends Biochem Sci 37:106–117
Bernal-Lugo I, Leopold A (1998) The dynamics of seed mortality. J Exp Bot 49:1455–1461
Bouchard RA (1990) Characterization of expressed meiotic prophase repeat transcript clones of *Lilium*: meiosis-specific expression, relatedness, and affinities to small heat shock protein genes. Genome 33(1):68–79
Bouchard RA, Walden DB (1990) Stage-specific expression of small hsp gene RNA in anthers. Maize Gene Coop News Lett 64:122
Buitink J, Claessens MMAE, Hemminga MA, Hoekstra FA (1998) Influence of water content and temperature on molecular mobility and intracellular glasses in seed and pollen. Plant Physiol 118:531–541
Cabiscol E, Bellí G, Tamarit J, Echave P, Herrero E, Ros J (2002) Mitochondrial Hsp60, resistance to oxidative stress, and the labile iron pool are closely connected in *Saccharomyces cerevisiae*. J Biol Chem 277(46):44531–44538
Carranco R, Almoguera C, Jordano J (1997) A plant small heat shock protein gene expressed during zygotic embryogenesis but non inducible by heat stress. J Biol Chem 272:27470–27475
Carranco R, Almoguera C, Jordano J (1999) An imperfect heat shock element and different upstream sequences are required for the seed specific expression of a small heat shock protein gene. Plant Physiol 121:723–730
Charng YY, Liu HC, Liu NY, Hsu FC, Ko SS (2006) Arabidopsis Hsa32, a novel heat shock protein, is essential for acquired thermotolerance during long recovery after acclimation. Plant Physiol 140:1297–1305
Chauhan H, Khurana N, Nijhavan A, Khurana JP, Khurana P (2012) The wheat chloroplastic small heat shock protein (sHSP26) is involved in seed maturation and germination and imparts tolerance to heat stress. Plant Cell Environ 35(11):1912–1931
Coca MA, Almoguera C, Jordano J (1994) Expression of sunflower low molecular weight heat shock proteins during embryogenesis and persistence after germination: localization and possible functional implications. Plant Mol Biol 25:479–492
Coca MA, Almoguera C, Thomas T, Jordano J (1996) Differential regulation of small heat-shock genes in plants: analysis of a water-stress-inducible and developmentally activated sunflower promoter. Plant Mol Biol 31:863–876
Crowe JH, Crowe LM, Petrelski S, Hoekstra FA, Araujo PD, Panek AD (1997) Anhydrobiosis: cellular adaptation to extreme dehydration. In: Dantzler WH (ed) Handbook of physiology, vol II. Oxford University Press, New York, pp 1445–1477
DeRocher AE, Vierling E (1994) Developmental control of small heat shock protein expression during pea seed maturation. Plant J 5:93–102
DeRocher AE, Vierling E (1995) Cytoplasmic HSP70 homologues of pea: differential expression in vegetative and embryonic organs. Plant Mol Biol 27:441–456
DeRocher AE, Helm KW, Lauzon LM, Vierling E (1991) Expression of a conserved family of cytoplasmic low molecular weight heat shock proteins during heat stress and recovery. Plant Physiol 96:1038–1047
Dietrich PS, Bouchard RA, Casey ES, Sinibaldi RM (1991) Isolation and characterization of a small heat shock protein gene from maize. Plant Physiol 96:1268–1276
Ehrnsperger M, Gräber S, Gaestel M, Buchner J (1997) Binding of non-native protein to Hsp25 during heat shock creates a reservoir of folding intermediates for reactivation. EMBO J 16:221–229
Fray RG, Lycett GW, Grierson D (1990) Nucleotide sequence of a heat-shock and ripening-related cDNA from tomato. Nucleic Acids Res 18(23):7148
Goldenberg CJ, Luo Y, Fenna M, Baler R, Weinmann R, Voellmy R (1988) Purified human factor activates heat shock promoter in a HeLa cell-free transcription system. J Biol Chem 2(63):19734–19739

Guo SJ, Zhou HY, Zhang XS, Li XG, Meng QW (2007) Overexpression of CaHSP26 in transgenic tobacco alleviates photoinhibition of PSII and PSI during chilling stress under low irradiance. J Plant Physiol 164:126–136
Haslbeck M, Franzmann T, Weinfurtner D, Buchner J (2005) Some like it hot: the structure and function of small heat-shock proteins. Nat Struct Mol Biol 12:842–846
Helm KW, Abernethy RH (1990) Heat shock proteins and their mRNAs in dry and early imbibing embryos of wheat. Plant Physiol 93:1626–1633
Hendrick JP, Hartl FU (1995) The role of molecular chaperones in protein folding. FASEB J 9:1559–1569
Hernandez LD, Vierling E (1993) Expression of low molecular weight heat shock proteins under field conditions. Plant Physiol 101:1209–1216
Hoekstra FA, Golovina EA, Buitink J (2001) Mechanisms of plant desiccation tolerance. Trends Plant Sci 6(9):431–438
Kalemba EM, Pukacka S (2008) Changes in late embryogenesis abundant proteins and a small heat shock protein during storage of beech (*Fagus sylvatica* L.) seeds. Environ Exp Bot 63:274–280
Kaur H, Petla BP, Kamble NU, Singh A, Rao V, Salvi P, Ghosh S, Majee M (2015) Differentially expressed seed aging responsive heat shock protein OsHSP18.2 implicates in seed vigor, longevity and improves germination and seedling establishment under abiotic stress. Front Plant Sci 6:713
Koo HJ, Park SM, Kim KP, Suh MC, Lee MO, Lee SK, Xinli X, Hong CB (2015) Small heat shock proteins can release light dependence of tobacco seed during germination. Plant Physiol 167(3):1030–1038
Kotak S, Vierling E, Baumlein H, von Koskull-Doring P (2007) A novel transcriptional cascade regulating expression of heat stress proteins during seed development of Arabidopsis. Plant Cell 19:182–195
Kruse E, Liu Z, Kloppstech K (1993) Expression of heat shock proteins during development of barley. Plant Mol Biol 23:111–122
Kurtz S, Rossi J, Petko L, Lindquist S (1986) An ancient development induction: heat-shock proteins induced in sporulation and oogenesis. Science 231:1154–1157
Lee GJ, Vierling E (2000) A small heat shock protein cooperates with heat shock protein 70 systems to reactivate a heat-denatured protein. Plant Physiol 122:189–197
Lee GJ, Pokala N, Vierling E (1995) Structural and in vitro molecular chaperone activity of cytosolic small heat shock proteins from pea. J Biol Chem 270:10432–10438
Lee GJ, Roseman AM, Saibil HR, Vierling E (1997) A small heat shock protein stably binds heat-denatured model substrates and can maintain a substrate in a folding-competent state. EMBO J 16:659–671
Leopold AC, Sun WQ, Bernal-Lugo I (1994) The glassy state in seeds: analysis and function. Seed Sci Res 4:267–274
Lindquist S, Craig EA (1988) The heat shock proteins. Annu Rev Genet 22:631–677
Lubaretz O, zur Nieden U (2002) Accumulation of plant small heat-stress proteins in storage organs. Planta 215:220–228
Luo YT, Amin J, Voellmy R (1991) Ecdysterone receptor is a sequence-specific transcription factor involved in the developmental regulation of heat shock genes. Mol Cell Biol 7:3660–3675
Macherel D, Benamar A, Avelange-Macherel MH, Tolleter D (2007) Function and stress tolerance of seed mitochondria. Physiol Plant 129:233–241
Maimbo M, Ohnishi K, Hikichi Y, Yoshioka H, Kiba A (2007) Induction of a small heat shock protein and its functional roles in *Nicotiana* plants in the defense response against *Ralstonia solanacearum*. Plant Physiol 145:1588–1599
Mtwisha L, Brandt W, McCready S, Lindsey GG (1998) HSP 12 is a LEA-like protein in *Saccharomyces cerevisiae*. Plant Mol Biol 37:513–521
Mu C, Zhang S, Yu G, Chen N, Li X, Liu H (2013) Overexpression of small heat shock protein LimHSP16. 45 in *Arabidopsis* enhances tolerance to abiotic stresses. PLoS One 8(12):e82264

Parker CS, Topol J (1984) A *Drosophila* RNA polymerase II transcription factor contains a promoter-region-specific DNA binding activity. Cell 36:357–369
Perez DE, Hoyer JS, Johnson AI, Moody ZR, Lopez J, Kaplinsky NJ (2009) BOBBER1 is a noncanonical Arabidopsis small heat shock protein required for both development and thermotolerance. Plant Physiol 151(1):241–252
Prändl R, Schöffl F (1996) Heat shock elements are involved in heat shock promoter activation during tobacco seed maturation. Plant Mol Biol 31:157–162
Prändl R, Kloske E, Schöffl F (1995) Developmental regulation and tissue-specific differences of heat-shock gene-expression in transgenic tobacco and Arabidopsis plants. Plant Mol Biol 28:73–82
Prieto-Dapena P, Castano R, Almoguera C, Jordano J (2006) Improved resistance to controlled deterioration in transgenic seeds. Plant Physiol 142:1102–1112
Rajjou L, Lovigny Y, Groot SPC, Belghazi M, Job C, Job D (2008) Proteome-wide characterization of seed aging in *Arabidopsis*: a comparison between artificial and natural aging protocols. Plant Physiol 148(1):620–641
Sanmiya K, Suzuki K, Egawa Y, Shono M (2004) Mitochondrial small heat-shock protein enhances thermotolerance in tobacco plants. FEBS Lett 557(1–3):265–268
Schöffl F, Key JL (1982) An analysis of mRNAs for a group of heat shock proteins of soybean using cloned cDNAs. J Mol Appl Genet 1:301–314
Shen-Miller J, Mudgett MB, Schopf JW, Clarke S, Berger R (1995) Exceptional seed longevity and robust growth: ancient sacred lotus from China. Am J Bot 82:1367–1380
Shen-Miller J, Lindner P, Xie Y, Villa S, Wooding K, Clarke SG, Loo RR, Loo JA (2013) Thermal-stable proteins of fruit of long-living Sacred Lotus Gaertn var. China Antique. Trop Plant Biol 6(2-3):69–84
Sun W, Bernard C, van de Cotte B, Van Montagu M, Verbruggen N (2001) *At-HSP17.6A*, encoding a small heat-shock protein in *Arabidopsis*, can enhance osmotolerance upon overexpression. Plant J 27(5):407–415
Sun W, Van Montagu M, Verbruggen N (2002) Small heat shock proteins and stress tolerance in plants. Biochim Biophys Acta 1577:1–9
Ukaji N, Kuwabara C, Takezawa D, Arakawa K, Yoshida S, Fujikawa S (1999) Accumulation of small heat-shock protein homologs in the endoplasmic reticulum of cortical parenchyma cells in mulberry in association with seasonal cold acclimation. Plant Physiol 120(2):481–489
Vierling E (1997) The small heat shock proteins in plants are members of an ancient family of heat induced proteins. Acta Physiol Plant 19:539–547
Vierling E, Sun A (1989) Developmental expression of heat shock proteins in higher plants. In: Cherry J (ed) Environmental stress in plants. Springer, Berlin, pp 343–354
Volkov RA, Panchuk II, Schöffl F (2005) Small heat shock proteins are differentially regulated during pollen development and following heat stress in tobacco. Plant Mol Biol 57:487–502
Wang W, Vinocur B, Shoseyov O, Altman A (2004) Role of plant heat-shock proteins and molecular chaperones in the abiotic stress response. Trends Plant Sci 9(5):244–252
Wang L, Zhao CM, Wang YJ, Liu J (2005) Overexpression of chloroplast-localized small molecular heat-shock protein enhances chilling tolerance in tomato plant. J Plant Physiol Mol Biol 31:167–174
Waters E, Lee G, Vierling E (1996) Evolution, structure and function of the small heat shock proteins in plants. J Exp Bot 47:325–338
Wehmeyer N, Vieling E (2000) The expression of small heat shock proteins in seeds responds to discrete developmental signals and suggests a general protective role in desiccation tolerance. Plant Physiol 122(4):1099–1108
Wehmeyer N, Hernandez LD, Finkelstein RR, Vicrling E (1996) Synthesis of small heat-shock proteins is part of the developmental program of late seed maturation. Plant Physiol 112(2):747–757

Winter J, Sinibaldi R (1991) The expression of heat shock proteins and cognate genes during plant development. In: Hightower L, Nover L (eds) Heat shock and development. Springer, Berlin, pp 85–105

Wu C (1984a) Two protein-binding sites in chromatin implicated in the activation of heat shock genes. Nature 309:229–234

Wu C (1984b) Activating protein factor binds in vitro to upstream control sequences in heat shock gene chromatin. Nature 311:81–84

Zarsky V, Garrido D, Eller N, Tupy J, Vicente O, Schöffl F, Heberle-Bors E (1995) The expression of a small heat shock gene is activated during induction of tobacco pollen embryogenesis by starvation. Plant Cell Environ 18:139–147

Zhou Y, Chen H, Chu P, Li Y, Tan B, Ding Y, Tsang EW, Jiang L, Wu K, Huang S (2012) *NnHSP17.5*, a cytosolic class II small heat shock protein gene from *Nelumbo nucifera*, contributes to seed germination vigor and seedling thermotolerance in transgenic *Arabidopsis*. Plant Cell Rep 31(2):379–389

Zou J, Liu C, Liu A, Zou D, Chen X (2012) Overexpression of *OsHsp17. 0* and *OsHsp23. 7* enhances drought and salt tolerance in rice. J Plant Physiol 169(6):628–635

Chapter 2
Plant Small Heat Shock Proteins and Its Interactions with Biotic Stress

Valeria Stefania Lopes-Caitar, Suellen Mika Hishinuma Silva, and Francismar Correa Marcelino-Guimaraes

Abstract Small heat shock proteins were first identified during heat shock stress, but currently have been often associated to plant biotic stresses. Considered stress defense proteins, HSP20s functions are especially related to interact with unfolded model substrate proteins, in ATP-independent manner, and keep them in a folding-competent state for subsequent refolding. These proteins have been reported to serve a role in plant-response against the most important crop phytopathogens in the world, such as nematodes and fungi. Their activation role during biotic stress is not completely elucidated, but some researches have demonstrated that in some occurrences of plant response to biotic stresses, there is a crosstalk between the abiotic stress responses. Some genetic evidence has revealed that the chaperones play a critical role in plant immunity. One hypothesis is the chaperone activity can provide the stability and accumulation of R proteins, and thus for the entire defense signaling cascade coordination. However, the researches about this issue still need to better elucidate HSP20 pathways and function in plant biotic stress.

Keywords HSP family • HSP20 • Plant defense mechanism • Pathogens

V.S. Lopes-Caitar (✉)
Department of Genetics and Molecular Biology, Universidade Estadual de Londrina, Londrina, Paraná, Brazil

Department of Plant Biotechnology, Brazilian Agricultural Research Corporation's – EMBRAPA Soybean, Road. Carlos João Strass, Postal Box 231, Londrina, Paraná 86001-970, Brazil
e-mail: valeria.caitar@colaborador.embrapa.br

S.M.H. Silva
Department of Biochemistry and Biotechnology, Universidade Estadual de Londrina, Londrina, Paraná, Brazil

Department of Genetics and Molecular Biology, Universidade Estadual de Londrina, Londrina, Paraná, Brazil

F.C. Marcelino-Guimaraes
Department of Plant Biotechnology, Brazilian Agricultural Research Corporation's – EMBRAPA Soybean, Road. Carlos João Strass, Postal Box 231, Londrina, Paraná 86001-970, Brazil

A.A.A. Asea et al. (eds.), *Heat Shock Proteins and Plants*, Heat Shock Proteins 10, DOI 10.1007/978-3-319-46340-7_2

Abbreviations

ACD	conserved α-crystallin domain
Avr gene	avirulence gene
bp	base pairs
ESTs	expressed sequence tags ETI effector-triggered immunity
GUS	β-glucuronidase
HMM	hidden markov model
HSE	heat shock element
HSFs	heat transcription factors
HSP	heat shock protein
MAMPs	microbe-associated molecular patterns
NB-ARC- LRR	nucleotide binding site-leucine-rich regions
NB-LRR	nucleotide binding domain and leucine-rich repeat
NIL-R	resistance near-isogenic lines
NIL-S	susceptible near-isogenic lines
PAMPs	pathogen associated molecular patterns
PCR	polymerase chain reaction
PRRs	plant pattern-recognition receptors
PTI	Physical association of pattern-triggered immunity
QTL	Quantitative trait locus
RdRp	RNA-dependent RNA polymerase
R genes	resistance genes
RLKs	receptor-like kinases
RLPs	receptor-like proteins
ROS	reactive oxygen species
RT-qPCR	reverse transcription quantitative polymerase chain reaction
SCN	soybean cyst nematode
sHSPs	small heat shock proteins
SSR	simple sequence repeats
Y2H	yeast two-hybrid

2.1 Introduction

In the course of evolution, organisms have developed strategies to overcome the different types of abiotic and biotic stress, including a network of proteins to protect cells (Shanker and Venkateswarlu 2011; Richter et al. 2010). An important group of these stress responsive proteins is composed by chaperones, or heat shock proteins (HSP), and their presence has been demonstrated in all living organisms (Tkáčová and Angelovičová 2012). HSPs comprise several families classified according to their molecular weight: HSP100, HSP90, HSP70/DnaK and HSP60/GroE and HSP20 or small heat shock proteins (sHSPs), representing sizes of 15–42 kDa (Waters 2013).

During their lifetime, due to their sessile nature, plants are continually exposed to adverse effects but may supplant them through the evolution of various morphological and physiological mechanisms that enable them to survive stress (Al-Whaibi 2011). In higher plants, HSP20 are encoded by nuclear multigenic families and have undergone great functional diversification (Waters et al. 1996). Plants usually contain around 20 HSP20, or more, but this number can be four times lower in animal organisms (Waters et al. 1996; Franck et al. 2004).

Furthermore, it is noteworthy that no HSP20 proteins located in plant organelles have homologs in animals, fungi, or even in green algae (Waters et al. 2008); these the HSP20 appear to be localized only in the cytoplasm/nucleus while plants have several subfamilies, which are distributed in different cellular compartments. In *Arabidopsis*, 19 genes encoding HSP20 were classified into 12 subfamilies, according their cellular localization and phylogeny (Scharf et al. 2001), while 36 Hsp20 were identified in *Populus trichocarpa* (Waters et al. 2008), 23 in *Oryza sativa* (Sarkar et al. 2009), 51 in *Glycine max* (Lopes-Caitar et al. 2013), 35 in *Capsicum annuum* L., 27 in *Triticum aestivum* and 13 in *Hordeum vulagre* (Pandey et al. 2014).

HSP20 are often the most abundant plant stress responsive class among the heat shock proteins group (Heckathorn et al. 1999). They also have been largely studied in human health and diseases (Bakthisaran et al. 2015). The primary structures of HSP20s are characterized by a conserved α-crystallin domain (ACD) from 80 to 100 amino acids at the C-terminal region (Cashikar et al. 2005). This domain is divided into two consensus regions designated I and II, separated by a hydrophobic region of variable size. The consensus region I (~27 amino acids) N-terminal constitutes conserved sequence Pro-$X_{(14)}$-Gly-Val-Leu; and the consensus region II (~29 amino acids) C-terminal, has the conserved sequence Pro-$X_{(14)}$/Val/Leu/Ile-Val/Leu/Ile. ACD is preceded by an N-terminal variable size region and considerable diversity in their sequence (Cashikar et al. 2005; Waters et al. 2008). Within HSP20-ACD domain there is a typical secondary structure formed, usually with the formation 6–8 β-sheet configuration. The β settings are numbered along the amino acids chain in the amino-caboxiterminal sense, being found within the ACD, usually the $\beta 2$ chains $\beta 9$ (Siddique et al. 2008) (Fig. 2.1).

Each class of HSPs has a characteristic function upon specific spectrum (broad or narrow) of substrate proteins in assisting their folding, refolding, oligomeric assembly, translocation, and/or degradation (Hartl et al. 2011). The HSP20 are known to act as ATP-independent molecular chaperones that bind to the newly synthesized proteins and denatured proteins to prevent them from forming irreversible aggregations, under stress conditions, also can be called "holdase" (Zhang et al. 2015b). However, they have been classified within the group of molecular chaperones for the ability to recognize and bind to denatured proteins or inappropriate folding and thus prevent improper interactions that lead to irreversible aggregate formation, precipitation, and degradation thereof by proteases (Sun et al. 2002).

The HSP20s are ubiquitin that form large oligomeric hetero-complexes ranging in size 200–800 kDa and have also been found to suppress protein aggregation in an ATP-independent manner, stabilize stress-damaged cell membranes, tag denatured proteins aggregates, or avoid inappropriate folding, to further action of proteases in

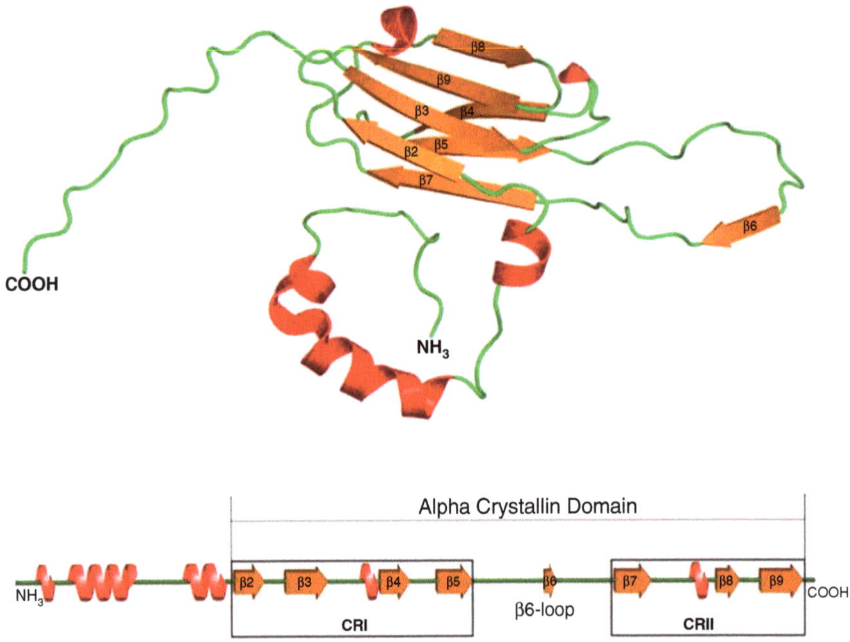

Fig. 2.1 Structural organization of a typical sHSP protein. The picture shows the X-ray structure of a Triticum aestivum sHSP dimer (protein data bank in Europe: 1GME) and primary structure (protein data bank PDBsum: 1GME) of the protein that are constituted by C-terminal region, N-terminal region and a α-crystallin domain (ACD) containing the conserved regions I (CRI) and II (CRII) and the β6-loop (Modified from Bondino et al. 2012)

both, prokaryotic and eukaryotic cells (Fu 2014). The main HSP20s activity is to interact, while still possible, with unfolded model substrate proteins. HSP20 keeps another protein in a folding-competent shape for subsequent refolding, by ATP-dependent chaperones as HSP70s and HSP100s, or labels them and allows the proteinases action, if necessary (Al-Whaibi 2011; Fu 2014).

In this chapter, we will discuss how plant small heat shock proteins are related to response against biotic stress and how the knowledge about its function and involvement in response to different types of stress can be explored as an important tool in agricultural sciences.

2.2 Plant HSP Family

The plant heat shock proteins have several functions related to stress response and it can be verified by each subclass. The HSP100 family usually located in the cytosol, mitochondria and chloroplast, exhibit weight around 97–114 kDa (Singla et al. 1998; Park and Seo 2015). This protein family has been detected in various plant species, such as in Glycine max, where the HSP103 kDa protein was present in soybean cell subjected to heat stress. Similarly, the presence of the protein HSP100 and 110 was detected in *Nicotiana tabacum* under heat stress (Barnett et al. 1980). Besides these species, already was noted the HSP100 protein in *Gossypium hirsutum, Gossypium hirsutum, Opuntia ficus indiea, Prosopis chilensis, Saccharum. Officinarum, Secale cereate, Triticum aestivum, T. aestivum, T. durum, Vigna radiate* and *Zea mays* (Singla et al. 1998).

In the model plant *Arabidopsis thaliana*, the family HSP90 includes 7 members: At*Hsp*90-1 to At*Hsp*90-4 genes, which constitute the cytoplasmic subfamily, AtHsp90-5, AtHsp90-6, and AtHsp90-7 genes, which are predicted to be within the plastidial, mitochondrial, and endoplasmic reticulum compartments, respectively (Krishna, 2001). Therefore, HSP90 are located in several cellular compartments, such as cytosol, mitochondria, chloroplast, nucleus and endoplasmic reticulum (Park and Seo 2015). Additionally, HSP90 can be found in *Chlamydomonas reinhardtii* and *Oryza sativa* species (Chen et al. 2006).

Regarding HSP70, *A. thaliana* genome contains 12 subfamilies, with five belonging to cytosol compartment (Hsc70-1, Hsc70-2, Hsc70-3, Hsp70 and Hsp70b), three belonging to endoplasmic reticulum lumen (BiP-1, BiP-2 and BiP-3), two belonging to mitochondrion matrix (mtHsc70-1 and mtHsc70-2), and two belonging to plastid stroma (cpHsc70-1 and cpHsc70-2) (Sung et al. 2001). The HSP70 family was also detected in other plants species, such as *Capsicum annuum L., Cucurbita maxima, Cucumis sativus, Nicotiana tabacum, Gossypium hirsutum, Populus trichocarpa, Solanum lycopersicum, Pisum sativum, Pisum sativum, Theobroma cacao, Hevea brasiliensis, Spinacia oleracea,* and *Oryza sativa* (Guo et al. 2014).

A phylogenetic analysis of HSP60 gene family from *Populus trichocarpa, A. thaliana* and *Oryza sativa*, suggested that these plant species contend approximately 28, 18 and 20 HSP60, respectively, categorized into four subfamilies Zhang et al. (2015a). This group is cytosol-localized Cpn60 (18 genes), mitochondrion-localized Hsp60 (3 genes), and chloroplast-localized Cpn60-a (4 genes) and Cpn60-b (3 genes) (Zhang et al. 2015a). Additionally, the HSP60 family was also identified in several other plant species, such as Glycine max, *Medicago truncatula, Physcomitrella patens, Ricinus communis, Solanum tuberosum, Sorghum bicolor, Triticum aestivum, Vitis vinífera* and *Zea mays* (Ratheesh Kumar et al. 2012).

Like other proteins, HSP20 was found in several plant species, such as *Arabidopsis thaliana, Petunia hybrid, Pisum sativium, Triticum aestivum, Zea mays, Chenopodium rubrum, Glycine max, Daucus carota, Helianthus annuus, Lycopericoscon esculentum, Medicago sativa, Ipomea nil* and *Lilium longijlorum* (Waters 1995).

The genome-wide analysis of the HSP20 protein family was performed with at least 6 species: *Oryza sativa, G. max, T. aestivum, Hordeum vulgare* L., *Capsicum annuum* and *Arabidopsis*. (Guo et al. 2015; Lopes-Caitar et al. 2013; Pandey et al. 2015; Sarkar et al. 2009; Scharf et al. 2001).

Pandey et al. (2015) reported in their studies HSP20 genes in wheat (TaHSP20) and barley (HvHSP20). Using Hidden Markov Model (HMM) and Blast algorithm, the authors identified 27 newly TaHSP20 candidate genes in wheat and 13 HvHSP20 in barley. Interestingly, the HSP20 protein predicted were located in both mitochondria and the nucleus, however, a large number of TaHSP20 and HVHSP20 protein was classified into mitochondria subfamilies (MT-sHSPs). Pandey et al. (2015)) also generated a phylogenetic tree of HSP20 revealing that HSP 16.9 from different plant species such as rice, barley, sorghum, and brachypodium show highly conserved protein sequence, moreover the HSP26 from *Tritricum durum, Triticum monococum, Oryza sativa*, and *Brachypodium* ssp. illustrated evolutionary relatedness close.

Similarly, the genome-wide analysis of the HSP20 gene family member was carried out in pepper specie. Through HMM analysis, 35 sequences were annotated as being pepper HSP20 genes. Among these, 8 genes are predict localized in cytosol, 1 gene (cytosol, mitochondrial), 1 gene (cytosol, extracellular), 5 gene (cytosol, nucleus), 1 gene (cytosol, peroxisomal), 1 gene (cytosol, extracellular, endoplasmic reticulum), 2 gene (cytosol, mitochondrial, chloroplast), 2 gene (cytosol, nucleus, golgi), 1 gene (mitochondrial), 1 gene (mitochondrial, chloroplast), 3 gene (chloroplast), 1 gene (chloroplast, nucleus), 1 gene (chloroplast, extracellular) and 7 gene (secretory pathway) (Guo et al. 2015).

Guo et al. (2015) analyzed the evolutionary relationship of the HSP20 gene families in pepper, tomato, *Arabidopsis*, rice, and maize. Interestingly, the HSP20 members from the five species were more closely to those in the same class in different species than to other in the same species. The author also addresses that three species of dicotyledon, forming 11 pairs orthologs (pepper/tomato) and 1 pair ortholog (pepper/Arabidopsis) were identified, suggesting that the ancestral genes of Hsp20 family might have existed before differentiation of *Arabidopsis*, tomato and pepper species.

The *Arabidopsis* genome revealed 19 genes encoding HSP20, grouped into 6 classes; class CI (At17.6B-CI, At17.8-CI, At17.6A-CI, At18.1-CI, At17.4-CI and At17.6C-CI), class CII (At17.7-CII and At17.6-CII), class CIII (At17.4-CIII), class Mitochondrial (At23.6-M and At23.5-M), class plastids (At25.3-P) and class endoplasmic reticulum (At22.0-ER). The other six sHSPs (At14.2-P(r), At15.4-CI(r), At15.7-CI(r), At18.5-CI(r), At21.7-CI(r) and At26.5-P(r)) are more distantly related, but also appear to be members of this protein family (Scharf et al. 2001).

The genome-wide analysis of HSP20 in rice performed by Sarkar et al. (2009)), demonstrated the organelle localization of rice HSP20 gene family. Thus, the 23 HSP20 of rice was identified, consisting of 14 subfamilies distributed to several compartments cellular, such as 16 nucleo-cytoplasmic (9 subfamilies), 3 mitochondrial (2 subfamilies), 2 endoplasmic reticulum, 1 plastidial and 1 peroxisomal having one subfamily each.

The analysis of the HSP20 gene family in soybean revealed 51 gene models as potential GmHSP20 candidates. Based on the phylogenetic tree constructed, these genes are divided into 13 subfamilies, these being, 37 nucleo-cytoplasmic (8 subfamilies and 3 orphan genes), 3 mitochondrial (2 subfamilies), 4 endoplasmic reticulum, 5 plastidial and 2 peroxisomal having 1 subfamily each (Lopes-Caitar et al. 2013). The large number of GmHSP20 proteins classified in the cytoplasmic subfamilies indicates that the cytoplasm may be the primary site of HSP20s' action. It is in the cytoplasm where proteins assembling reside after synthesis by the translation process. Thus, it is likely that this concentration of HSP20 in cytoplasmic subfamilies is related to its function to prevent these proteins from performing improper folding or inappropriate interactions forming aggregates.

In this study, Lopes-Caitar et al. (2013) also performed some sequence and phylogenetic analysis and it was identified that those GmHSP20 responsible to heat stress showed a minimum sequence identity of 17.39 % and a maximum of 98.05 %, particularly evident in the region C terminus, including the α-crystalline domain region. The evolutionary relationship in plant HSP20 subfamilies was evidenced when members from different species can be grouped at the same subfamily, instead of grouping by plant species (Fig. 2.2).

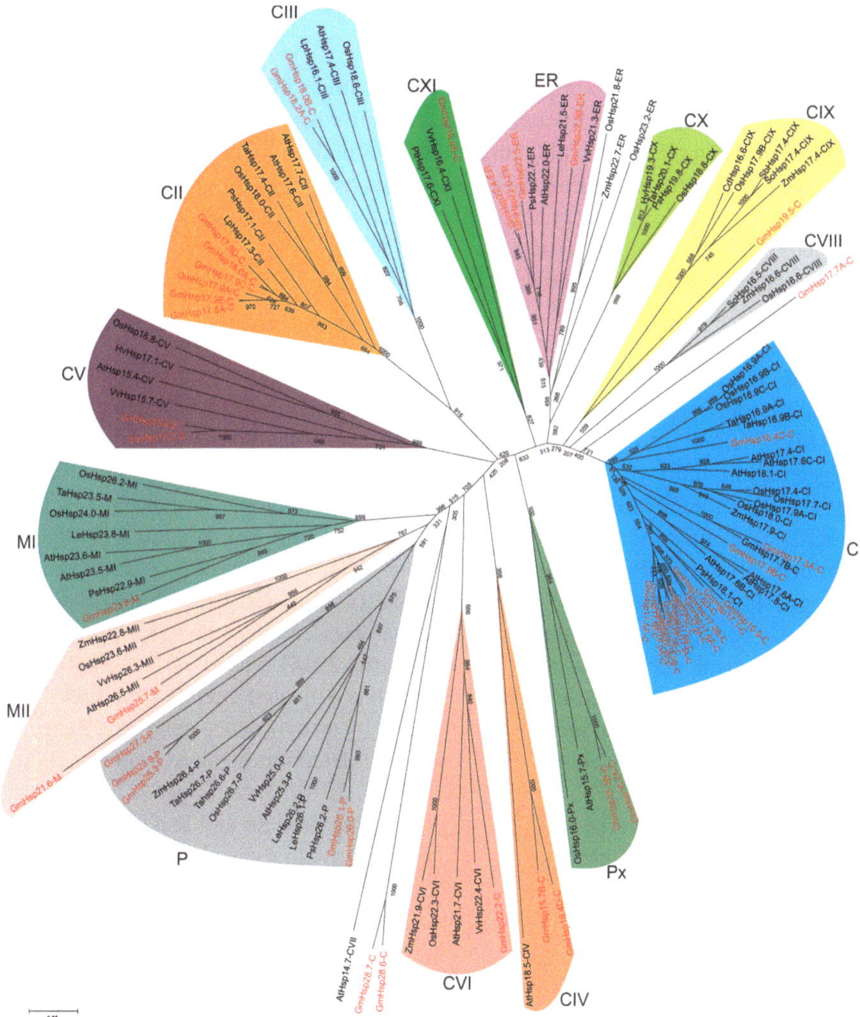

Fig. 2.2 Phylogenetic evidence of evolutionary relationship in plant HSP20 subfamilies. HSP20 genes from different plant species (At Arabidopsis thaliana, the Oryza sativa, Vv Vitis vinifera, Ta Triticum aestivum, Ps pea, Le Solanum lycopersicum, Zm Zea mays, Lp Lycopersicon peruvianum, Cd Cynodon dactylon, So Saccharum officinarum, Sb Sorghum bicolor, Hv Hordeum vulgare) are grouped in subfamilies, according their subcellular localization. Subfamilies are described by nucleus/cytoplasmic CVI, CVII, CVIII, CIX and CX; ER (endoplasmic reticulum); P (plastids/chloroplasts); Px (peroxisome); MI and MII (mitochondria) (Lopes-Caitar et al. 2013)

This evidence indicates that there is a synteny between soybean, rice, *Arabidopsis* and other species containing HSP20s. The HSP20 genes probably had a common ancestor that gave rise to different subfamilies before diversification among these species, and this was conserved during evolution (Ouyang et al. 2009).

2.3 General Role of HSP20 in Plant Defense Mechanism Against Biotic Stresses

HSP20 are commonly associated with abiotic stress (Wang et al. 2004; Efeoğlu 2009), but they have been frequently described in response to biotic stresses. Some HSPs have been reported to play important roles in the immune response in animals and their presence in the circulation serves as a warning mechanism to the host (Evans et al. 2015; Dühring et al. 2015). As well as other organisms, plants are challenged by a variety of fungi, bacteria, or viral infections. They respond to pathogens using elaborate networks of genetic interactions (Mandadi and Scholthof 2013), and recently, evidence has revealed that the chaperones play a critical role in plant immunity (Park and Seo 2015; Shirasu 2009).

The current knowledge about the plant innate immune system proposes two-branched defense responses, pathogen (microbe)-associated molecular patterns (PAMPs/MAMPs)-triggered immunity (PTI), or basal resistance, and effector-triggered immunity (ETI), or race-specific resistance. PAMPs, which are initiated by the direct recognition pathogen through plant pattern-recognition receptors (PRRs), constitute the first mode of defense against pathogen infection (Dodds and Rathjen 2010). PRRs are plasma membrane proteins, most are characterized as receptor-like kinases (RLKs) or receptor-like proteins (RLPs), which recognizes relatively conserved molecules within a class of essential proteins for microbe fitness or survival, such as flagellin (Vidhyasekaran 2014). The other branch provides a remarkable level of disease resistance, which is activated directly or indirectly by pathogen-specific protein recognition encoded by an avirulence gene (Avr gene), or elicitor/effector (resistance gene-by-gene). The detection involves genes encoding resistance proteins (R genes) (Van Ooijen et al. 2010). Most R genes are intracellular proteins, characterized as nucleotide binding domain and leucine-rich repeat (NB-LRR)-containing protein family. After Avr protein recognition by R genes, the defense response is activated and requires the interaction of several auxiliary proteins, as those involved in ROS generation, kinases and heat shock proteins (Van Ooijen et al. 2010; Buchanan et al. 2000). The chaperone activity during biotic stress has been shown to be important for the stability and accumulation of R proteins, and thus for all defense signaling cascade coordination (Septiningsih et al. 2013; Boter et al. 2007).

The activation of plant HSP20 under heat stress is directly related to Ca^{2+} concentration in the cell. High Ca^{2+} concentration activates calmodulins and/or MAP kinases and followed by HSFs phosphorylation (Heat Transcription Factors), which are able to recognize certain cis elements present in the heat shock genes promoter sequence, inducing its expression (Swindell et al. 2007). How plant HSP20 activation occurs during biotic stress is something still not clear. However, many studies have demonstrated that gene expression patterns in response to different stresses (Table 2.1), even when activated by different stresses, are regulated at any time by an overlapping of signals (Fraire-Velázquez et al. 2011; Singh 2002).

Table 2.1 Summary of different studies of plant sHSP under biotic stress conditions

Phatogen	Species	Plant	Gene	Reference
Bacteria	*Ralstonia solanacearum*	*Nicotiana tabacum*	NtHsp17	Maimbo et al. (2007)
	Xanthomonas axonopodis pv. citri	Orange	Hsp15.5	Garofalo et al. (2009)
	Xanthomonas campestris pv. vesicatoria	Pepper	Hsp16	
	Pseudomonas syringae	*Arabidopsis thaliana*	HSP17, HSP17.6II, HSP17.6A-CI, HSP17.6, HSP21, HSP23.6-MITO	Bricchi et al. (2012)
Fungi	*Magnaporthe grisea*	Rice	Hsp17.4-CI, Hsp18.0-CI, Hsp16.9A-CI, Hsp18.0-CII, Hsp22.3-CVI, Hsp16.6-CVIII, Hsp18.8- CX, Hsp17.8-CXI	Sarkar et al. (2009)
	Fusarium oxysporum	Tomato	I-2	Van Ooijen et al. (2010)
	Rhizoctonia solani	*Arabidopsis thaliana*	HSP17.4 and HSP17.6A	Foley et al. (2013)
	Rhizopus nigricans Ehrenb	*Solanum lycopersicum*	HSP17.6	Pan et al. (2013a)
	Venturia inaequalis	*Malus domestica*	HSP21	Hüsselmann (2014)
	Blumeria graminis f. sp. hordei	*Hordeum vulgare*	Hsp16.9 and Hsp17.5.	Ahmed et al. (2015)
	Phytophthora infestans	*Solanum tuberosum*	HSP17.8	Yogendra et al. (2015)

(continued)

Table 2.1 (continued)

Phatogen	Species	Plant	Gene	Reference
Nematodes	*Meloidogyne incognita*	Sunflower	HaHsp17.7G4, HaHsp18.6, HaHsp17.6	Escobar et al. (2003)
	Meloidogyne javanica	Soybean	Gm*Hsp*17.6-L	Fuganti et al. (2004, 2010)
	Meloidogyne incognita	Sunflower	Hahsp17.6G1 (G1), Hahsp18.6G2 (G2), Hahsp17.7G4 (G4)	Barcala et al. (2008)
	Heterodera glycines	Soybean	GmHsp14g06910, GmHsp10g32000, GmHsp08g07340, GmHsp04g05720, GmHsp17g08020, GmHsp16g29750, GmHsp11g14950	Kandoth et al. (2011)
	Meloidogyne javanica	Soybean	GmHsp22.4, GmHsp17.9B, GmHsp17.9A, GmHsp17.4, GmHsp22.4, GmHsp17.6B	Lopes-Caitar et al. (2013)
	Rotylenchulus reniformis	*Gossypium hirsutum*	HSP20	Li et al. (2015b)
Virus	*Turnip vein-clearing tobamovirus* (TVCV), O*ilseed rape tobamovirus* (ORMV), P*otato virus X potexvirus* (PVX), *Cucumber mosaic cucumovirus* (CMV), T*urnip mosaic potyvirus* (TuMV)	*Arabidopsis thaliana*	HSP17.6A, HSP17.4	Whitham et al. (2003)
	SYNV (*Sonchus yellow net virus*), NSV (*Impatiens necrotic spot virus*)	*Solanum tuberosum*	HSP18, HSP20	Senthil (2005)
	Rice stripe virus (RSV)	Rice	OsHSP20	Li et al. (2015a)
		Nicotiana benthamiana	NbHSP20	

2.4 Plant HSP20 in Immune Response During Bacteria Infection

In nature, plant-bacteria interaction, in most cases, have no influence in plant growth or development; some of them may be beneficial, especially considering crop yield increase, as occurs between soybean and Rhizobia bacteria (Dudeja and Giri 2014). However, bacteria pathogenic for plants is also responsible for devastating losses in agriculture. There are around 100 phytopathogenic bacteria species already described (Stavrinides 2009). Recently, a list was published of the top 10 phytopathogenic bacteria in a rank order (Bull et al. 2014; Mansfield et al. 2012).

The gram-negative plant pathogenic bacteria, *Xanthomonas* spp, have caused important agricultural losses. In Taiwanaise cabbage crops, *Xanthomonas campestris* pv. *campestris* is the disease responsible of black rot in crucifers (Li et al. 2014); in Uganda, *Xanthomonas* Wilt of Banana is the cause of around 50 % yield losses and is becoming a threat the food security of about 70 million people owing to its impact on an important staple crop (Vurro et al. 2010).

To exploit plant nutrients, phytopathogenic bacteria have evolved sophisticated infection strategies. Some of most important phytopathogenic bacteria, as *Pseudomonas syringae*, *Ralstonia solanacearum* and *Xanthomonas* spp., are able to cause diseases in plants through a type III secretion system (Maejima et al. 2014).

As in most plant-pathogen interaction, plant response against bacteria can occur by pathogen-associated molecular patterns (PAMPs), as a basal defense response (Thomma et al. 2011). However, throughout evolution bacteria have overcame this basal level of recognition by evolving effectors to suppress basal resistance; some plants genotypes could also evolve and aquire a resistance response. In effector-triggered immunity (ETI), the pathogen effector protein can be recognized by R genes (resistance gene) and present a fast and effective defense response against the infection (Dodds and Rathjen 2010).

Although, there are not many studies describing the association between plant-HSP20 during bacterial infection, some reports are demonstrating that it is an important issue to be elucidated.

Maimbo et al. (2007) studied the compatible and incompatible interactions of *Nicotiana tabacum* plants with pathogenic strains and non-pathogenic bacterium *Ralstonia solanacearum*, for induction of hypersensitive response. Through the technique of PCR differential display, it was possible to isolate and functionally characterize genes related to tobacco interaction with *R. solanacearum*. Among the fragments differentially regulated in inoculated tobacco leaves, sequences similar were found in *Medicago truncatula Hsp*20, *Cucumis melo* and a hypothetical protein from *Arabidopsis*. In tobacco, this isolated fragment shows a corresponding sequence to the gene encoding the protein NtHSP17 with proven response to induction by heat stress. The results obtained by real-time PCR in tobacco, immediately after the silencing of this gene followed by pathogen inoculation, demonstrated that the expression of some defense related genes, such as ethylene-response element-binding protein and genes related to pathogenesis PR1 (pathogen-related) and PR4, were compromised. The silencing of the Nt*Hsp*17 gene, coding for an NtHSP20, also resulted in increased virulence of *R. solanacearum*, including non-pathogenic strain. These results suggest that HSP20 may have an important role in the immune response of tobacco plants when attacked by *R. solanacearum*.

Similar results were found by Garofalo et al. (2009), who analyzed the expression of Hsp20 proteins in orange and pepper plants during compatible and incompatible interactions with the bacterial pathogens *Xanthomonas axonopodis* pv. *citri* and *Xanthomonas campestris* pv. *vesicatoria*, respectively. Hsp20 expression profiles in different situations were related to the time after inoculation treatment and the compatibility interaction, suggesting that these proteins are related to basal defense response in plants.

Bricchi et al. (2012) performed *Arabidopsis* expression analysis by microarray technology during *Pseudomonas syringae* infection. The results showed that six small heat shock proteins were down-regulated (HSP17, HSP17.6II, HSP17.6A-CI, HSP17.6, HSP21, HSP23.6-MITO). According to Pavlova et al. (2009), the down-regulation of HSPs during a biotic stress is quite rare, although gene expression of HSP20s was verified as down-regulated by SA in *Arabidopsis*.

2.5 HSP20 Activity in Plant Response Against Fungi

Around 10 % of the 100,000 known fungal species are able to colonize plants, and a fraction of these are able to cause disease (Epton 1989). Recently, a list of the top 10 phytopathogenic fungi and oomycete class were published (Kamoun et al. 2015; Dean et al. 2012). There are several ways that fungi infect its hosts, depending on their lifestyle. In general, the fungi attack is initiated by spore germination on plant surface, followed by short tube formation which differs in appressorium structure and plant-cell penetration with the primary hyphae (Mendgen and Hahn 2002). As in all pathogen attack, plants can present different behaviors during fungi infection, going from susceptibility to resistance.

Some plant-HSP20 has been also associated with defense response against fungi. Sarkar et al. (2009) performed Hsp20 gene family characterization in rice and its expression profile under stress conditions and developmental stages. An opposite pattern of expression was observed for different family members. Four rice Hsp20 genes were induced after 4 days under *Magnaporthe grisea* infection (Hsp17.4-CI, Hsp18.0-CI, Hsp16.9A-CI, Hsp18.0-CII), while another four genes, all belonging to mitochondrial subfamily, were repressed (Hsp22.3-CVI, Hsp16.6-CVIII, Hsp18.8- CX and Hsp17.8-CXI).

Van Ooijen et al. (2010) described a specific interaction between HSP20 and the tomato (*Solanum lycopersicum*) R protein I-2 that confers resistance to *Fusarium oxysporum*. They demonstrated that the resistance against *Fusarium oxysporum* mediated by hypersensitivity response is strongly reduced when a specific class of Hsp20 genes is silenced in tobacco. In this case, Hsp20 proteins interact specifically with proteins encoded by gene I-2 from tomato, triggering the hypersensitive response. This is the unique report showing the physical interaction between a HSP20 and an R gene type NB-ARC-LRR (nucleotide binding site-leucine-rich regions) demonstrating its evident involvement in plant immunity.

Foley et al. (2013) performed a global gene expression analyses in *Arabidopsis* at 7 days post infection by two different isolates of *Rhizoctonia solani.* This pathogen is an important soil-borne necrotrophic fungal, with a wide host range and little effective resistance in crops (Wang et al. 2014). *Arabidopsis* is resistant to *R. solani* AG8, but susceptible to *R. solani* AG2-1. HSP20s seem to be associated with resistance phenotype, and the class of HSPs showed the most striking response to AG8, with five out of the seven HSPs induced greater than tenfold by AG8, but no HSP genes were expressed to a higher degree in AG2-1 infected tissue. After quantitative RT-PCR experiments were performed to validate the response of HSP17.4 and HSP17.6A, which had the highest fold induction among HSP genes following AG8 infestation, both genes were silenced and overexpressed to check their function during the response to *R. solani* infection. The silent mutants showed a response against AG8 or AG2-1 similar to wild type. The overexpressed mutants for HSP 17.4 and 17.6A were not affected in their resistance or susceptibility to *R. solani* AG8 or AG2-1, respectively. The authors concluded that, even though this gene family presented the largest transcriptional difference between AG2-1 and AG8 infected tissue, it seems that individual HSP20s do not have a direct role in the plant's response to *R. solani.* However, considering the large number of HSP20s induced specifically by AG8, HSP20s may be part of the effect rather than the cause of the plant's resistance to *R. solani.*

In a *Solanum Lycopersicum* (tomato) proteomic profile analysis during *Rhizopus nigricans* infection, Pan et al. (2013b) identified, among others, a HSP20 related to this interaction. *Rhizopus nigricans* Ehrenb is a major necrotrophic pathogenic fungus that causes serious decay on fruit during development and postharvest storage (Pan et al. 2013a). In this study, the translation of HSP17.6 was reduced in the *R. nigricans* infected tissue, a trend validated by the steady state level of its transcript. The authors considered these molecular regulations as a host defense mechanism against *R. nigricans.* A similar result was reported in the thesis written by Hüsselmann (2014); this study performed the *Malus domestica* Borkh proteomic analysis during the infection by fungus *Venturia inaequalis.* The results showed one apple HSP21 was identified and found to be absent or degraded in the stressed proteome.

Ahmed et al. (2015) were able to demonstrate two barley HSP20s being targeted by pathogen effectors. The authors described, by yeast two-hybrid assay, the interaction between two *Blumeria graminis f. sp. hordei* [Bgh] secrete effector proteins with *Hordeum vulgare* small heat shock proteins, Hsp16.9 and Hsp17.5. They found that two barley powdery mildew effector candidates, CSEP0105 and CSEP0162, contributed to Bgh infection success, and also two HSP20 were identified as host targets for these two CSEPs. Notably, Hsp16.9 showed in vitro chaperone activity, which was specifically suppressed by CSEP0105. The data showed that barley Hsp16.9 and Hsp17.5 have high amino acid sequence similarity with the tomato HSP20 previously described by Maimbo et al. (2007), relating to *R. solanacearum* infection. These data suggest that CSEP0105 promotes virulence by interfering with the chaperone activity of a barley small heat shock protein essential for defense and stress responses.

Yogendra et al. (2015) provided compelling evidence on the functionality of StWRKY1 transcription factor, in a compatible interaction of *Solanum tuberosum –Phytophthora infestans*. Based on potato metabolic profiling and reverse genetics, they validated the transcriptional activation of StWRKY1 following *P. infestans* invasion. The data allowed identifying resistance-related metabolites belonging to the phenylpropanoid pathway and their biosynthetic genes regulated by WRKY1. Moreover, the results suggest a transcriptional activation of WRKY1 by StHSP17.8 only in the resistant potato genotype. The results showed a simultaneous induction a small heat shock protein and WRKY1 during the fungi infection. The presence of a heat shock element in the resistant genotype WRKY1 promoter region is related with this simultaneous induction. In addition, the regulation of WRKY1 by the HSP17.8 was confirmed by LUC transient expression assay. These results suggest that this interaction with HSP20 in the resistant genotype is responsible for high transcriptional activity of WRKY1, probably by the chaperone activity in keep the stability and accumulation of this protein during the stress condition.

2.6 Plant HSP20 Associated to Nematodes Attack

Different from other organisms, so far no direct interaction is considered beneficial between plants and any nematode species (Usta 2013). Among the top 10 plant-parasitic nematodes in molecular plant pathology, root-knot nematodes (*Meloidogyne* spp.); cyst nematodes (*Heterodera* and *Globodera* spp.); and root lesion nematodes (*Pratylenchus* spp.) are the most harmful species for agriculture (Jones et al. 2013).

Some plant-HSP20 has also been associated with defense response against nematode. Fuganti et al. (2004), working with microsatellite markers, identified a QTL linked to *Meloidogyne javanica* infection in a segregating population for resistance in soybean. The QTL mapped (SOYHSP 176) in soybean F linkage group (chromosome 13) between SSR molecular markers Satt114 and Satt423, in the promoter region of Glyma13G176000.1. A differential expression profile of this gene was demonstrated in resistant and susceptible genotypes under *M. javanica* infection (Fuganti et al. 2010). The authors showed the soybean Gm*Hps*17.6-L gene induction on resistant genotypes (PI595099), but no changes in gene expression levels on susceptible genotype (BRS133). These data indicate that this protein may have an important role on complex soybean resistance response against phytonematodes.

Kandoth et al. (2011)) studied the soybean resistance mechanisms against the nematode *Heterodera glycines*, and compared the gene expression profiles of syncytia in two soybean lines, differentiated only by the presence of Rhg1 gene, a resistance gene first described in Peking for resistance to soybean cyst nematode (SCN) (Caldwell et al. 1960). Among the 1447 differentially expressed genes found between both lines, resistant near-isogenic lines (NIL-R) and susceptible near-isogenic lines (NIL-S), 4 genes corresponded to HSP20 proteins; and other enzymes that lead to reactive oxygen species increased levels. These results indicate that a complex network of molecular events occur during Rhg1 mediated resistance, leading to a defense response against a soybean root pathogen and the possibility that the Hsp20 plays an important role in resistance.

Other studies also carried out on soybeans, have suggested that the cis elements positioning of heat shock response in the Hsp gene promoters is directly related to its activation during biotic stress. Escobar et al. (2003) studied the influence of position HSE (Heat Shock Element) changes in the sunflower gene HaHsp17.7G4 promoter region under its expression in the giant cells formed after infection with *Meloidogyne incognita* in transgenic tobacco plants. The results obtained by the authors demonstrated that HaHsp17.7 responds to infection by the nematode and that the presence of a single element HSE is sufficient for induction, since it occurs until – 83 base pairs (bp) from the site initiation of transcription. When the HSE element is positioned at distances greater than – 83 bp from the transcription start site, the Hsp20 gene ceases to be induced in response to nematode infection.

Subsequently, the comparative analysis of Ha*Hsp*17.7 promoter sequence (- 83 bp G4) and two other sunflower Hsp20 genes, which also respond to infection by nematodes, HaHsp18.6 (G2 contains two HSE a proximal – 49 bp) and HaHsp17.6 (G1 contains an HSE, furthest – 108 bp), in *Arabidopsis* plants transformed by the floral deep technique, corroborated the data obtained by Escobar et al. (2003). In this experiment, Barcala et al. (2008) confirmed the importance of the single core element HSE presence, distant to – 83 bp of the transcription start site. These analyzes used a vector with GUS gene marker, under regulation of different promoter configurations by HSE elements position and number. In addition, in G1 mutant, the distance between the HSE and TATAbox were much larger than in first HSE in G2 or G4 (compared to 49 bp 12 bp 21 bp in G2 and G4). CAATbox elements have also been predicted and found between the matrix HSE of G2 promoter (CAAT box position – 177 bp and – 72 bp) and G4 (CAATbox position – 141 bp). In contrast, the CAATbox been identified in the promoter positioned – 67 bp, before the HSE element.

The results showed that the G2 promoter was activated after nematode infection in transgenic tobacco plants, while G1 was not induced upon the same stress condition. Furthermore, it seems the G4 promoter responds only to root-knot nematodes, because it was not activated in plants infected by cyst nematode (Barcala et al. 2008). The data indicate primarily that the HSE elements arrangement in promoter region, and thus far it's setting in association with other elements, such as TATAbox and CAATbox, is an important influence on *Hsp*20 genes response under biotic stress. Moreover, this configuration appears to have influence on these genes' expression in different pathogens species.

Lopes-Caitar et al. (2013) performed the genome-wide analysis of the Hsp20 gene family in soybean and gene its expression analysis under abiotic and biotic stresses. The results indicated six soybean Hsp20 genes had some significant expression profile changes during nematode infection. GmHsp22.4, GmHsp17.9B, GmHsp17.9A and GmHsp17.4 genes were induced by *M. javanica* in the susceptible genotype and have been already described by Kandoth et al. (2011) as induced under cyst nematode infection. Two soybean Hsp20 genes, GmHsp22.4 and GmHsp17.6B, showed a divergent expression pattern between the examined resistant and susceptible genotypes under *M. javanica* infection. Furthermore, the promoter analysis showed that Hsp20 genes responsive to biotic stress follows the same cis element arrangement, especially for HSE, that have been described for other plant species under nematode infection.

Li et al. (2015b) performed a RNA-seq with three different *Gossypium hirsutum* phenotypes under 3, 9 and 12 days post *Rotylenchulus reniformis* infection. In this study, the results indicated a total of 52 HSP20 genes showed significant expression profile changes. Susceptible (DP90 & SG747) and hypersensitive (LONREN-1) genotypes showed the largest number of repressed Hsp20 pathway; 23 and 41 genes, respectively.

2.7 HSP20 May Protect Plant Cells During Viral Infection

The virus infectious cycle depends largely on the nature of their genetic material, which most often in plants are polo-positive viral RNA. Upon viral particle entry in the plant cell, the RNA genome is translated and the coating removed; both processes are highly coordinated. It seems a link among viral protein synthesis, some of these proteins assembly and the host factor to form a replication complex. After the virus moves to the neighboring cells, the infection is quickly spread throughout the plant (Pallas and Garcia 2011). A complete review of the top 10 plant viruses in molecular plant pathology and of economic importance was published recently (Scholthof et al. 2011; Rybicki 2015), while Mandadi and Scholthof (2013) presented the current state of knowledge of plant-virus interactions.

There are several studies that have described an association of plant HSP70 and HSP90 during infection by viruses (Jiang et al. 2014; Mine et al. 2012; Chen et al. 2008), and just few reports of plant small heat shock proteins involved in viral response. However, for the science community, it is clear that plant small heat shock proteins (HSP20s) have a fundamental role in protecting cells during viral infection, but their biological function remains unknown (Piotrowska et al. 2010).

Whitham et al. (2003) performed an expression profile analysis, by microarray technology, using Arabidopsis, under five different virus infection, *turnip vein-clearing tobamovirus* (TVCV), *oilseed rape tobamovirus* (ORMV), *potato virus X potexvirus* (PVX), *cucumber mosaic cucumovirus* (CMV), and *turnip mosaic potyvirus* (TuMV), at 1, 2, 4, and 5 days after inoculation (DAI). Two small heat shock proteins, HSP17.6A, and HSP17.4, were coordinately induced in response to two tobamoviruses, ORMV and TVCV, at 1 DAI. Both, genes were also lately induced in Arabidopsis during the infection by CMV, PVX and TuMV. The authors suggested that the coordinated induction of HSPs, plant defense proteins, suggests that might be a common mechanism that controls their expression in response to viral infection.

Similiarly, Senthil (2005) studied how SYNV (*Sonchus yellow net virus*) e INSV (*Impatiens necrotic spot virus*) could affect the *Solanum tuberosum* transcriptional profile. Applying the model plant *Nicotiana benthamiana*, they inoculated the virus in two different experiments and performed and hererologous microarray with potato expressed sequence tags (ESTs). Interestingly, a higher expression fold was identified for two small heat shock proteins, HSP 18 and HSP 20, at 4 dpi., with a decrease in expression at 5 d.p.i. during INSV infection. However, these genes were not induced by SYNV infection. According to the authors, the expression induction of the HSP20, during virus attack, is related to general plant response to stress.

Li et al. (2015a) investigated the interactions between plant HSP20 and Rice stripe virus (RSV) protein, RdRp, by yeast two-hybrid (Y2H). They found that the expression and sub-cellular localization of a host small heat shock protein was significantly affected by RSV infection. RSC is the causal agent of chlorotic stripes or mottling and necrotic streaks in the newly expanded leaves and growth is stunted; it is a devastating viral disease of rice in East Asia. Also Li et al. (2015a) showed the HSP20 features changes in the host that is a consequence of its interaction with the viral RdRp, a large protein that presents multiple functional motifs and domains. Apparently, the N-terminus (amino acid residues 1–296) of the RdRp was crucial for the interaction between the HSP20s and viral RdRp and responsible for the alteration of the sub-cellular localization and distribution pattern of HSP20s in protoplasts of rice and epidermal cells of *Nicotiana benthamiana.* This may suggest that interactions between HSP20 and the viral RdRp play an important role in viral replication. These findings are the first report that a plant virus or a viral protein alters the expression pattern or sub-cellular distribution of HSP20 and may be therefore a step forward in understanding a virus that causes a seriously damaging disease of one of the most important crop plants in the world.

2.8 Conclusion

Plant small heat shock proteins are currently associated with abiotic stress responses; however, it is clear they are involved in the response to most important phytopathogens, especially for fungi. Currently, it is still not clear how these special defense proteins act during plant biotic stresses, but all of the evidence suggests that HSP20s can be a promising target to plant breeding issues. In addition, this chapter is an incentive to update the researches related to plant-HSP20 and better elucidate their functions under biotic stress.

Acknowledgements The authors thank the support from Soybeans Embrapa and institutes State University of Londrina. We also thank many lab members and colleagues for critical and productive collaborations over many years.

References

Ahmed AA, Pedersen C, Schultz-Larsen T, Kawaaitaal M, Jørgensen HJL, Thrdal-Christensen H (2015) The barley powdery mildew candidate secreted effector protein csep0105 inhibits the chaperone activity of a small heat shock protein. Plant Physiol 168:321–333

Al-Whaibi MH (2011) Plant heat-shock proteins: a mini review. J King Saud Univ Sci 23:139–150

Bakthisaran R, Tangirala R, Rao CM (2015) Small heat shock proteins: role in cellular functions and pathology. Biochim Biophys Acta – Proteins Proteomics 1854:291–319

Barcala M, García A, Cubas P et al (2008) Distinct heat-shock element arrangements that mediate the heat shock, but not the late-embryogenesis induction of small heat-shock proteins, correlate with promoter activation in root-knot nematode feeding cells. Plant Mol Biol 66:151–164

Barnett T, Altschuler M, McDaniel CN, Mascarenhas JP (1980) Heat shock induced proteins in plant cells. Dev. Genet. 1, 331–340

Bondino HG, Valle EM, Ten Have A (2012) Evolution and functional diversification of the small heat shock protein/α-crystallin family in higher plants. Planta 235:1299–1313

Boter M, Amigues B, Peart J et al (2007) Structural and functional analysis of SGT1 reveals that its interaction with HSP90 is required for the accumulation of Rx, an R protein involved in plant immunity. Plant Cell 19:3791–3804

Bricchi I, Bertea CM, Occhipinti A, Paponov IA, Maffei ME (2012) Dynamics of membrane potential variation and gene expression induced by Spodoptera littoralis, Myzus persicae, and Pseudomonas syringae in Arabidopsis. PLoS One 7:e46673

Buchanan BB, Gruissem W, Jones RL (2000) Biochemistry and molecular biology of plants. Am Soc Plant Physiol 2:1280

Bull CT, Coutinho TA, Denny TP et al (2014) List of new names of plant pathogenic bacteria (2011–2012). J Plant Pathol 96:223–226

Caldwell BE, Brim CA, Ross JP (1960) Inheritance of resistance of soybeans to the cyst nematode. Heterodera glycines Agron J 1960(52):635–636

Cashikar AG, Duennwald M, Lindquist SL (2005) A chaperone pathway in protein disaggregation. J Biol Chem 280:23869–23875

Chen B, Zhong D, Monteiro A (2006) Comparative genomics and evolution of the HSP90 family of genes across all kingdoms of organism. BCM genomics 7:156

Chen Z, Zhou T, Wu X, Hong Y, Li H (2008) Influence of cytoplasmic heat shock protein 70 on viral infection of Nicotiana benthamiana. Mol Plant Pathol 9:809–817

Dean R, Van Kan JA, Pretorius ZA et al (2012) The top 10 fungal pathogens in molecular plant pathology. Mol Plant Pathol 13:414–430

Dodds PN, Rathjen JP (2010) Plant immunity: towards an integrated view of plant-pathogen interactions. Nat Rev Genet 11:539–548

Dudeja SS, Giri R (2014) Beneficial properties, colonization, establishment and molecular diversity of endophytic bacteria in legumes and non legumes. Afr J Microbiol Res 8:1562–1572

Dühring S, Germerodt S, Skerka C et al (2015) Host-pathogen interactions between the human innate immune system and Candida albicans – understanding and modeling defense and evasion strategies. Front Microbiol 6:625

Efeoğlu B (2009) Heat shock proteins and heat shock response in plants. Science 22:67–75

Epton HA (1989) Fungal infection of plants. Endeavour 13:93

Escobar C, Barcala M, Portillo M, Almoguera C, Jordano J, Fenoll C (2003) Induction of the Hahsp17.7G4 promoter by root-knot nematodes: involvement of heat-shock elements in promoter activity in giant cells. Mol Plant Microbe Interact 16:1062–1068

Evans SS, Repasky EA, Fisher DT (2015) Fever and the thermal regulation of immunity: the immune system feels the heat. Nat Rev Immunol 15:335–349

Foley RC, Gleason CA, Anderson JP, Hamann T, Singh K (2013) Genetic and genomic analysis of Rhizoctonia solani interactions with Arabidopsis; evidence of resistance mediated through NADPH oxidases. PLoS One 8:e56814

Fraire-Velázquez S, Rodríguez-Guerra R, Sánchez-Calderón L (2011) Abiotic and biotic stress response crosstalk in plants. InTech 346
Franck E, Madsen O, van Rheede T, Ricard G, Huynen MA, de Jong WW (2004) Evolutionary diversity of vertebrate small heat shock proteins. J Mol Evol 59:792–805
Fu X (2014) Chaperone function and mechanism of small heat-shock proteins. Acta Biochim Biophys Sin Shanghai 46:347–356
Fuganti R, Benevent MA, Silva JFV et al (2004) Identificação de marcadores moleculares de microssatélites para seleção de genótipos de soja resistentes a Meloidogyne javanica. Nematol Bras 28:125–130
Fuganti R, Machado M de FP da S, Lopes VS et al (2010) Size of AT(n) insertions in promoter region modulates Gmhsp17.6-L mRNA transcript levels. J Biomed Biotechnol 2010:1–9
Garofalo CG, Garavaglia BS, Dunger G, Gotting N, Orelhano EG, Ottado J (2009) Expression analysis of small heat shock proteins during compatible and incompatible plant–pathogen interactions. Adv Stud Biol 1:197–205
Guo M, Zhai Y, Lu J et al (2014) Characterization of CaHsp70-1, a pepper heat-shock protein gene in response to heat stress and some regulation exogenous substances in Capsicum annuum L. Int J Mol Sci 15(11):19741–19759
Guo M, Liu J-H, Lu J-P et al (2015) Genome-wide analysis of the CaHsp20 gene family in pepper: comprehensive sequence and expression profile analysis under heat stress. Front plant sci 6:806
Hartl FU, Bracher A, Hayer-Hartl M (2011) Molecular chaperones in protein folding and proteostasis. Nature 475:324–332
Heckathorn SA, Downs CA, Coleman JS (1999) Small heat shock proteins protect electron transport in chloroplasts and mitochondria during stress. Am Zool 39:865–876
Hüsselmann LHH (2014) Analysis of the early events in the interaction between *Venturia inaequalis* and the susceptible golden delicious apple (*Malus X Domestica* Borkh). University of the Western Cape, 278
Jiang S, Lu Y, Li K et al (2014) Heat shock protein 70 is necessary for Rice stripe virus infection in plants. Mol Plant Pathol 15:907–917
Jones JT, Haegeman A, Danchin EGJ et al (2013) Top 10 plant-parasitic nematodes in molecular plant pathology. Mol Plant Pathol 14:946–961
Kamoun S, Furzer O, Jones JDG et al (2015) The Top 10 oomycete pathogens in molecular plant pathology. Mol Plant Pathol 16:413–434
Kandoth P, Ithal N, Recknor J (2011) The soybean Rhg1 locus for resistance to the soybean cyst nematode Heterodera glycines regulates the expression of a large number of stress-and defense-related. Plant Physiol 155:1960–1975
Krishna P, Gloor G (2001) The Hsp90 family of proteins in Arabidopsis thaliana. Cell Stress Chaperones 6, 3:238–46
Li CT, Liao CT, Du SC, Hsiao YT, Lo HH, Hsiao YM (2014) Functional characterization and transcriptional analysis of galE gene encoding a UDP-galactose 4-epimerase in Xanthomonas campestris pv. Campestris. Microbiol Res 169:441–452
Li J, Xiang C, Yang J, Chen J-P, Zhang HM (2015a) Interaction of HSP20 with a viral RdRp changes its sub-cellular localization and distribution pattern in plants. Nat Publ Gr 1–15
Li R, Rashotte AM, Singh NK, Lawrence KS, Waver DB, Locy RD (2015b) Transcriptome analysis of cotton (Gossypium hirsutum L.) genotypes that are susceptible, resistant, and hypersensitive to reniform nematode (Rotylenchulus reniformis). PLoS One 10:e0143261
Lopes-Caitar VS, de Carvalho MCCG, Darben LM et al (2013) Genome-wide analysis of the Hsp20 gene family in soybean: comprehensive sequence, genomic organization and expression profile analysis under abiotic and biotic stresses. BMC Genomics 14:1–17
Maejima K, Oshima K, Namba S (2014) Exploring the phytoplasmas, plant pathogenic bacteria. J Gen Plant Pathol 80:210–221
Maimbo M, Ohnishi K, Hikichi Y, Yoshioka H, Kiba A (2007) Induction of a small heat shock protein and its functional roles in Nicotiana plants in the defense response against Ralstonia solanacearum. Plant Physiol 145:1588–1599

Mandadi KK, Scholthof K-BG (2013) Plant immune responses against viruses: how does a virus cause disease? Plant Cell 25:1489–1505
Mansfield J, Genin S, Magori S et al (2012) Top 10 plant pathogenic bacteria in molecular plant pathology. Mol Plant Pathol 13:614–629
Mendgen K, Hahn M (2002) Plant infection and the establishment of fungal biotrophy. Trends Plant Sci 7:352–356
Mine A, Hyodo K, Tajima Y et al (2012) Differential roles of Hsp70 and Hsp90 in the assembly of the replicase complex of a positive-strand RNA plant virus. J Virol 86:12091–12104
Ouyang Y, Chen J, Xie W, Wang L, Zhang Q (2009) Comprehensive sequence and expression profile analysis of Hsp20 gene family in rice. Plant Mol Biol 70, 3:341–57
Pallas V, Garcia JA (2011) How do plant viruses induce disease? Interactions and interference with host components. J Gen Virol 92:2691–2705
Pan X, Zhu B, Luo Y, Fu D (2013a) Unraveling the protein network of tomato fruit in response to necrotrophic phytopathogenic Rhizopus nigricans. PLoS One 8:e73034
Pan X-Q, Fu D-Q, Zhu B-Z, Lu C-W, Lou Y-B (2013b) Overexpression of the ethylene response factor SlERF1 gene enhances resistance of tomato fruit to Rhizopus nigricans. Postharvest Biol Technol 75:28–36
Pandey B, Kaur A, Gupta OP, Sharma I, Sharma P (2014) Identification of HSP20 gene family in wheat and barley and their differential expression profiling under heat stress. Appl Biochem Biotechnol 175:2427–2446
Pandey B, Kaur A, Gupta OP, Sharma I, Sharma P (2015) Identification of HSP20 gene family in wheat and barley and their differential expression profiling under heat stress. Appl Biochem Biotechnol 175:2427–2446
Park C-J, Seo Y-S (2015) Heat shock proteins: a review of the molecular chaperones for plant immunity. Plant Pathol 31:323–333
Pavlova EL, Rikhvanov EG, Tauson EL et al (2009) effect of salicylic acid on the development of induced thermotolerance and induction of heat shock protein synthesis in the arabidopsis thaliana cell culture. Russ J Plant Physiol 56:68–73
Piotrowska J, Hansen SJ, Park N, Jamka K, Sarnow P, Gustin KE (2010) Stable formation of compositionally unique stress granules in virus-infected cells. J Virol 84:3654–3665
Ratheesh Kumar R, Nagarajan NS, Arunraj SP et al (2012) HSPIR: a manually annotated heat shock protein information resource. Bioinformatics 28:2853–2855
Richter K, Haslbeck M, Buchner J (2010) The heat shock response: life on the verge of death. Mol Cell 40:253–266
Rybicki EP (2015) A Top Ten list for economically important plant viruses. Arch Virol 160:17–20
Sarkar NK, Kim Y-K, Grover A (2009) Rice sHsp genes: genomic organization and expression profiling under stress and development. BMC Genomics 10:393
Scharf K-D, Siddique M, Vierling E (2001) The expanding family of Arabidopsis thaliana small heat stress proteins and a new family of proteins containing α-crystallin domains (Acd proteins). Cell Stress Chaperones 6:225–237
Scholthof K-BG, Adkins S, Czosnek H et al (2011) Top 10 plant viruses in molecular plant pathology. Mol Plant Pathol 12:938–954
Senthil G (2005) Specific and common changes in Nicotiana benthamiana gene expression in response to infection by enveloped viruses. J Gen Virol 86:2615–2625
Septiningsih EM, Collard BCY, Heuer S, Bailey-Serres J, Ismail AM, Mackill DJ (2013) Applying genomics tools for breeding submergence tolerance in rice. Translational genomics for crop breeding: abiotic stress, yield and quality 2:9–30
Shanker AK, Venkateswarlu B (2011) Abiotic stress response in plants – physiological biochemical and genetic perspectives. InTech, 358
Shirasu K (2009) The HSP90-SGT1 chaperone complex for NLR immune sensors. Annu Rev Plant Biol 60:139–164

Siddique M, Gernhard S, Koskull-Döring P, Vierling E, Scharf K-D (2008) The plant sHSP superfamily: five new members in Arabidopsis thaliana with unexpected properties. Cell Stress Chaperones 13:183–197

Singh K (2002) Transcription factors in plant defense and stress responses. Curr Opin Plant Biol 5:430–436

Singla SL, Pareek A, Grover A (1998) Plant Hsp100 family with special reference to rice. J Biosci 23:337–345

Stavrinides J (2009) Origin and evolution of phytopathogenic bacteria. In: Plant pathogenic bacteria: genomics and molecular biology 4. Caister Academic, Norfolk, pp 1497–1497

Sun W, Van Montagu M, Verbruggen N (2002) Small heat shock proteins and stress tolerance in plants. Biochim Biophys Acta Gene Struct Expr 1577:1–9

Sung DY, Vierling E, Guy CL (2001) Comprehensive expression profile analysis of the Arabidopsis HSP70 gene family. Plant Physiol 126:789–800

Swindell WR, Huebner M, Weber AP (2007) Transcriptional profiling of Arabidopsis heat shock proteins and transcription factors reveals extensive overlap between heat and non-heat stress response pathways. BMC Genomics 8:125

Thomma BPHJ, Nurnberger T, Joosten MHAJ (2011) Of PAMPs and effectors: The blurred PTI-ETI dichotomy. Plant Cell 23:4–15

Tkáčová J, Angelovičová M (2012) Heat shock proteins (HSPs): a review. Sci Pap Anim Sci Biotechnol 45:349–353

Usta C (2013) Microorganisms in biological pest control – a review (bacterial toxin application and effect of environmental factors). Intech 288–317

Van Ooijen G, Lukasik E, Van Den Burg HA, Vossen JH, Cornelissen BJC, Takken FLW (2010) The small heat shock protein 20 RSI2 interacts with and is required for stability and function of tomato resistance protein I-2. Plant J 63:563–572

Vidhyasekaran P (2014) PAMP signals in plant innate immunity. PAMP Signals in Plant Innate Immun 21:442

Vurro M, Bonciani B, Vannacci G (2010) Emerging infectious diseases of crop plants in developing countries: impact on agriculture and socio-economic consequences. Food Secur 2:113–132

Wang W, Vinocur B, Shoseyov O, Altman A (2004) Role of plant heat-shock proteins and molecular chaperones in the abiotic stress response. Trends Plant Sci 9:244–252

Wang X, Jiang N, Liu J, Liu W, Wang G-L (2014) The role of effectors and host immunity in plant–necrotrophic fungal interactions. Virulence 5:722–732

Waters ER (1995) The molecular evolution of the small heat-shock proteins in plants. Genetics 141:785–795

Waters ER (2013) The evolution, function, structure, and expression of the plant sHSPs. J Exp Bot 64:391–403

Waters ER, Lee GJ, Vierling E (1996) Evolution, structure and function of the small heat shock proteins in plants. J Exp Bot 47:325–338

Waters ER, Aevermann BD, Sanders-Reed Z (2008) Comparative analysis of the small heat shock proteins in three angiosperm genomes identifies new subfamilies and reveals diverse evolutionary patterns. Cell Stress Chaperones 13:127–142

Whitham SA, Quan S, Chang HS et al (2003) Diverse RNA viruses elicit the expression of common sets of genes in susceptible Arabidopsis thaliana plants. Plant J 33:271–283

Yogendra KN, Kumar A, Sarkar K et al (2015) Transcription factor StWRKY1 regulates phenylpropanoid metabolites conferring late blight resistance in potato. J Exp Bot 66:7377–7389

Zhang J, Liu B, Li J et al (2015a) Hsf and Hsp gene families in Populus: genome-wide identification, organization and correlated expression during development and in stress responses. BCM genomics 16:1–19

Zhang K, Ezemaduka AN, Wang Z et al (2015b) A novel mechanism for small heat shock proteins to function as molecular chaperones. Sci Rep 5:8811

Chapter 3
Small Heat Shock Proteins, a Key Player in Grass Plant Thermotolerance

Kehua Wang, Xunzhong Zhang, and Erik H. Ervin

Abstract Small heat shock proteins (sHSPs) are by far the most complex group of HSPs that function to protect practically all cellular compartments in plants under stress due to their unusual abundance and diversity. Recent advances in proteomics, genomics, and other cellular and molecular techniques have facilitated the identification and characterization of sHSPs in higher plants, especially grass plants from the family of Poaceae that are economically important as crops and grasslands. Here we introduce the structure and function of plant sHSPs, and then summarize recent research progress on the role of sHSPs in grass tolerance to heat stress.

Keywords Grass plant • Heat stress • Small heat shock proteins • Thermo-tolerance

Abbreviations

ATP	adenosine triphosphate
ER	endoplasmic reticulum
GST	glutathione s-transferase
HPLC	high performance liquid chromatography
HSF	heat shock factor
LMW	low molecular weight
MS	mass spectrometry
PAGE	polyacrylamide gel electrophoresis
Q-TRAP	quadrupole linear ion trap

K. Wang (✉)
Department of Grassland Science, College of Animal Science and Technology, China Agricultural University, 2 Yuanmingyuan West Road, Beijing 100193, China
e-mail: kehwang@cau.edu.cn

X. Zhang • E.H. Ervin
Department of Crop & Soil Environmental Sciences, College of Agriculture and Life Sciences, Virginia Polytechnic Institute and State University, Blacksburg 24061, VA, USA

A.A.A. Asea et al. (eds.), *Heat Shock Proteins and Plants*,
Heat Shock Proteins 10, DOI 10.1007/978-3-319-46340-7_3

sHSPs	small heat shock proteins
2-DE	two-dimensional gel electrophoreses
UV-B	Ultraviolet B
ZR	zeatin riboside

3.1 Introduction

Heat shock proteins (HSPs) are a group of proteins that were first described in relation to heat shock in *Drosophila* cells over a half century ago (Ritossa 1962). They are now known to also be induced under many other stresses including drought, salinity, cold, UV light, and oxidation (Cho and Choi 2009; He et al. 2015; Liu et al. 2010; Rizhsky et al. 2002). Heat-shock proteins occur under non-stressful conditions as well, simply playing a role as a "monitor" for old proteins in cells (Haslbeck et al. 2015). HSPs are thought to be important in thermo-tolerance due to the following observations: (1) coincident and universal induction in response to temperature stress in all investigated organisms, and (2) fast and intensive biosynthesis (Wahid et al. 2007).

In higher plants heat stress redirects protein synthesis and results in a nearly exclusive synthesis of stress proteins, particularly heat shock proteins (Schulze et al. 2005). This increase in expression is transcriptionally regulated. The dramatic up-regulation of the HSPs is a key part of the heat shock response and is regulated primarily by heat shock factor (HSF) (Pirkkala et al. 2001; Wu 1995). HSPs have been assigned to 11 multi-protein families across different organisms based on a new eukaryotic/*Escherichia coli* nomenclature system, including HSP100/ClpB, HSP90/HtpG, HSP70/DnaK, HSP60/GroEL, HSP40/CbpA and small HSP (sHSP) proteins (Bharti and Nover 2002; Kotak et al. 2007; Wang et al. 2004). Among them, sHSPs are particularly interesting due to their high abundance and being the only ATP-independent HSPs among the 11 molecular chaperone families (Santhanagopalan et al. 2015). Our review will mainly focus on the current progress of plant sHSP research, particularly in grass species.

3.2 Definition of sHSPS

Small heat shock proteins (sHSPs) are one of the most ubiquitous HSP subgroups, whose molecular weights were originally recognized to be from 17 to 28 kDa and are now expanded from ~12 to 42 kDa (Santhanagopalan et al. 2015; Vierling 1991). Small heat shock proteins were first discovered as proteins that were up-regulated after a heat shock treatment together with several other HSPs, including HSP70s and HSP40s over 40 years ago (Tissiere et al. 1974). Several years later the first observations of sHSP synthesis in higher plants were reported for soybean (*Glycine max*) and tobacco (*Nicotiana tabacum*) tissue culture cells (Barnett et al. 1980). Soon afterwards sHSPs were reported in heat shocked maize (*Zea mays*) seedlings, by Baszczynski et al. (1982). Although not all prokaryotes, but nearly all eukaryotic organisms have two or more genes encoding sHSP. For example, there are two in yeast, 10 in humans, and over 19 genes in *Arabidopsis thaliana*. Even more have been reported in other higher plant species, such as the grasses (Kappe et al. 2003; Kriehuber et al. 2010; Ouyang et al. 2009; Pandey et al. 2015; Santhanagopalan et al. 2015). In wheat (*Triticum aestivum*), there are 50 known/predicted sHSP genes; rice (*Oryza sativa*) has been reported to have 23 (Waters et al. 2008) or 39 genes of sHSP (Ouyang et al. 2009).

The majority of sHSPs in plants are induced by heat stress and some are developmentally expressed, demonstrating their crucial role in the heat stress response of plants (Sun et al. 2002; Waters 2013). The most abundant plant sHSPs induced are class I cytoplasm sHSPs that can account for over 1 % of total leaf or root water soluble protein (Hsieh et al. 1992). Other sHSPs, such as the chloroplast-located sHSPs can amount to only about 0.02 % of total leaf protein (Chen et al. 1990). Studies using most recently developed quantitative proteomic techniques combined with electron cryo-tomography found sHsp concentration increasing from 0.06 μM in non-stimulated cells of *Leptospira interrogans* to 30 μM (500-fold) in heat-stressed cells (Beck et al. 2009). In plants, not only the protein of sHSPs were found to be numerous, but the corresponding mRNAs were also produced at very high levels during heat stress (nearly >20,000 copies per cell) (Santhanagopalan et al. 2015).

Higher plant sHSPs have evolved independently of animal and bacterial homologs and comprise multiple paralogous sHSPs in different cell compartments, including the nucleus, endoplasmic reticulum, peroxisomes, mitochondria, and chloroplasts (van Montfort et al. 2002; Waters et al. 1996). The cytosolic sHSPs of angiosperms are further grouped into five or more families (11 or 13 families) that originated hundreds of millions of years ago and show evidence of continued diversification (Lopes-Caitar et al. 2013; Ouyang et al. 2009; Pandey et al. 2015; Siddique et al. 2003, 2008; Waters et al. 2008; Waters and Vierling 1999b). Further, the chloroplast-localized sHSPs evolved when land-plants developed, through gene duplication from cytosolic sHSPs (Chen and Vierling 1991; Waters and Vierling 1999a). The chloroplast sHSP subfamily may need to be further expanded to two subfamilies as it was done for the mitochondria sHSP subfamily based on comparative analysis (Waters et al. 2008) (Fig. 3.1). The relationship of different families of plant sHSPs is illustrated for selected sHSPs from *Arabidopsis* and some grass plants from the Poaceae family (Fig. 3.2). The evolution of diverse land plant sHSPs has been generally thought to be driven by plants' sessile nature as they are incapable of escaping environmental stresses that are potentially counteracted by the function of sHSPs (Eisenhardt 2013; Waters 2013). The abundance and diversity of plant sHSPs has made their function and evolution of more interest (Bondino et al. 2012).

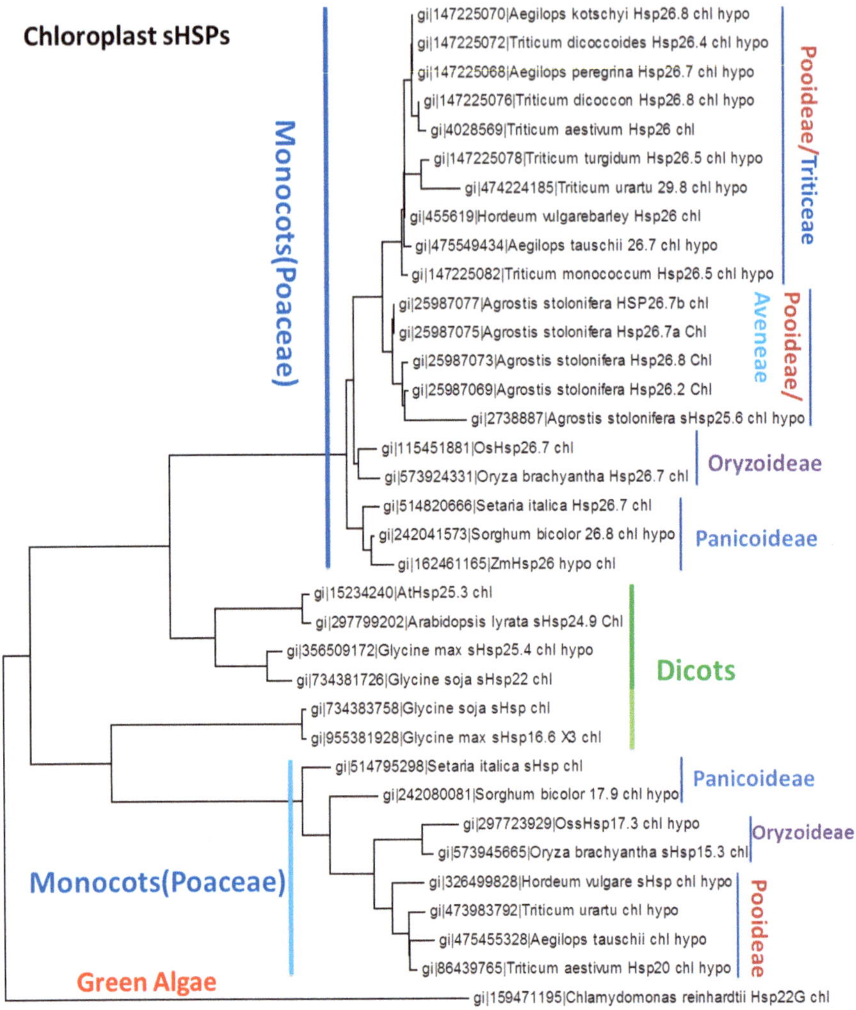

Fig. 3.1 Phylogenetic analysis of chloroplast sHSPs of monocots (grass species) with dicots (Arabidopsis, Glycine max, and Glycine soja). The tree was derived by Neighbor- joining method with MEGA 5 from alignment of amino acid sequences of chloroplast sHSPs/hypothetic chloroplast sHSPs using ClustalW (Tamura et al. 2011; Thompson et al. 2002)

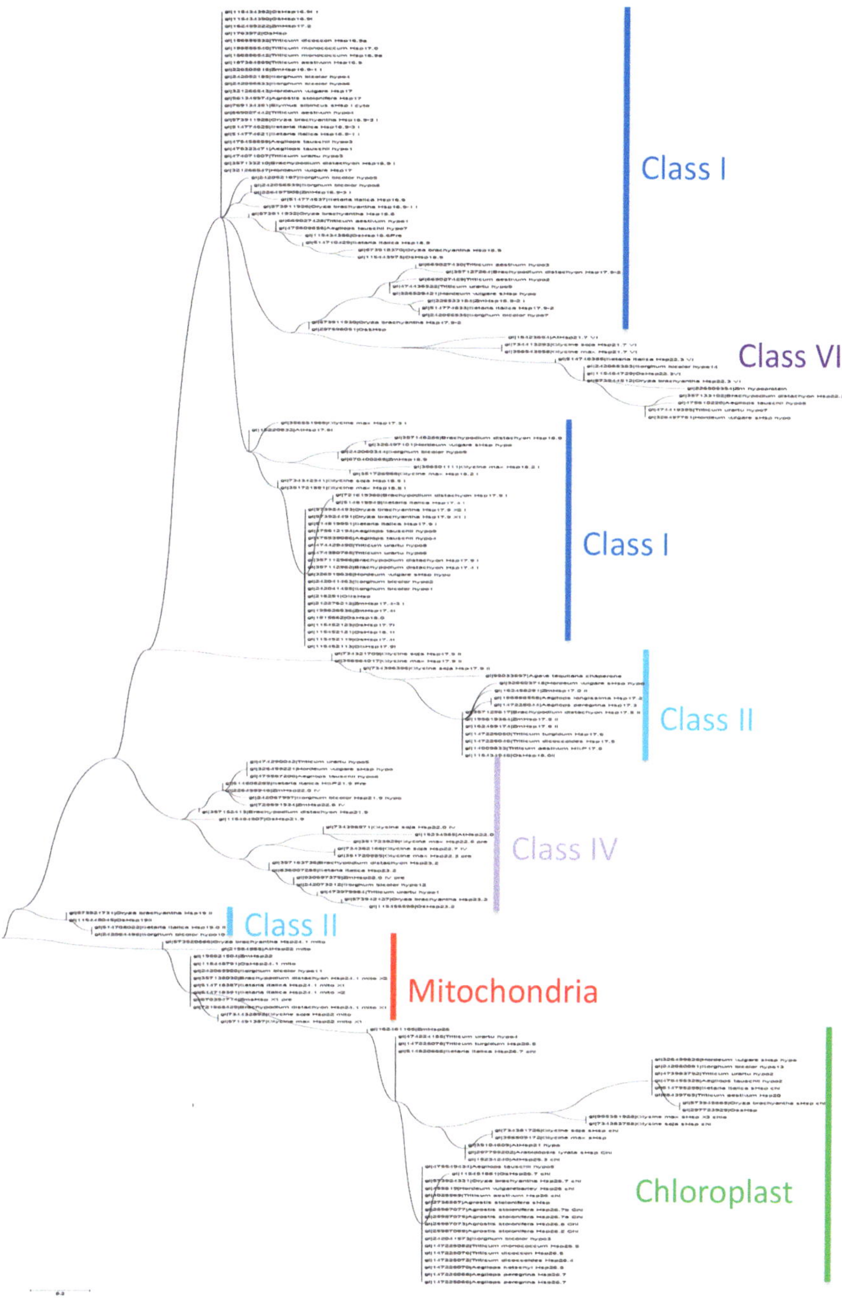

Fig. 3.2 Phylogenetic relationship of sHSPs of different grass species with Arabidopsis, Glycine max, and Glycine soja. The tree was derived by Neighbor- joining method with MEGA 5 from alignment of amino acid sequences of sHSPs/predicted sHSPs using ClustalW (Tamura et al. 2011; Thompson et al. 2002)

3.3 Structure, Function and Regulation of SHSP

The common characteristic of all sHSPs is the presence of three functionally important regions, a conserved α-crystallin domain of 90–100 amino acids (Dejong et al. 1993, 1998; Waters et al. 1996), an evolutionarily divergent N-terminal region and a short flexible C-terminal region (Carver and Lindner 1998; Carver et al. 1997; Cheng et al. 2008; Kim et al. 1998; van Montfort et al. 2002) (Fig. 3.3a). The α-crystallin domain is the most representative motif of sHSPs, but multiple sequence alignments of sHSPs of different organisms revealed very few consensus residues. Only the A-x-x-x-n-G-v-L motif at the end of the α-crystallin domain is conserved, and it is also the most significant indicator of the domain (Haslbeck et al. 2015; Lentze et al. 2003; Studer et al. 2002) (Fig. 3.3a). The length and structure of the α-crystallin domain, rather than the sequence, are well conserved with an average of 94 amino acids forming a compact β-sheet sandwich (Kriehuber et al. 2010). The N-terminal part preceding the α-crystallin domain is highly variable and even more poorly conserved compared to the α-crystallin domain. And the C-terminal is a short extension following the α-crystallin domain (Sun and MacRae 2005). For example, the entire extension is only 14–21 (average 15) amino acids in Ta16.9 and other plant class I proteins. A majority of all sHSPs contains a notably conserved I/V/L-x-I/V/L motif (IxI motif) in the C-terminal extension (de Jong et al. 1998) (Fig. 3.3a, b).

The α-crystallin domain is necessary for assembling the basic building block of dimer formation, but the α-crystallin domain by itself is not sufficient for the formation of the higher-order structures or oligomers (Bagneris et al. 2009; Baranova et al. 2011; Clark 2016; Clark et al. 2011; Ghosh et al. 2006; Hilton et al. 2013a; Shi

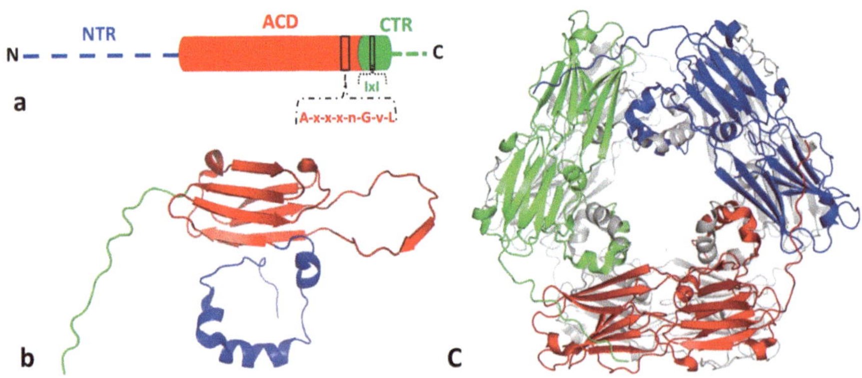

Fig. 3.3 (**a**) Domain organization of sHsps. N-terminal domain (NTR) (blue), α-crystallin domain (ACD) (*red*), C-terminal extension (CTR) (*green*). The motifs common to all sHsps are indicated. (**b**) Monomer structure of wheat HSP16.9. The domains are color coded according to the domain organization shown in (**a**). (**c**) The crystal structure of wheat HSP16.9 reported by van Montfort et al. (2001). (PDB ID: 1GME). Wheat HSP16.9 is a homo-dodecameric protein consisting of two disk-like layers. The three HSP16.9 dimers in the upper disk are colored in *green*, *blue* and *red*, respectively. The lower disk is colored in *gray*

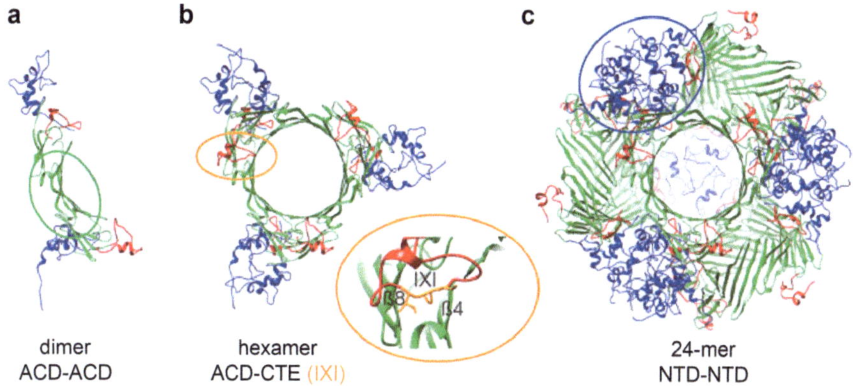

Fig. 3.4 The hierarchy of the human αB-crystallin 24-meric assembly (Braun et al. 2011) (PDB: 2YGD). (**a**) The dimer assembled through the interactions between the ß -strands located within the ACDs of the adjacent monomers. (**b**) The sub-assembly of a hexamer formed by inter-dimer interactions through the binding of C-terminal IXI motifs to the hydrophobic groove of the adjacent monomer (*inset*). (**c**) 24-mer assembled from hexameric blocks through contacts within N-terminal regions (Haslbeck et al. 2015)

et al. 2013; Weeks et al. 2014). Both the N-terminal and C-terminal flanking regions facilitate the oligomer formation. The N-terminal arms are major substrate binding domains, and contacts within N-terminal regions are important for the higher order oligomers (Fu et al. 2005; McDonald et al. 2012; Mani et al. 2015; McHaourab et al. 2011; Santhanagopalan et al. 2015; Takeda et al. 2011; Tiroli and Ramos 2007). In addition, IxI motif in the C-terminal extension makes an important contact linking sHSP dimers into higher order oligomers (Haslbeck et al. 2015). Other than that, the C-terminal extension can also adopt different angles in relation to the α-crystallin domain, which facilitates the assembly of oligomers with different geometries (Hanazono et al. 2013; van Montfort et al. 2001). In summary, the three parts contribute to the oligomer assembly process in a hierarchical way (Fig. 3.4, adopted from Haslbeck et al. 2015).

The ability to assemble into large oligomers of 12–40 subunits (Hanazono et al. 2013; Kim et al. 1998; van Montfort et al. 2001) is the key to the function of many sHSPs, and also the most striking feature; although a few sHSPs are present as small oligomers of 2–4 subunits (Basha et al. 2013; Garrido et al. 2012; Kokke et al. 1998) (Fig. 3.3b, c). The canonic sHSPs function as a buffer system to bind unfolding proteins, forming stable complexes upon stress and to protect them from irreversible aggregation (Fig. 3.5, adopted from Santhanagopalan et al. 2015). In vitro experiments revealed that the non-native protein trapped in sHSP/substrate complexes remain folding-competent (Buchner et al. 1998; Lee et al. 1997) for subsequent refolding in cooperation with ATP-dependent chaperones, such as HSP70s and HSP100s (Cashikar et al. 2005; Goloubinoff et al. 1999; Mogk et al. 2003; Veinger et al. 1998) (Fig. 3.4).

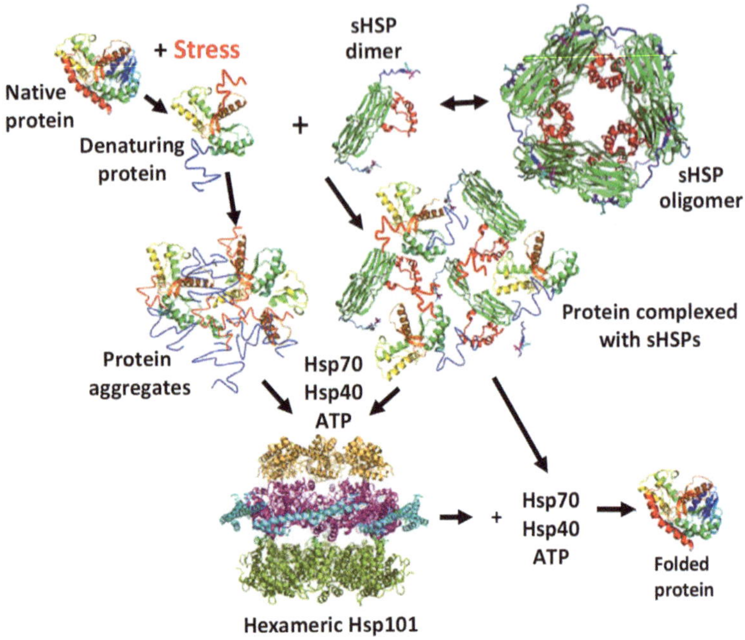

Fig. 3.5 Mechanism of the chaperone function of sHSPs. Under heat stress sHSPs bind non-native/partially unfolded proteins in an energy-independent way and keep them in a folding-competent state in sHSPs/substrate complexes. Subsequently, the substrates are refolded by downstream ATP-dependent chaperone systems, such as HSP70/40, and HSP100 (Santhanagopalan et al. 2015)

The current model for sHSP chaperone activity was defined based on studies of plant class I sHSPs of wheat Ta16.9 and pea (*Pisum sativum*) homologous protein Ps18.1. Both Ta16.9 and Ps18.1 are dodecamers in solution at room temperature, and most likely other plant class I sHSPs as well, such as tobacco, and the perennial grass, sugarcane (*Saccharum* spp.) (Basha et al. 2010; Tiroli-Cepeda and Ramos 2010; Tiroli and Ramos 2007; Yoon et al. 2005). It has been suggested that sHSP oligomers act as reservoirs of the active dimeric units of the chaperone (Hilton et al. 2013b; Stengel et al. 2010; van Montfort et al. 2001). Generally, there are four different regulatory stimuli (1) the presence of unfolded or partially folded substrates; (2) changes in the environmental temperature or stresses; (3) phosphorylation or more general post- translational modifications and, (4) the formation of hetero-oligomers, which affect the association/dissociation equilibrium of sHSPs and lead to their activation (Haslbeck et al. 2015). Stress can activate the sHSPs by shifting the equilibrium to the dimeric form, which can then bind partially unfolded/misfolded proteins. This activation mechanism appears to be largely dependent on the ratio of sHSP to substrate and the Hsp70/Hsp40 system alone is effective in refolding the substrate proteins only if sHSPs are present at proper or excess concentrations, where soluble and well defined sHSP/substrate complexes form. At excess levels of the substrate protein, sHSPs become incorporated into aggregates of the substrate protein. For the refolding of substrates from these aggregate-like sHSP/substrate complexes, a mechanism involving Hsp70/Hsp40 and members of the Hsp100 family is necessary (Haslbeck et al. 2005; Lee and Vierling 2000; Mogk et al. 2003; Patel et al. 2014). This process requires neither binding nor hydrolysis of ATP, indicating that the energy required for the sHSPs to perform any conformational changes necessary for their function comes from the ambient environment (McHaourab et al. 2009). This is different from the mechanism of transferring energy used by the majority of the molecular chaperones, and reveals that the sHSPs populate a relatively shallow free-energy surface (Papoian 2008; Zheng et al. 2012). Recently a super-transformer model illustrating both the dynamic structures of sHSPs and the variable sHSP-substrate interactions was proposed. In this model, each sHSP monomer, like a transformer, is able to adapt to different conformations, and easily assembles into various transformable and dynamic oligomers (super-transformer) (Fu 2015). In summary, sHSPs are an excellent illustration of the idea that protein dynamics are able to regulate and control protein function. This mechanism seems to be conserved from bacteria to the eukaryotes.

In summary, research on function and regulation of sHSPs has been advanced considerably during the past ten to 15 years along with successful resolution of the representative eukaryotic sHSP crystal structure of wheat Hsp16.9 (Ta16.9) and *Schizosaccharomyces pombe* Hsp16.0. However, more high resolution, and especially complete full length sHSP structures are needed, particularly for higher plants with independently evolved sHSPs from metazoan and bacterial homologs. In addition, exactly how sHSPs function and what they protect remains undetermined.

3.4 The Grass Family and the Mechanisms for Heat Tolerance

The grass family Poaceae, which is comprised of more 10,000 species, is the single most important plant family in agriculture. It includes annual species cultivated as major grain crops, such as maize, rice, wheat, barley (*Hordeum vulgare*), and oat (*Avena sativa*), along with other cereals accounting for a major portion of food consumed worldwide, and perennial species cultivated as forage grasses for livestock, turfgrasses for lawns, parks, sports fields, and golf courses, and cash crop for bioenergy, sugar production. In addition, it also surpasses all other flowering plant families in ecological importance as grasses cover >20 % of the earth's land surface and often dominate temperate and tropical habitats (Gaut 2002).

Grasses are classified into two categories (warm-season and cool-season) based on the ranges of temperature and precipitation at which they are adapted. Cool-season and warm-season grasses have distinctive photosynthetic pathways, known as C3 and C4, respectively. Cool-season or C3 grass species have an optimum growth temperature in the range from 18–24 °C while C4 or warm-season grass species grow the best between 30 and 35 °C (Turgeon 2011). High temperature is particularly detrimental to cool-season grass species, particularly in transition zone and semi-tropical climate regions. Temperature rises of 5 °C or more above optimal can cause alteration of metabolism or even damages to plant function and development, resulting in growth reductions and yield loss (Kotak et al. 2007; Schulze et al. 2005; Wahid et al. 2007). The predicted 2–5 °C higher global temperature due to global warming in the twenty-first century, and heat waves as the early warning signs of global warming happening globally, are one of the most serious challenges facing agriculture worldwide (Porter 2005). Better understanding the mechanism of grass species thermo-tolerance and improving grass species tolerance to high temperature becomes a more urgent topic for the academies globally.

Similar to other land plants, grasses may survive heat stress through heat-avoidance or heat-tolerance mechanisms (Fry and Huang 2004; Schulze et al. 2005; Wahid et al. 2007). Heat avoidance is the ability of plants to maintain internal temperatures below lethal stress levels; plant tolerance to high temperature may be achieved through various mechanisms, including changes at the molecular, cellular, biochemical, physiological, and whole-plant levels (Bita and Gerats 2013; Kotak et al. 2007). Heat stress redirects protein synthesis in higher plants, including degradation of proteins, inhibition of protein accumulation, and induction of certain groups of protein synthesis, particularly heat shock proteins (Krishna 2004; Monjardino et al. 2005; Schulze et al. 2005; Vierling 1991). Small HSPs are of particular importance as they are among the most dominant proteins induced in higher plants upon heat stress (Sun et al. 2002).

3.5 Identification and Characterization of SHSPS Associated with Heat Tolerance in Grasses

The involvement and role of sHSPs in thermo-tolerance has been mainly studied in various annual grasses cultivated as grain crops, including wheat, rye (*Secale cereale*), barley, and oat belonging to the Pooideae subfamily (all C3 grass species), and maize, and sorghum (*Sorghum bicolor*) belonging to the Panicoideae subfamily (many are C4 grass species). For perennial grass species, the function and role of sHSPs in thermo-tolerance has been only examined in a few, mainly cool season grasses, such as creeping bentgrass (*Agrostis stolonifera*), fescues (*Festuca* spp.), and perennial ryegrass (*Lolium perenne*). Several studies have been done on sugarcane, a warm-season perennial grass. Here we treat it as an annual grass since it is cultivated as an annual crop. Below we will give a more detailed review on how these sHSPs are regulated in the adaptive and tolerant mechanisms of annual crops, perennial forage and turfgrasses, and other grass plants under heat stress.

3.5.1 SHSPS Identified in Annual Species Cultivated as Grain Crops

In the early 1980s, expression of an 18 kDa sHSP was first observed in a study examining gene expression of 5-d plumules of corn by heat shock (Baszczynski et al. 1982). A year later, the same group reported similar findings *in vitro* by isolating and expressing the corn mRNA in both the rabbit reticulocyte and the wheat germ systems (Baszczynski et al. 1983). Since then, more sHSPs have been identified by various studies on different annual grass plants using one-dimensional protein electrophoresis and other methods (Cooper and Ho 1983; Craig 1985; Krishnan et al. 1989; Necchi et al. 1987). For example, Necchi et al. (1987) observed inductions of several low molecular HSPs (14–17 kDa) along with other HSPs in five cereal species (common wheat, drurum wheat (*T. durum*), barley, rye, and triticale (x *Triticosecale*)) responding to heat shock at 40 °C for four to 12 h. A subsequent study reported that distinct levels of acquired thermo-tolerance between wheat varieties were associated with significant quantitative differences in the synthesis of multiple HSPs, particularly sHSPs (Krishnan et al. 1989). In addition, in response to heat shock, maize seedlings have four more mitochondrial LMW-HSPs expressed compared to seedlings of wheat and rye, which was suggested to be the possible reason for higher heat tolerance in maize than in wheat and rye (Korotaeva et al. 2001).

Since the 1990s, genes of sHSPs from some crop species, such as maize, barley, and rice began to be cloned and their characteristics and function were further analyzed (Boston et al. 1996; Chen and Vierling 1991; Dietrich et al. 1991; Frappier et al. 1998; Goping et al. 1991; Lee et al. 2000; Ham et al. 2013; Klein et al. 2014; Marmiroli et al. 1993; Yeh et al. 1995). For example, a rice class I 16.9-kDa HSP gene, pTS1, was cloned and it was shown that the recombinant rice 16.9-kDa sHSP can provide thermo-protection *in vitro*. Moreover, the 16.9-kDa sHSP, obtained after a cleavage of the recombinant portion of the glutathione S-transferase (GST) protein, was also shown to form a protein complex of approximately 310 kDa under native conditions as the native class I sHSPs from heat-shocked rice seedlings can, indicating that the formation of the oligomers are also important for *in vitro* thermo-protection. Also in rice, an Oshsp26 gene, encoding a chloroplast-localized sHSP, was isolated. The Oshsp26 gene was expressed following heat stress, and also following oxidative stress even in the absence of heat stress, suggesting an important role in protecting the chloroplast against damage caused by oxidative stress as well as by heat stress (Lee et al. 2000). One of the most important breakthroughs of sHSP research is the resolution of the crystallographic structure of a wheat class I cytosolic sHSP, TaHsp16.9-CI (wHSP16.9, PDB Id: 1GME) (van Montfort et al. 2001). Years later, the same group further studied the chaperone activity of wheat TaHsp16.9-CI, along with the related wheat protein TaHsp17.8-CII, a class II plant cytosolic sHSP. The data indicate that sHSPs show substrate specificity and suggest that N-terminal residues contribute to substrate interactions, which may reflect the evolutionary distinction between the two classes, class I and class II (Basha et al.

2004). Similarly, the recombinant proteins of two sugarcane cytosolic class I sHSPs, SsHsp17.2 and SsHsp17.9, were used to characterize their function under high temperature *in vitro*. Data showed that the stability and their chaperone activity of the two sHSPs were not equivalent, indicating a substrate specificity involved in different sites at the N-terminus. In addition, high temperature affected the oligomeric state of the two sHSPs, resulting in enhanced chaperone activity by dissociating dodecamers to dimers (Cagliari et al. 2005; Tiroli-Cepeda and Ramos 2010; Tiroli and Ramos 2007). Based on the available structure data and other results, the standard model of how sHSPs function was proposed (see part 4.3 for details) (Haslbeck et al. 2005; Lee and Vierling 2000; Mogk et al. 2003).

More recently advances in "omics" technologies have provided new opportunities for the identification and characterization of heat-responsive proteins including sHSPs (Lee et al. 2007; Majoul et al. 2004; Skylas et al. 2001; Wang et al. 2015). Using classical two-dimensional gel electrophoresis (2-DE) coupled with mass spectrometry (MS), Thouraya et al. (2004) detected up-regulation of 5 sHSPs in a study analyzing the effect of heat stress on wheat grain proteome. Lee et al. (2007) identified 48 proteins in a study investigating rice leaf proteome in response to heat stress, including seven newly induced or highly up-regulated sHSPs, with one of the sHSPs being further confirmed by western blot as a mitochondrial sHSP.

Other than the studies that simply observed/identified sHSPs under heat stress, studies also found positive correlation between sHSPs expression and plant thermo-tolerance. Skylas et al. (2001)) observed a sHSP16.9 showing an increase in protein abundance in mature wheat grain of a heat tolerant wheat cultivar. The authors determined this protein as a possible marker protein for heat tolerance. Similarly, in a study of sHSPs in rice cultivars of different heat stress tolerance, Chen et al. (2014) found five sHSPs were up-regulated, with all being significantly higher in the heat-tolerant rice cultivar "Co39" than those in the heat-sensitive cultivar "Azucena". The results indicated that the expression levels of these five sHSPs were positively related to the ability of rice plants to tolerate heat stress. Further, the assembled large hetero-oligomeric complex by some of these five sHSPs was thought to be a key step for rice to tolerate heat stress.

In maize, the chloroplastic sHSP26 was found to be able to interact strongly with specific chloroplast proteins and to stabilize their abundance under heat stress, including ATP synthase subunit beta, chlorophyll a/b binding protein, oxygen-evolving enhancer protein 1 and photosystem I reaction center subunit IV, which improve chloroplast performance and in turn enhance plant thermo-tolerance (Hu et al. 2015b). Sule et al. (2004) were able to distinguish six isoforms of a 16.9 kDa sHSP in one single spot of a gel in a proteomic study of barley heat response by using an electrospray ionization-quadrupole linear ion trap (Q-TRAP) spectrometer combined with nano-HPLC. Interestingly, different isoforms of the same sHSP seems to play different roles in heat tolerance because they were present in a different pattern in the heat tolerant and sensitive cultivars. An in-depth analysis of those proteins and their isoforms is warranted in order to better understand their specific role in relation to plant heat-tolerance and susceptibility.

Genome-wide or transcriptome-wide data analysis proves to be a powerful way to identify genes, including sHSPs. Guan et al. (2004) characterized the genomic structures of two chromosomes in rice and identified nine cytosolic class I small heat shock protein genes, namely Oshsp16.9A, Oshsp16.9B, Oshsp16.9C, Oshsp16.9D, Oshsp17.9B, Oshsp17.3, Oshsp17.7, Oshsp17.9A and Oshsp18.0, which were further confirmed by 2-DE combined with MS. This work showed that all sHSP-CI genes except Oshsp17.9B were induced dramatically after heat shock for 2 h. A later comparative analysis of the sHSPs in three angiosperm genomes (*Arabidopsis thaliana, Populus strichocarpa, Oryza sativa*) identified a total of 23 rice sHSPs, including eight cytosolic class I, two for class II and mitochondria, one each for class III, IV, V, peroxisome, chloroplast, and endoplasmic reticulum (ER), and five "orphan" ones without close homologues in dicots (Waters et al. 2008). Similarly, Sarkar et al. (2009) identified 23 rice sHSPs through database search and phylogenetic analysis. However, they were categorized slightly differently as 16 cytoplasmic/nuclear, one plastid one peroxisomal, two ER, and three mitochondrial genes. Three of the 16 nuclear/cytoplasmic *sHSP* genes were found only in monocots. In addition, expression analysis based on microarray and RT-PCR showed that 19 *sHSP* genes were up-regulated by high temperature stress, and two of them, *Hsp16.6-CVIII and Hsp17.9B-CIX* were monocot specific. Interestingly, another study (Ouyang et al. 2009) on comprehensive sequence and expression profile analysis of the Hsp20 gene family in rice reported finding a total of 39 OsHsp20 genes, with the majority of the genes up-regulated at temperatures of 35 and 42 °C. In wheat, other than the previously described 14 sHSPs, another 27 new candidate HSP20 family genes were identified; and 13 in barley were identified as well. The expression analysis indicated an up-regulation of all the transcripts under heat stress condition except TaCBM38894 candidate (Pandey et al. 2015). Further data from experiments instead of digital analysis will be necessary to show the detailed role of each individual sHSP, and also to clarify the exact number of sHSPs in rice, and probably other grass plants.

In addition to identifying the sHSPs in the common annual crop grass species, a few studies have further tested the role of sHSPs in thermo-tolerance using transgenic method, which is also an effective strategy of breeding heat-tolerant annual crop and perennial grasses. For example, overexpression of both a rice chloroplast sHSP (*Oshsp26*) gene and a cytosolic class I sHSP (*Os16.9*) conferred better tolerance to heat stress in *Escherichia coli*, and the chloroplast sHSP (*Oshsp26*) gene also conferred better tolerance to oxidative stress in *E. coli* (Lee et al. 2000; Yeh et al. 1997, 2002). Overexpression of the rice chloroplast-localized *OsHSP26* gene enhanced tolerance against oxidative and heat stresses as well in tall fescue (*Festuca arundinacea*) (Kim et al. 2012). A later study of overexpression of rice *sHSP17.7* found both the transformed *E. coli* and the transgenic rice plants had better tolerance to heat and UV-B stresses (Murakami et al. 2004). In addition to rice sHSPs, constitutive expression of sHSP genes from wheat and maize also improved plant heat resistance. Chauhan et al. (2012) found wheat chloroplastic sHSP (*HSP26*) was highly inducible by heat stress. Transgenic Arabidopsis plants were substantially tolerant under continuous high temperature regimen than wild-type plants.

Correspondingly, antisense Arabidopsis plants showed negligible tolerance even for non-lethal heat shock (Chauhan et al. 2012). Similarly, overexpression of a maize cytosolic class I small heat shock protein *ZmHSP16.9* enhanced tolerance of tobacco plants to heat and oxidative stress (Sun et al. 2012).

3.6 SHSPS Identified in Perennial Species Cultivated as Forage or Turf Grasses

Because it is more of a problem for cool-season grasses than warm-season ones, most heat stress related researches have been done on cool-season grass species, especially the major species used for golf courses, such as creeping bentgrass. In this species, a LMW-HSP (HSP25) was reported to be genetically involved in heat tolerance (Luthe et al. 2000; Park et al. 1996, 1997). The researchers found the heat-tolerant variants synthesized two to three additional sHSPs (25 kDa) than the heat-sensitive variants. A subsequent study from this group showed that heat sensitivity was associated with reduced capacity of the susceptible bentgrass variants to accumulate a chloroplastic sHSP (ApHsp26.2) under heat stress, which is due to a point mutation that generated a premature stop codon (Fig. 3.4) (Wang and Luthe 2003). In order to better understand the roles of HSPs in heat tolerance of perennial grass species, rough bentgrass (*Agrostis scabra*) identified in Yellowstone National Park, USA, has been investigated in comparison with creeping bentgrass. Rough bentgrass is a unique C3 perennial grass species growing actively in hot soils associated with geothermal activity, which may have evolved with different heat resistant strategies (Stout and Al-Niemi 2002; Tercek et al. 2003). Heat-induced changes in one-dimensional protein profiles of rough bentgrass were compared to those of creeping bentgrass "L-93" (heat tolerant cultivar) and "Penncross" (heat sensitive cultivar). In the shoots, expression of three sHPS and a 66 kDa HSP were induced or enhanced at 35–45 °C in "L-93" and rough bentgrass, but only at 40–45 °C in "Penncross". The results indicated the better heat tolerance of "L-93" and rough bentgrass may be related to an earlier induction of the related sHSPs as well as maintenance of them under elevated temperature (Xu et al. 2011). The findings were further confirmed by several studies at the gene level (Tian et al. 2009; Xu et al. 2011). Tian et al. (2009) identified a heat-responsive gene of HSP20-like chaperone in rough bentgrass using a suppression subtractive hybridization (SSH) library. The HSP20-like chaperone is highly homologous to an HSP20-like chaperone from clover (*Medicago truncatula*) that contains the p23 domain. P23 is one of the co-chaperones of HSP90 and stabilizes the HSP90 hetero-complex (Cha et al. 2009). The data here suggested that the p23 sHSP along with HSP90 could be important for heat tolerance in perennial grasses. In another study, the expression levels of a HSP70 gene and a sHSP (*HSP16*) gene were compared between heat-sensitive creeping bentgrass and thermo-tolerant rough bentgrass (Xu et al. 2011). The expression of HSP16 was highly induced in both species at 45 °C after 24 h, and the abundance

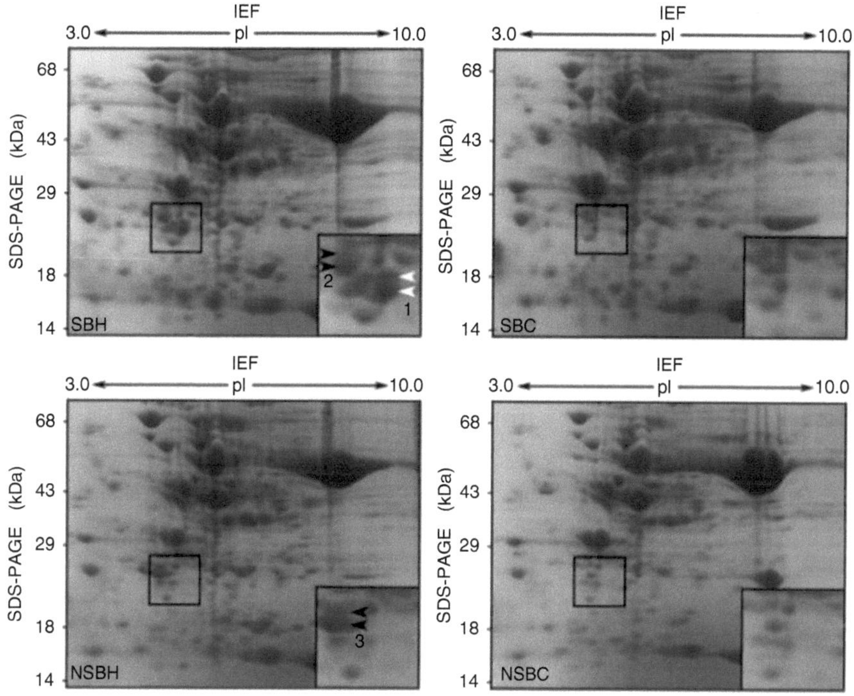

Fig. 3.6 Two-dimensional gel electrophoresis of total leaf protein from heat-shocked (37 °C for 12 h) SB (tolerant variant) (SBH), heat-shocked NSB (sensitive variant) (NSBH), control SB (SBC), and control NSB (NSBC). *Black arrows*, CP-sHSPs isoforms present in both SB and NSB. *White arrows*, CP-sHSP isoforms present only in SB (Wang and Luthe 2003)

was more substantial in the thermal species; whereas HSP70 gene was constitutively expressed at optimum temperature with only slight up-regulation at elevated temperatures in both species. Most likely the HSP16 plays a more important role in thermo-tolerance in the two perennial grasses than HSP70 does (Xu et al. 2011) (Fig. 3.6).

It is well known that heat acclimation can enhance plant resistance to more severe heat stress later on, and it has also been found to induce HSPs in various plant species (Kotak et al. 2007; Vierling 1991). A study examining the effects of heat acclimation and sudden heat stress on protein changes in creeping bentgrass "Penncross" found that both heat treatments led to the accumulation of HSPs, but heat acclimation induced more HSPs than sudden heat, including a 23 kDa sHSP. These new HSPs during heat acclimation might be associated with the enhanced thermo-tolerance of creeping bentgrass after heat acclimation, but direct correlation of these two factors is yet to be determined (He et al. 2005).

Both cytokinins and N have been also reported to enhance plant performance under high temperature, including creeping bentgrass (Liu and Huang 2002; Liu et al. 2002; Wang et al. 2012, 2013). Veerasamy et al. (2007) further investigated the

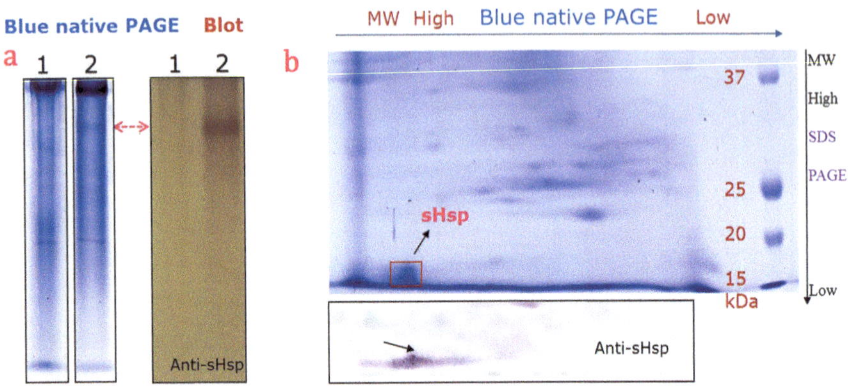

Fig. 3.7 Blue native PAGE (Polyacrylamide gel electrophoresis) and immunoblots of sHSPs in creeping bentgrass under heat stress (28/38 °C, d/n) for 1 day. (**a**) Native one-dimensional electrophoresis gel and immunoblots of sHSPs; (**b**) Two-dimensional electrophoresis gel and immunoblots of sHSPs. Equal amounts of shoot total protein (300 μg) from creeping bentgrass under optimum temperature (1) and high temperature (2) were loaded in each lane

effects of exogenous applied zeatin riboside (ZR), a synthetic cytokinin, on protein metabolism associated with heat tolerance in creeping bentgrass "Penncross". The up-regulated expression levels of a few HSPs, including a 32 kDa sHSP were more prominent in ZR-treated plants than in the control plants under heat stress. The results suggest that some HSPs including the sHSP are among the primary targets in cytokinin regulation of heat tolerance in cool-season perennial grass species. In a study on the role of HSPs in heat stress tolerance of creeping bentgrass at different N levels, Wang et al. (2014) found all the investigated HSPs (HSP101, HSP90, HSP70, and sHSPs) were up-regulated by heat stress, and their expression patterns indicated cooperation between different HSPs and their roles in bentgrass thermo-tolerance. Interestingly, the amount of sHSP kept relatively stable during a period of 7-week heat stress, which may indicate their important role as "paramedics" of the cell in monitoring denatured proteins for plants under stress. Moreover, the sHSPs identified by immunoblot were found to form an oligomer as expected by blue native gel. Further mass spectrometry identified at least three sHSPs (sHSP18.0, sHSP16.8a, and sHSP16.8b) were in the oligomer, suggesting a hetero-oligomer formation during the dynamic heat stress response (Fig. 3.7).

In other forage and turfgrass species, such as fescues, Hu et al. (2014) observed an activation of a low molecular weight HSPs (LMW-HSP, HSP20) in both high-temperature stressed genotypes of tall fescue through a RNA sequencing transcriptome analysis. Later the same group (Hu et al. 2015a) reported that two tall fescue heat shock protein (HSP) genes, including the LMW-HSP/HSP20, exhibited transcriptional memory for their higher transcript abundance during one or more subsequent stresses (S2, S3, S4) relative to the first stress (S1). They concluded that the activated transcriptional memory from two trainable genes could help to induce higher thermo-tolerance in tall fescue. Wang et al. (2015) conducted a comparative leave transcriptome analyses of two grass species (tall fescue, relatively more heat tolerant; perennial ryegrass, heat sensitive) exposed to heat and found 52 sHSPs in tall fescue, with 46 of them being up-regulated under heat stress. Meanwhile, in perennial ryegrass, a grass species closely related to tall fescue, only 40 sHSPs were found, 35 of them were up-regulated by heat. A greater number of activated sHSPs were thought to play a role in the better heat tolerance of tall fescue than perennial ryegrass.

As described above, most studies observed a positive relationship between sHSP expression and perennial grass thermo-tolerance. However, a recent study on a sHSP (AsHSP17) discovered in creeping bentgrass by transcriptome analysis found that overexpression of *AsHSP17* led to hypersensitivity to heat stress. The results suggested that AsHSP17 may function as a protein chaperone to negatively regulate plant responses to adverse environmental stresses, including heat stress (Sun et al. 2016). Similar results were found in a study of tall fescue under heat stress. Zhang et al. (2005) cloned differentially expressed HSPs belonging to the four classes, HSP100, HSP90, HSP70, and sHSPs, between the two genotypes (heat tolerant PI297901 and heat sensitive PI283316) of tall fescues under heat stress. Notably, several LMW-HSP genes identified by SSH or RT-PCR methods either only presented in the sensitive genotype or expressed at a higher level in the sensitive genotype. This may imply that the repair and protection mechanisms were highly activated in heat-sensitive plants for survival under stressful conditions or these sHSPs functioned as negative regulators during heat stress.

3.7 Conclusion

In conclusion, there is accumulating evidence showing that sHSPs are important in plant thermo-tolerance, including both annual and perennial grass plants. With the advancement of new methods and "omics" techniques, more sHSPs in grass plants have been identified. Recent progress in genetic manipulation of plants also opens up opportunities of incorporating important heat stress genes, such as sHSPs, into grass species for crop improvement, although the success of genetic transformation approach has been thus far limited.

Acknowledgements We thank the National Science Foundation of China (31472140) and the Fundamental Research Funds for the Central Universities of China (2014RC007) for financial support.

References

Bagneris C, Bateman OA, Naylor CE, Cronin N, Boelens WC, Keep NH, Slingsby C (2009) Crystal structures of alpha-crystallin domain dimers of alpha b-crystallin and hsp20. J Mol Biol 392:1242–1252

Baranova EV, Weeks SD, Beelen S, Bukach OV, Gusev NB, Strelkov SV (2011) Three-dimensional structure of alpha-crystallin domain dimers of human small heat shock proteins HSPB1 and HSPB6. J Mol Biol 411:110–122

Barnett T, Altschuler M, Mcdaniel CN, Mascarenhas JP (1980) Heat shock-induced proteins in plant-cells. Dev Genet 1:331–340

Basha E, Lee GJ, Demeler B, Vierling E (2004) Chaperone activity of cytosolic small heat shock proteins from wheat. Eur J Biochem 271:1426–1436

Basha E, Jones C, Wysocki V, Vierling E (2010) Mechanistic differences between two conserved classes of small heat shock proteins found in the plant cytosol. J Biol Chem 285:11489–11497

Basha E, Jones C, Blackwell AE, Cheng G, Waters ER, Samsel KA, Siddique M, Pett V, Wysocki V, Vierling E (2013) An unusual dimeric small heat shock protein provides insight into the mechanism of this class of chaperones. J Mol Biol 425:1683–1696

Baszczynski CL, Walden DB, Atkinson BG (1982) Regulation of gene expression in corn *(Zea Mays* L.) by heat shock. Can J Biochem 60:569–579

Baszczynski CL, Walden DB, Atkinson BG (1983) Regulation of gene expression in corn (*Zea mays* L.) by heat shock. II. In vitro analysis of RNAs from heat-shocked seedlings. Can J Biochem Cell Biol 61:395–403

Beck M, Malmstrom JA, Lange V, Schmidt A, Deutsch EW, Aebersold R (2009) Visual proteomics of the human pathogen Leptospira interrogans. Nat Methods 6:817–823

Bharti K, Nover L (2002) Heat stress-induced signalling. In: Scheel D, Wasternack C (eds) Plant signal transduction: frontiers in molecular biology. Oxford University Press, Oxford, pp 74–115

Bita C, Gerats T (2013) Plant tolerance to high temperature in a changing environment: scientific fundamentals and production of heat stress tolerant crops. Front Plant Sci 4:273. doi:10.3389/fpls.2013.00273

Bondino HG, Valle EM, Ten Have A (2012) Evolution and functional diversification of the small heat shock protein/alpha-crystallin family in higher plants. Planta 235:1299–1313

Boston RS, Viitanen PV, Vierling E (1996) Molecular chaperones and protein folding in plants. Plant Mol Biol 32:191–222

Braun N, Zacharias M, Peschek J, Kastenmuller A, Zou J, Hanzlik M, Haslbeck M, Rappsilber J, Buchner J, Weinkauf S (2011) Multiple molecular architectures of the eye lens chaperone alpha B-crystallin elucidated by a triple hybrid approach. Proc Natl Acad Sci U S A 108:20491–20496

Buchner J, Ehrnsperger M, Gaestel M, Walke S (1998) Purification and characterization of small heat shock proteins. Mol Chaperon 290:339–349

Cagliari TC, Tiroli AO, Borges JC, Ramos CHI (2005) Identification and in silico expression pattern analysis of Eucalyptus expressed sequencing tags (ESTs) encoding molecular chaperones. Genet Mol Biol 28:520–528

Carver JA, Lindner RA (1998) NMR spectroscopy of alpha-crystallin. Insights into the structure, interactions and chaperone action of small heat-shock proteins. Int J Biol Macromol 22:197–209

Carver JA, Lindner RA, Kapur A, Gaestel M, Esposito G, Smulders RHPH, De Jong WW (1997) How do small heat-shock proteins stabilize unfolding proteins? Some insights from NMR spectroscopic studies. Congenit Anom 37:376

Cashikar AG, Duennwald M, Lindquist SL (2005) A chaperone pathway in protein disaggregation - Hsp26 alters the nature of protein aggregates to facilitate reactivation by Hsp104. J Biol Chem 280:23869–23875

Cha J-Y, Ermawati N, Jung MH, Su'udi M, Kim K-Y, Kim J-Y, Han C-D, Lee KH, Son D (2009) Characterization of orchardgrass p23, a flowering plant Hsp90 cohort protein. Cell Stress Chaperones 14:233–243

Chauhan H, Khurana N, Nijhavan A, Khurana JP, Khurana P (2012) The wheat chloroplastic small heat shock protein (sHSP26) is involved in seed maturation and germination and imparts tolerance to heat stress. Plant Cell Environ 35:1912–1931

Chen Q, Vierling E (1991) Analysis of conserved domains identifies a unique structural feature of a chloroplast heat-shock protein. Mol Gen Genet 226:425–431
Chen Q, Lauzon LM, Derocher AE, Vierling E (1990) Accumulation, stability, and localization of a major chloroplast heat-shock protein. J Cell Biol 110:1873–1883
Chen X, Lin S, Liu Q, Huang J, Zhang W, Lin J, Wang Y, Ke Y, He H (2014) Expression and interaction of small heat shock proteins (sHsps) in rice in response to heat stress. Biochim Biophys Acta 1844:818–828
Cheng G, Basha E, Wysocki VH, Vierling E (2008) Insights into small heat shock protein and substrate structure during chaperone action derived from hydrogen/deuterium exchange and mass spectrometry. J Biol Chem 283:26634–26642
Cho E, Choi Y (2009) A nuclear-localized HSP70 confers thermoprotective activity and drought-stress tolerance on plants. Biotech Lett 31:597–606
Clark JI (2016) Functional sequences in human alphaB crystallin. BBA Gen Subj 1860:240–245
Clark JI, Ghosh J, Houck S (2011) Identification of interactive sites in the M Tuberculosis (Mtb) Small Heat Shock Protein (sHSP), MTB HSP16.3. FASEB J 25
Cooper P, Ho TH (1983) Heat shock proteins in maize. Plant Physiol 71:215–222
Craig EA (1985) The heat-shock response. Crc Cr Rev Biochem 18:239–280
de Jong WW, Caspers GJ, Leunissen JAM (1998) Genealogy of the alpha-crystallin – small heat-shock protein superfamily. Int J Biol Macromo 22:151–162
de Jong WW, Leunissen JAM, Voorter CEM (1993) Evolution of the alpha-crystallin small heat-shock protein family. Mol Biol Evol 10:103–126
Dietrich PS, Bouchard RA, Casey ES, Sinibaldi RM (1991) Isolation and characterization of a small heat shock protein gene from maize. Plant Physiol 96:1268–1276
Eisenhardt BD (2013) Small heat shock proteins: recent developments. Biomol concepts 4:583–595
Frappier JR, Walden DB, Atkinson BG (1998) Translation of some maize small heat shock proteins is initiated from internal in-frame AUGs. Genetics 148:471–477
Fry JD, Huang B (2004) Applied turfgrass science and physiology. Wiley, Hoboken
Fu X (2015) Insights into how small heat shock proteins bind a great diversity of substrate proteins: a super-transformer model. In: Tanguay MR, Hightower EL (eds) The big book on small heat shock proteins. Springer International Publishing, Cham, pp 101–117
Fu X, Zhang H, Zhang X, Cao Y, Jiao WW, Liu C, Song Y, Abulimiti A, Chang ZY (2005) A dual role for the N-terminal region of Mycobacterium tuberculosis Hsp16.3 in self-oligomerization and binding denaturing substrate proteins. J Biol Chem 280:6337–6348
Garrido C, Paul C, Seigneuric R, Kampinga HH (2012) The small heat shock proteins family: the long forgotten chaperones. Int J Biochem Cell Biol 44:1588–1592
Gaut BS (2002) Evolutionary dynamics of grass genomes. New Phytol 154:15–28
Ghosh JG, Shenoy AK Jr, Clark JI (2006) N- and C-terminal motifs in human alpha B crystallin play an important role in the recognition, selection, and solubilization of substrates. Biochemistry 45:13847–13854
Goloubinoff P, Mogk A, Zvi APB, Tomoyasu T, Bukau B (1999) Sequential mechanism of solubilization and refolding of stable protein aggregates by a bichaperone network. PNAS 96:13732–13737
Goping IS, Frappier JR, Walden DB, Atkinson BG (1991) Sequence, identification and characterization of cDNAs encoding two different members of the 18 kDa heat shock family of Zea mays L. Plant Mol Biol 16:699–711
Guan JC, Jinn TL, Yeh CH, Feng SP, Chen YM, Lin CY (2004) Characterization of the genomic structures and selective expression profiles of nine class I small heat shock protein genes clustered on two chromosomes in rice (Oryza sativa L.). Plant Mol Biol 56:795–809
Ham DJ, Moon JC, Hwang SG, Jang CS (2013) Molecular characterization of two small heat shock protein genes in rice: their expression patterns, localizations, networks, and heterogeneous overexpressions. Mol Biol Rep 40:6709–6720

Hanazono Y, Takeda K, Oka T, Abe T, Tomonari T, Akiyama N, Aikawa Y, Yohda M, Miki K (2013) Nonequivalence observed for the 16-meric structure of a small heat shock protein, SpHsp16.0, from schizosaccharomyces pombe. Structure 21:220–228
Haslbeck M, Franzmann T, Weinfurtner D, Buchner J (2005) Some like it hot: the structure and function of small heat-shock proteins. Nat Struct Mol Biol 12:842–846
Haslbeck M, Weinkauf S, Buchner J (2015) Regulation of the chaperone function of small Hsps. In: Tanguay MR, Hightower EL (eds) The big book on small heat shock proteins. Springer International Publishing, Cham, pp 155–178
He YL, Liu XZ, Huang BR (2005) Protein changes in response to heat stress in acclimated and nonacclimated creeping bentgrass. JASHS 130:521–526
He W, Zhuang H, Fu Y, Guo L, Guo B, Guo L, Zhang X, Wei Y (2015) De novo transcriptome assembly of a Chinese locoweed (Oxytropis ochrocephala) species provides insights into genes associated with drought, salinity and cold tolerance. Front Plant Sci 6:1086
Hilton GR, Hochberg GKA, Laganowsky A, McGinnigle SI, Baldwin AJ, Benesch JLP (2013a) C-terminal interactions mediate the quaternary dynamics of alpha B-crystallin. Philos Trans R Soc B 368(1617):20110405
Hilton GR, Lioe H, Stengel F, Baldwin AJ, Benesch JLP (2013b) Small heat-shock proteins: paramedics of the cell. Top Curr Chem 328:69–98
Hsieh MH, Chen JT, Jinn TL, Chen YM, Lin CY (1992) A class of soybean low-molecular-weight heat-shock proteins - immunological study and quantitation. Plant Physiol 99:1279–1284
Hu T, Sun X, Zhang X, Nevo E, Fu J (2014) An RNA sequencing transcriptome analysis of the high-temperature stressed tall fescue reveals novel insights into plant thermotolerance. BMC Genomics 15:1–13
Hu T, Liu S-Q, Amombo E, Fu J-M (2015a) Stress memory induced rearrangements of HSP transcription, photosystem II photochemistry and metabolism of tall fescue (Festuca arundinacea Schreb.) in response to high-temperature stress. Front Plant Sci 6:403. doi:10.3389/fpls.2015.00403
Hu X, Yang Y, Gong F, Zhang D, Zhang L, Wu L, Li C, Wang W (2015b) Protein sHSP26 improves chloroplast performance under heat stress by interacting with specific chloroplast proteins in maize (Zea mays). J Proteomics 115:81–92
Kappe G, Franck E, Verschuure P, Boelens WC, Leunissen JAM, de Jong WW (2003) The human genome encodes 10 alpha-crystallin-related small heat shock proteins: HspB1-10. Cell Stress Chaperone 8:53–61
Kim KK, Kim R, Kim S-H (1998) Crystal structure of a small heat-shock protein. Nature 394:595–599
Kim KH, Alam I, Kim YG, Sharmin SA, Lee KW, Lee SH, Lee BH (2012) Overexpression of a chloroplast-localized small heat shock protein OsHSP26 confers enhanced tolerance against oxidative and heat stresses in tall fescue. Biotechnol Lett 34:371–377
Klein RD, Chidawanyika T, Tims HS, Meulia T, Bouchard RA, Pett VB (2014) Chaperone function of two small heat shock proteins from maize. Plant Sci 221:48–58
Kokke BPA, Leroux MR, Candido EPM, Boelens WC, de Jong WW (1998) Caenorhabditis elegans small heat-shock proteins Hsp12.2 and Hsp12.3 form tetramers and have no chaperone-like activity. FEBS Lett 433:228–232
Korotaeva NE, Antipina AI, Grabelnykh OI, Varakina NN, Borovskii GB, Voinikov VK (2001) Mitochondrial Low-molecular-weight heat-shock proteins and the tolerance of cereal mitochondria to hyperthermia. Russ J Plant Physiol 48:798–803
Kotak S, Larkindale J, Lee U, von Koskull-Döring P, Vierling E, Scharf K-D (2007) Complexity of the heat stress response in plants. Curr Opin Plant Biol 10:310–316
Kriehuber T, Rattei T, Weinmaier T, Bepperling A, Haslbeck M, Buchner J (2010) Independent evolution of the core domain and its flanking sequences in small heat shock proteins. FASEB J 24:3633–3642
Krishna P (2004) Plant responses to heat stress. In: Hirt H, Shinozaki K (eds) Plant responses to abiotic stress. Springer, Berlin/Heidelberg, pp 73–101

Krishnan M, Nguyen HT, Burke JJ (1989) Heat shock protein synthesis and thermal tolerance in wheat. Plant Physiol 90:140–145
Lee GJ, Vierling E (2000) A small heat shock protein cooperates with heat shock protein 70 systems to reactivate a heat-denatured protein. Plant Physiol 122:189–197
Lee GJ, Roseman AM, Saibil HR, Vierling E (1997) A small heat shock protein stably binds heat-denatured model substrates and can maintain a substrate in a folding-competent state. EMBO J 16:659–671
Lee BH, Won SH, Lee HS, Miyao M, Chung WI, Kim IJ, Jo J (2000) Expression of the chloroplast-localized small heat shock protein by oxidative stress in rice. Gene 245:283–290
Lee DG, Ahsan N, Lee SH, Kang KY, Bahk JD, Lee IJ, Lee BH (2007) A proteomic approach in analyzing heat-responsive proteins in rice leaves. Proteomics 7:3369–3383
Lentze N, Studer S, Narberhaus F (2003) Structural and functional defects caused by point mutations in the alpha-crystallin domain of a bacterial alpha-heat shock protein. J Mol Biol 328:927–937
Liu XH, Huang BR (2002) Cytokinin effects on creeping bentgrass response to heat stress: II. Leaf senescence and antioxidant metabolism. Crop Sci 42:466–472
Liu XH, Huang BR, Banowetz G (2002) Cytokinin effects on creeping bentgrass responses to heat stress: I. Shoot and root growth. Crop Sci 42:457–465
Liu S, Zhang P, Cong B, Liu C, Lin X, Shen J, Huang X (2010) Molecular cloning and expression analysis of a cytosolic Hsp70 gene from Antarctic ice algae Chlamydomonas sp. ICE-L. Extremophiles 14:329–337
Lopes-Caitar VS, de Carvalho MCCG, Darben LM, Kuwahara MK, Nepomuceno AL, Dias WP, Abdelnoor RV, Marcelino-Guimaraes FC (2013) Genome-wide analysis of the Hsp20 gene family in soybean: comprehensive sequence, genomic organization and expression profile analysis under abiotic and biotic stresses. BMC Genomics 14:577. doi:10.1186/1471-2164-14-577
Luthe D, Krans J, Park S-Y, Wang D (2000) The presence and role of heat shock proteins in creeping bentgrass. In: Wilkinson R (ed) Plant environment interactions. Marcel Dekker, New York, pp 283–319
Majoul T, Bancel E, Triboi E, Ben Hamida J, Branlard G (2004) Proteomic analysis of the effect of heat stress on hexaploid wheat grain: characterization of heat-responsive proteins from non-prolamins fraction. Proteomics 4:505–513
Mani N, Ramakrishna K, Suguna K (2015) Characterization of rice small heat shock proteins targeted to different cellular organelles. Cell Stress Chaperone 20:451–460
Marmiroli N, Pavesi A, Di Cola G, Hartings H, Raho G, Conte MR, Perrotta C (1993) Identification, characterization, and analysis of cDNA and genomic sequences encoding two different small heat shock proteins in Hordeum vulgare. Genome 36:1111–1118
McDonald ET, Bortolus M, Koteiche HA, McHaourab HS (2012) Sequence, structure, and dynamic determinants of Hsp27 (HspB1) equilibrium dissociation Are encoded by the N-terminal domain. Biochemistry 51:1257–1268
McHaourab HS, Godar JA, Stewart PL (2009) Structure and mechanism of protein stability sensors: chaperone activity of small heat shock proteins. Biochemistry 48:3828–3837
McHaourab HS, Mishra S, Koteiche HA, Shi J, Williams D, Stewart PL (2011) Role of the N-terminal domain in polydispersity and substrate binding by small heat shock proteins: lessons from Hsp16.5. ARVO Annu Meet Abstr 2011:1612
Mogk A, Deuerling E, Vorderwulbecke S, Vierling E, Bukau B (2003) Small heat shock proteins, ClpB and the DnaK system form a functional triade in reversing protein aggregation. Mol Microbiol 50:585–595
Monjardino P, Smith AG, Jones RJ (2005) Heat stress effects on protein accumulation of maize endosperm. Crop Sci 45:1203–1210
Murakami T, Matsuba S, Funatsuki H, Kawaguchi K, Saruyama H, Tanida M, Sato Y (2004) Over-expression of a small heat shock protein, sHSP17.7, confers both heat tolerance and UV-B resistance to rice plants. Mol Breed 13:165–175

Necchi A, Pogna NE, Mapelli S (1987) Early and late heat-shock proteins in wheats and other cereal species. Plant Physiol 84:1378–1384
Ouyang Y, Chen J, Xie W, Wang L, Zhang Q (2009) Comprehensive sequence and expression profile analysis of Hsp20 gene family in rice. Plant Mol Biol 70:341–357
Pandey B, Kaur A, Gupta OP, Sharma I, Sharma P (2015) Identification of HSP20 gene family in wheat and barley and their differential expression profiling under heat stress. Appl Biochem Biotechnol 175:2427–2446
Papoian GA (2008) Proteins with weakly funneled energy landscapes challenge the classical structure-function paradigm. PNAS 105:14237–14238
Park SY, Shivaji R, Krans JV, Luthe DS (1996) Heat-shock response in heat-tolerant and nontolerant variants of Agrostis palustris Huds. Plant Physiol 111:515–524
Park SY, Chang KC, Shivaji R, Luthe DS (1997) Recovery from heat shock in heat-tolerant and nontolerant variants of creeping bentgrass. Plant Physiol 115:229–240
Patel S, Vierling E, Tama F (2014) Replica exchange molecular dynamics simulations provide insight into substrate recognition by small heat shock proteins. Biophys J 106:2644–2655
Pirkkala L, Nykänen P, Sistonen L (2001) Roles of the heat shock transcription factors in regulation of the heat shock response and beyond. FASEB J 15:1118–1131
Porter JR (2005) Rising temperatures are likely to reduce crop yields. Nature 436:174
Ritossa F (1962) A new puffing pattern induced by temperature shock and DNP in drosophila. Experientia 18:571–573
Rizhsky I, Liang H, Mittler R (2002) The combined effect of drought stress and heat shock on gene expression in tobacco. Plant Physiol 130(3):1143–1151
Santhanagopalan I, Basha E, Ballard KN, Bopp NE, Vierling E (2015) Model chaperones: small heat shock proteins from plants. In: Tanguay MR, Hightower EL (eds) The big book on small heat shock proteins. Springer International Publishing, Cham, pp 119–153
Sarkar NK, Kim YK, Grover A (2009) Rice sHsp genes: genomic organization and expression profiling under stress and development. BMC Genomics 10:393
Schulze E-D, Beck E, Muller-Hohenstein K (2005) Plant ecology. Springer, Berlin
Shi J, Koteiche HA, McDonald ET, Fox TL, Stewart PL, McHaourab HS (2013) Cryoelectron microscopy analysis of small heat shock protein 16.5 (Hsp16.5) complexes with T4 lysozyme reveals the structural basis of multimode binding. J Biol Chem 288:4819–4830
Siddique M, Port M, Tripp J, Weber C, Zielinski D, Calligaris R, Winkelhaus S, Scharf KD (2003) Tomato heat stress protein Hsp16.1-CIII represents a member of a new class of nucleocytoplasmic small heat stress proteins in plants. Cell Stress Chaperone 8:381–394
Siddique M, Gernhard S, von Koskull-Doering P, Vierling E, Scharf K-D (2008) The plant sHSP superfamily: five new members in Arabidopsis thaliana with unexpected properties. Cell Stress Chaperone 13:183–197
Skylas DJ, Copeland L, Rathmell WG, Wrigley CW (2001) The wheat-grain proteome as a basis for more efficient cultivar identification. Proteomics 1:1542–1546
Stengel F, Baldwin AJ, Painter AJ, Jaya N, Basha E, Kay LE, Vierling E, Robinson CV, Benesch JLP (2010) Quaternary dynamics and plasticity underlie small heat shock protein chaperone function. PNAS 107:2007–2012
Stout RG, Al-Niemi TS (2002) Heat-tolerant flowering plants of active geothermal areas in Yellowstone National Park. Ann Bot 90:259–267
Studer S, Obrist M, Lentze N, Narberhaus F (2002) A critical motif for oligomerization and chaperone activity of bacterial alpha-heat shock proteins. Euro J Biochem 269:3578–3586
Sule A, Vanrobaeys F, Hajos G, Van Beeumen J, Devreese B (2004) Proteomic analysis of small heat shock protein isoforms in barley shoots. Phytochemistry 65:1853–1863
Sun Y, MacRae TH (2005) Small heat shock proteins: molecular structure and chaperone function. Cell Mol Life Sci 62:2460–2476
Sun WN, Van Montagu M, Verbruggen N (2002) Small heat shock proteins and stress tolerance in plants. BBA Gene Struct Express 1577:1–9

Sun L, Liu Y, Kong X, Zhang D, Pan J, Zhou Y, Wang L, Li D, Yang X (2012) ZmHSP16.9, a cytosolic class I small heat shock protein in maize (Zea mays), confers heat tolerance in transgenic tobacco. Plant Cell Rep 31:1473–1484
Sun X, Sun C, Li Z, Hu Q, Han L, Luo H (2016) AsHSP17, a creeping bentgrass small heat shock protein modulates plant photosynthesis and ABA-dependent and independent signalling to attenuate plant response to abiotic stress. Plant Cell Environ 39:1320–1337
Takeda K, Hayashi T, Abe T, Hirano Y, Hanazono Y, Yohda M, Miki K (2011) Dimer structure and conformational variability in the N-terminal region of an archaeal small heat shock protein, StHsp14.0. J Struct Biol 174:92–99
Tamura K, Peterson D, Peterson N, Stecher G, Nei M, Kumar S (2011) MEGA5: molecular evolutionary genetics analysis using maximum likelihood, evolutionary distance, and maximum parsimony methods. Mol Biol Evol 28:2731–2739
Tercek MT, Hauber DP, Darwin SP (2003) Genetic and historical relationships among geothermally adapted Agrostis (bentgrass) of North America and Kamchatka: evidence for a previously unrecognized, thermally adapted taxon. Am J Bot 90:1306–1312
Thompson JD, Gibson TJ, Higgins DG (2002) Multiple sequence alignment using ClustalW and ClustalX. Current protocols in bioinformatics/editoral board, Andreas D. Baxevanis … [et al]. Wiley, Chapter 2. doi:10.1002/0471250953.bi0203s00
Thouraya M, Bancel E, Triboi E, Hamida JB, Branlard G (2004) Proteomic analysis of the effect of heat stress on hexaploid wheat grain: characterization of heat-responsive proteins from non-prolamins fraction. Proteomics 4:505–513
Tian J, Belanger FC, Huang B (2009) Identification of heat stress-responsive genes in heat-adapted thermal Agrostis scabra by suppression subtractive hybridization. J Plant Physiol 166:588–601
Tiroli AO, Ramos CHI (2007) Biochemical and biophysical characterization of small heat shock proteins from sugarcane Involvement of a specific region located at the N-terminus with substrate specificity. Int J Biochem Cell Bio 39:818–831
Tiroli-Cepeda AO, Ramos CHI (2010) Heat causes oligomeric disassembly and increases the chaperone activity of small heat shock proteins from sugarcane. Plant Physiol Biochem 48:108–116
Tissiere A, Mitchell HK, Tracy UM (1974) Protein synthesis in salivary glands of drosophila melanogaster - relation to chromosome puffs. J Mol Bio 84:389–398
Turgeon AJ (2011) Turfgrass management. Prentice-Hall/Pearson Higher Education, Upper Saddle River
van Montfort RL, Basha E, Friedrich KL, Slingsby C, Vierling E (2001) Crystal structure and assembly of a eukaryotic small heat shock protein. Nat Struct Biol 8:1025–1030
van Montfort R, Slingsby C, Vierling E (2002) Structure and function of the small heat shock protein/alpha-crystallin family of molecular chaperones. Adv Protein Chem 59:105–156
Veerasamy M, He YL, Huang BR (2007) Leaf senescence and protein metabolism in creeping bentgrass exposed to heat stress and treated with cytokinins. JASHS 132:467–472
Veinger L, Diamant S, Buchner J, Goloubinoff P (1998) The small heat-shock protein IbpB from Escherichia coli stabilizes stress-denatured proteins for subsequent refolding by a multichaperone network. J Bio Chem 273:11032–11037
Vierling E (1991) The roles of heat-shock proteins in plants. Annu Rev Plant Physiol 42:579–620
Wahid A, Gelani S, Ashraf M, Foolad MR (2007) Heat tolerance in plants: an overview. Environ Exp Bot 61:199–223
Wang DF, Luthe DS (2003) Heat sensitivity in a bentgrass variant failure to accumulate a chloroplast heat shock protein isoform implicated in heat tolerance. Plant Physiol 133:319–327
Wang W, Vinocur B, Shoseyov O, Altman A (2004) Role of plant heat-shock proteins and molecular chaperones in the abiotic stress response. Trend plant sci 9:244–252
Wang K, Zhang X, Ervin E (2012) Antioxidative responses in roots and shoots of creeping bentgrass under high temperature: Effects of nitrogen and cytokinin. J Plant Physiol 169:492–500

Wang K, Zhang X, Ervin E (2013) Effects of nitrate and cytokinin on creeping bentgrass under supraoptimal temperatures. J Plant Nutr 36:1549–1564
Wang K, Zhang X, Goatley M, Ervin E (2014) Heat shock proteins in relation to heat stress tolerance of creeping bentgrass at different N levels. PLoS One 9:e102914. doi:10.1371/journal.pone.0102914
Wang Y, Dai Y, Tao X, Wang J-Z, Cheng H-Y, Yang H, Ma X-R (2015) Heat shock factor genes of tall fescue and perennial ryegrass in response to temperature stress by RNA-Seq analysis. Front Plant Sci 6:1226
Waters ER (2013) The evolution, function, structure, and expression of the plant sHSPs. J Exp Bot 64:391–403
Waters ER, Vierling E (1999a) The diversification of plant cytosolic small heat shock proteins preceded the divergence of mosses. Mol Biol Evol 16:127–139
Waters ER, Vierling E (1999b) Chloroplast small heat shock proteins: evidence for atypical evolution of an organelle-localized protein. PNAS 96:14394–14399
Waters ER, Lee GJ, Vierling E (1996) Evolution, structure and function of the small heat shock proteins in plants. J Exp Bot 47:325–338
Waters ER, Aevermann BD, Sanders-Reed Z (2008) Comparative analysis of the small heat shock proteins in three angiosperm genomes identifies new subfamilies and reveals diverse evolutionary patterns. Cell Stress Chaperone 13:127–142
Weeks SD, Baranova EV, Heirbaut M, Beelen S, Shkumatov AV, Gusev NB, Strelkov SV (2014) Molecular structure and dynamics of the dimeric human small heat shock protein HSPB6. J Struct Biol 185:342–354
Wu C (1995) Heat shock transcription factors: structure and regulation. Annu Rev Cell Dev Bio 11:441–469
Xu Y, Zhan C, Huang B (2011) Heat shock proteins in association with heat tolerance in grasses. Int J Proteom 2011:1–11. doi:10.1155/2011/529648
Yeh CH, Yeh KW, Wu SH, Chang PF, Chen YM, Lin CY (1995) A recombinant rice 16.9-kDa heat shock protein can provide thermoprotection in vitro. Plant Cell Physiol 36:1341–1348
Yeh CH, Chang PFL, Yeh KW, Lin WC, Chen YM, Lin CY (1997) Expression of a gene encoding a 16.9-kDa heat-shock protein, Oshsp16.9, in *Escherichia coli* enhances thermotolerance. PNAS 94:10967–10972
Yeh C-H, Chen Y-M, Lin C-Y (2002) Functional regions of rice heat shock protein, Oshsp16.9, required for conferring thermotolerance in *Escherichia coli*. Plant Physiol 128:661–668
Yoon HJ, Kim KP, Park SM, Hong CB (2005) Functional mode of NtHSP17.6. a cytosolic small heat-shock protein from Nicotiana tabacum. J Plant Bio 48:120–127
Zhang Y, Mian MAR, Chekhovskiy K, So S, Kupfer D, Lai H, Roe BA (2005) Differential gene expression in Festuca under heat stress conditions. J Exp Bot 56:897–907
Zheng W, Schafer NP, Davtyan A, Papoian GA, Wolynes PG (2012) Predictive energy landscapes for protein-protein association. PNAS 109:19244–19249

Chapter 4
Induction of Heat Shock Proteins During the Bud Dormancy Stage in Woody Fruit Plants

Yoshihiro Takemura and Fumio Tamura

Abstract In the autumn, bud endodormancy in deciduous fruit trees is induced by short days or low temperatures. Once established, endodormancy is broken by the accumulation of low temperature days, although the precise conditions for this transition vary by species and cultivar. By comparing the reactions of buds of the same cultivar to different chilling periods, these studies identified several genes involved in dormancy phase transitions such as stress response-, cell cycle- and phytohormone-related genes. The actions of heat shock proteins (HSP) may be one of the mechanisms that plants use to survive unfavorable conditions during the bud dormancy stage. More specifically, HSP have chaperone activity that maintain proteins in their functional conformation and thereby prevent the degradation of proteins exposed to cold stress. Small HSPs (sHSPs) function as membrane stabilizers and reactive oxygen species (ROS) scavengers, as well as act synergistically with the antioxidant system. These small proteins also play a key role in maintaining membrane quality attributes, including fluidity and permeability under chilling stress. Membrane damage and ROS production are multifaceted adverse effects of cold injury in plants. The production and accumulation of HSPs can produce greater chilling injury tolerance during endodormancy, which suggests that these proteins play a central role in acquired tolerance to cold conditions.

Keywords Budbreak • Chill units • Chilling requirements • Climatic change • Endodormancy • Global warming

---

Y. Takemura (✉) • F. Tamura
Faculty of Agriculture, Tottori University, Koyama, Tottori 680-8553, Japan
e-mail: takemura_yoshihiro67@yahoo.co.jp

A.A.A. Asea et al. (eds.), *Heat Shock Proteins and Plants*,
Heat Shock Proteins 10, DOI 10.1007/978-3-319-46340-7_4

Abbreviations

ABA	abscisic acid
ADH	alcohol dehydrogenase
APX	ascorbate peroxidase
APX	ascorbate peroxidase
CAT	catalase
CBF	C-repeat binding factor
CR	chilling requirement
DAM	dormancy-associated MADS-box
DREB	dehydration-responsive element binding protein
GDBRPK	grape dormancy breaking-related protein kinase
GR	glutathione reductase
GST	glutathione S-transferase
H_2O_2	hydrogen peroxide
HC	hydrogen cyanamide
HSP	heat shock proteins
HSTF	functional heat shock transcription factor
PDC	pyruvate decarboxylase
POD	peroxidase
ROS	reactive oxygen species
sHSPs	small HSPs
SNF-like protein	sucrose non-fermenting protein kinase
SuSy	sucrose synthase
TPM	transcripts per million
Trxh	thioredoxin h

4.1 Introduction

Perennial woody fruit plants cultivated in temperate zones synchronize their annual growth patterns with seasonal environmental changes. During unfavorable winter conditions, temperate-zone deciduous fruit trees use bud dormancy as a defensive mechanism (Faust et al. 1997). Bud dormancy is classified into three stages: paradormancy, endodormancy, and ecodormancy (Lang 1987). Paradormancy is equivalent to correlative inhibition or apical dominance; endodormancy refers to deep or winter dormancy; and ecodormancy occurs during late winter and spring and is imposed by temperatures unfavorable to growth (Campoy et al. 2011) (Fig. 4.1).

During endodormancy in the autumn, after trees stop growing and their leaves fall, buds enter an inactive state. At this time, trees cannot initiate bud growth, even if the environmental conditions are favorable. Endodormancy is broken by the accumulation of low temperatures, also known as the chilling requirement (CR), which varies by species and cultivar (Westwood 1978; Saure 1985). In recent years, climate

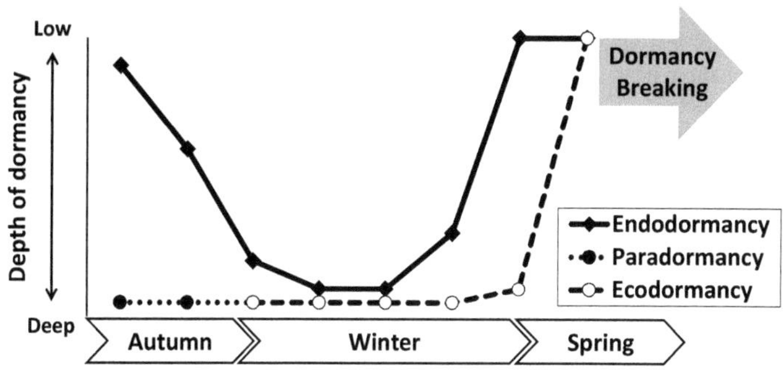

Fig. 4.1 Proposed schematic of dormancy phases regulated by temperature in woody plants. They classified the inhibitions as paradormancy, which is equivalent to correlative inhibition or apical dominance; endodormancy, which is deep dormancy or winter dormancy; and ecodormancy, which is found in late winter and spring and imposed by temperatures unfavourable to growth

change, and especially global warming, has precluded the adequate chilling needed to satisfy the CR for breaking endodormancy, which then inhibits the normal growth of new organs in the spring (Sugiura et al. 2007). This imperfect breaking of endodormancy subsequently has a deleterious impact upon vegetative growth and fruit production.

4.2 Research Findings in Bud Dormancy of Woody Fruit Plants

Several recent studies have focused on the breaking of dormancy in grapes (Or et al. 2000, 2002; Pang et al. 2007; Halaly et al. 2008) by using hydrogen cyanamide (HC) to modify the disruption of endodormancy. Treating grape buds with HC increased hydrogen peroxide (H_2O_2) concentration and inhibited catalase (CAT) activity (Pérez et al. 2008). These changes are the first steps in a cascade that upregulates several signaling proteins such as transcription factors, protein phosphatases, and protein kinases (Neill et al. 2002). The application of HC also results in transcriptional upregulation of the grape dormancy breaking-related protein kinase (GDBRPK), a sucrose non-fermenting protein kinase (SNF-like protein), pyruvate decarboxylase (PDC), alcohol dehydrogenase (ADH), thioredoxin h (Trxh), glutathione S-transferase (GST), ascorbate peroxidase (APX), glutathione reductase (GR), and sucrose synthase (SuSy) (Or et al. 2000; Pérez and Lira 2005; Keilin et al. 2007; Halaly et al. 2008; Pérez et al. 2008). High temperature treatments also broke endodormancy in grapevine buds (Tohbe et al. 1998). Similarly, endodormancy was terminated when buds of the Japanese pear were exposed to high temperature conditions (45 °C for 4 h), along with the accumulation of nine protein spots observed by the 2-D method (Tamura et al. 1998). In addition, the 19-kDa protein of these spots shifts to the basic side with both high and low temperature treatments (Tamura et al. 1998) (Fig. 4.2).

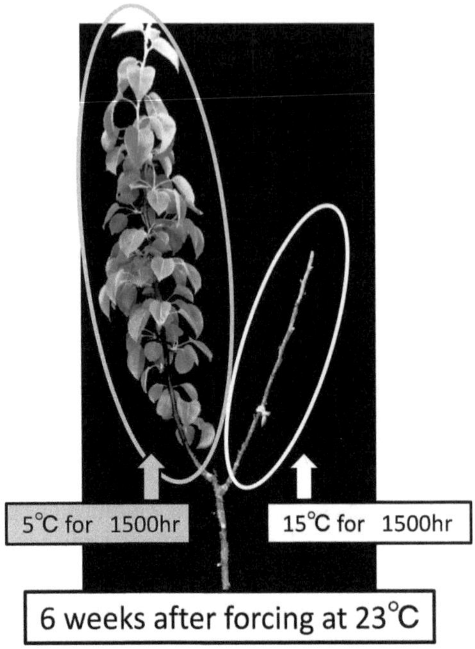

Fig. 4.2 The difference of growth situation of leaf buds in branch of Japanese pear treated by difference temperature (5 °C or 15 °C). After endodormancy induction, only buds in branch treated 5 °C was breaking in growing suitable condition

To address this issue, several studies have been conducted to examine endodormancy in various tree species. Water availability is one of the basic factors that determine bud development because dormancy is closely related to changes in water movement, as this is an essential event in the overwintering of woody plants (Welling and Palva 2006). For example, in two cultivars of peach buds with different CRs during endodormancy, the total soluble sugar and water accumulation periods were different, even though the species was the same (Yooyongwech et al. 2009). In addition, phytohormone fluctuations influence bud dormancy. For example, endogenous abscisic acid (ABA) levels induced by low temperature or drought stress increased with the establishment of endodormancy and decreased at the conclusion of endodormancy in apple (Seeley and Powell 1981) and pear (Tamura et al. 1992) buds. After dormancy was induced by short autumn days, ABA signal transduction components (*PP2C* or *AREB3*) were induced in poplar buds (Ruttink et al. 2007). A conceptual model developed for seasonal dormancy transitions in crown buds of leafy spurge based on microarray studies also highlighted the role of *DREB1A/CBF2*, *COP1*, *HY5*, *DELLA*s, *DAM*, and *FT* in the maintenance of well-defined dormancy phases (Doğramacı et al. 2010). The cold-induced expression of some dehydration-responsive element binding protein (*DREB*)/C-repeat binding factor (CBF)-family members (Fowler et al. 2005), and *DREB1A* has also been identified as a central regulator of molecular networks involved in endodormancy induction (Doğramacı et al. 2010) and termination (Doğramacı et al. 2011). A recent study of

Pyrus pyrifolia focused on determining the molecular levels of MIKC-type dormancy-associated MADS-box (*DAM*) genes, which may be candidate endodormancy-breaking genes (Ubi et al. 2010). Expression of dam genes decreased during the breaking of endodormancy in the Japanese pear 'Kosui' and was very low in the Taiwanese pear 'Hengshanli' (TP-85-119), which is a less dormant pear species (Ubi et al. 2010). A comparison between 'Kosui' and 'Hengshanli' identified two novel transcription factors (*NAC* and *PRR*) whose expression levels varied concomitantly with dormancy phase changes (Nishitani et al. 2012).

Recently, microarray analysis and RNA sequencing using next-generation sequencing technology (RNA-seq) have been widely used for the transcriptomic analysis of dormancy in plants such as the Japanese pear (Liu et al. 2012; Nishitani et al. 2012; Bai et al. 2013; Takemura et al. 2015a), grapevine (Díaz-Riquelme et al. 2012), and Japanese apricot (Habu et al. 2014). By comparing the responses of buds of the same species but different cultivars to varying chilling periods, these studies identified several genes involved in dormancy phase transitions such as stress response-, cell cycle- and phytohormone-related genes.

4.3 Heat-Shock Proteins Detected During Bud Dormancy Stage

Candidate genes related to endodormancy induction and breaking can be used as molecular markers for understanding the state of endodormancy or breeding low chilling cultivar. Transcriptomic analysis on a genome-wide basis can promote the identification of gene networks associated with the regulation or transition of bud dormancy. In addition, several research papers have reported the detection of HSP during the bud dormancy stage. RNA-seq on the bud of the 'Suli' pear (*P. pyrifolia* white pear group) during dormancy showed that expression levels of genes encoding proteins associated with HSP up-regulated from November 15 to December 15 in the endodormant phase, and down-regulated between January 15 and February 15 in the ecodormant phase (Liu et al. 2012). Similarly, RNA-seq of Japanese apricot flower buds indicated that the value of transcripts per million (TPM) clean tags of gene annotated HSP21 increased from the endodormancy to ecodormancy stages, and then decreased at the dormancy release stage (Zhong et al. 2013). In a study using transcriptional profiling of bud dormancy release in oaks by next-generation sequencing, HSP18.2 was identified as an ecodormancy gene that is up-regulated by ABA (Ueno et al. 2013). In this way, endogenous ABA levels are induced by low temperatures during endodormancy, and then decrease during endodormancy release, which suggests that both HSP and ABS have a similar pattern of activity (Table 4.1).

In addition to genetic research, the analysis of proteins expressed as a practical matter is important for the elucidation of the mechanisms involved in endodormancy induction or breaking. Cold acclimation in woody plants has been associated

Table 4.1 Heat-shock proteins detected during bud dormancy stage or cold treatment in woody plants

Reference	Plant name	Expression pattern or size of HSP
Renaut et al. (2004)	Poplar	Detected: Chilling at 4 °C (HSP70)
Zhang et al. (2008)	Grape	Up-regulated: Cold acclimation (HSP70)
Liu et al. (2012)	Chinese pear	Up-regulated: Endodormant phase
		Down-regulated: Ecodormant phase
Ueno et al. (2013)	Sessile oak	Detected: Ecodormancy stage (HSP18.2)
Zhong et al. (2013)	Japanese apricot	Up-regulated: Endodormancy to ecodormancy stages (HSP21): Paradormancy (before leaf fall) to endodormancy (heat shock cognate protein 70–1)
		Down-regulated: Dormancy release stage (HSP21): Ecodormancy to dormancy release (heat shock cognate protein 70–1)
Takemura et al. (2015a, b)	Japanese pear	Detected: Pre-breaking period of endodormancy (17.6 kDa class I HSP and 18.1 kDa class I HSP)

with distinct metabolic changes and protein changes that are largely related to variations in the expression of HSP (Wisniewski et al. 1996). In a study evaluating the responses of poplar to cold temperatures, molecular chaperone-like proteins including sHSP, a fragment of HSP70, and heat shock cognate 70, were identified by the 2-D method (Renaut et al. 2004). The results showed that these proteins were more abundant in leaves under stressful cold conditions (4 °C) than at the optimal growth temperature of 23 °C (Renaut et al. 2004). In comparison analysis among proteins associated with seasonal bud dormancy at four critical stages in Japanese apricot, the expression levels of heat shock cognate protein 70–1 increased from paradormancy (before leaf fall) to endodormancy, and then decreased from ecodormancy to dormancy release (Zhuang et al. 2013). In a quantitative proteomic analysis of the responses of the bark tissues of peach to short photoperiods and low-temperatures, two other HSP70 family members similar to the Arabidopsis chloroplast HSP70s were up-regulated at 5 °C (Renaut et al. 2008).

A short photoperiod also influences the regulation of dormancy and cold acclimation of woody plants (Kozlowski 2002). Interestingly, none of the HSP70 family members in proteomic analysis in the bark tissues of peach was identified in the suppression subtractive hybridization experiment in peach bark, and those HSP-like genes that were identified appeared to be predominantly regulated by short, rather than long, days (Bassett et al. 2006). In almost all organisms, HSP70 functions as a chaperone for newly synthesized proteins to prevent their accumulation (Sung et al. 2001). Furthermore, HSP70 plays a crucial role in the development of accumulation to temperature stress induced by cold acclimation pretreatment in grape plants (Zhang et al. 2008). A study of *Pisum sativum* reported cooperation in the activities of HSP70 and sHSP such as s HSP18.1 (Lee and Vierling 2000). HSP of molecular weights similar to those of HSP18.2 were expressed in the bud immediately before endodormancy breaking in Japanese pear (Takemura et al. 2015a, b). Among nine protein spots observed by the 2-D method in floral buds of Japanese pear in which endodormancy was broken by both high and low temperatures, the size of HSP spots that shifted to the basic side was approximately 19-kDa (Tamura et al. 1998).

Five principal HSPs that are distinguished by their molecular chaperone activity can be classified according to their approximate molecular weight: (1) HSP100, (2) HSP90, (3) HSP70, (4) HSP60, and (5) sHSP (Kotak et al. 2007). HSP display chaperone activity by maintaining proteins in their functional conformation, thereby preventing protein degradation during times of cold stress. This protective function may be one of the mechanisms that plants use to survive unfavorable conditions during the bud dormancy stage. Higher plants are characterized by the presence of at least 20 types of sHSPs (Vierling 1991). sHSPs are usually undetectable in plants under favorable conditions, but are induced by exposure to low temperatures (Zhang et al. 2008).

4.4 Role of Heat-Shock Proteins During the Bud Dormancy Stage

In Arabidopsis and other plant species, the synthesis of HSPs can be induced by low temperatures, salinity, desiccation, high intensity irradiations, wounding, as well as heavy metal, osmotic, and oxidative stressors (Swindell et al. 2007). Specifically, accumulation gene transcripts corresponding to HSPs observed in persimmon fruit stored at 1 °C as an acclimation pretreatment showed that these proteins prevented damage caused by low temperature conditions (Souza et al. 2011).

During induction or breaking periods of endodormancy, many biochemical or molecular biological changes occur in the buds of deciduous fruit trees. Previous research found that in many higher plants, cold-hardening during the winter is associated with changes in lipid composition within the cellular membrane (Uemura and Steponkus 1994; Uemura et al. 1995) or the accumulation of compatible solute within the cell (Koster and Lynch 1992; Wanner and Junttila 1999; Kamata and Uemura 2004). In peach buds of the same species but different cultivars with unique CRs, the accumulation periods of total soluble sugar and water contents during endodormancy were different (Yooyongwech et al. 2009). Moreover, Yooyongwech et al. (2009) showed that before the end of endodormancy, expression levels of the Pp-PIP1 and Pp-γTIP1 genes that encode aquaporins, which regulate water transport in tonoplasts and plasma membranes, increased more in the peach buds of high-chilling cultivars as compared to low-chilling varieties.

Proline plays a crucial role in the osmotic regulation between cytoplasm and vacuole, the redox regulation of the NAD+/NADH ratio, and the promotion of ROS scavenging systems (Sharp et al. 1990; Bohnert and Jensen 1996). The accumulation of endogenous proline in transgenic rice plants over-expressing OsHSP23.7 and OsHSP17.0 was significantly more than that in wild type plants (Zou et al. 2012).

In Arabidopsis overexpressing yeast HSP26, Xue et al. (2009) showed that there was an increase in endogenous proline content. Additionally, increased proline content has been observed in 'Superior Seedless' grapevine buds when HC treatment was used to induce dormancy release (Ben Mohamed et al. 2012). In this way, both HSP and proline contents increase from the endodormancy to ecodormancy stages.

Several recent studies have focused on redox reactions involved in endodormancy transition stages, and specifically, the decomposition of H_2O_2 by CAT or peroxidase (POD) (Or et al. 2002; Pérez and Burgos 2004; Pérez et al. 2008). In the flower bud of the Japanese pear 'Kosui,' H_2O_2 content increased between the endodormancy induction and endodormancy breaking periods, but then decreased after the ecodormancy stage (Kuroda et al. 2002). The transient peak of H_2O_2 preceding dormancy release could act as a signaling molecule to trigger the transition from dormancy to bud breaking (Halaly et al. 2008). In grapevines, a similar decrease in H_2O_2 before endodormancy breaking both under natural conditions and with HC treatment has been shown. The decrease in CAT activity in grapevine buds following HC treatment causes an increase in the level of H_2O_2, which may activate the pentose-phosphate pathway, thus leading to dormancy termination, bud bursting, and rapid growth (Simmonds and Simpson 1972; Nir et al. 1986). During stress, the pentose phosphate pathway serves as a key source of reduced NADPH for ROI removal (Pandolfi et al. 1995; Juhnke et al. 1996). Glutathione reductase, which transfers an electron from NADPH to glutathione, and ascorbate peroxidase (APX), which catalyses the reduction of H_2O_2 using ascorbate as an electron donor to prevent free radical toxicity, both belong to the protective enzyme system of the ascorbate–glutathione cycle (Chen 1991; Mittler 2002).

APX has also been identified by MALDI–TOF/TOF MS as the differentially expressed proteome of Japanese apricot flower buds, which have four critical stages that extend from paradormancy (before leaf fall) to dormancy release, with extremes of expression occurring at endodormancy and ecodormancy, and a decline in expression during dormancy release (Zhuang et al. 2013). In contrast, budbreak in the 'Anna' apple (*Malus domestica* Borkh.) has been induced by cold, heat, and allyl disulfide, with these treatments also increasing the ratios of ascorbic acid and the activities of ascorbate free-radical reductase, APX, dehydroascorbate reductase, and glutathione reductase in the buds (Wang and Faust 1994). Additionally, expression levels of genes encoding APX after 12 h of heat shock treatment as well as APX activity for 0–5 days after HC treatment increased rapidly in grape buds in the endodormancy stage (Ben Mohamed et al. 2012).

HSPs also assume the protective role of the secondary oxidative stress occurring under demanding conditions such as low temperatures. The expression of genes encoding HSP function as direct sensors of H_2O_2 in plants and the regulation of the expression of genes encoding for antioxidant proteins and raised resistance against oxidative stress (Panchuk et al. 2002). Additionally, in transgenic tobacco, accumulated sHSP enhanced APX, CAT, and POX activities (Li et al. 2012). The gene encoding for APX1 in peas contains a functional heat shock transcription factor (HSTF) binding motif in its promoter region (Mittler and Zilinskas 1992). HSTF-dependent APX1 gene expression in Arabidopsis also regulated the expression of genes involved in the functioning of the antioxidant system in order to enhance resistance to oxidative stress (Panchuk et al. 2002). In addition, the accumulation of HSP70 in grapes was related to an increase of gene expression and activity of antioxidant enzymes (Zhang et al. 2005).

Membrane damage and ROS production are multifaceted adverse effects of chilling stress in plants. However, the production and accumulation of HSPs contribute to higher cold injury tolerance during endodormancy, suggesting that HSPs play a central role in the acquired resistance to low temperature stress.

4.5 Conclusion

Endodormancy in temperate-zone deciduous fruit trees is an essential mechanism to defend buds from unfavorable conditions during winter. In order to accurately predict the time of bud break, it is important to regulate endodormancy in these woody plants. Failure to fulfill chilling requirements for breaking endodormancy can cause growth inhibition of the plants, thereby preventing germination during spring. Given the possibility of warmer temperatures and reduced potential for chill hours during winter, it would be useful to develop molecular markers that can help estimate whether plants have broken endodormancy. Up to this point, details of the mechanisms regulating the induction and breaking of endodormancy remain unknown. HSP plays a crucial role in the acclimation to low temperatures in plants. In addition, HSP is closely associated with physiological changes that occur during the bud dormancy stage in woody fruit plants. In the future, analysis of HSP-related endodormancy, in addition to these genetic approaches, will lead us to a better-developed understanding of the mechanisms of endodormancy.

Acknowledgements We gratefully acknowledge the work of past and present members of our laboratory. This work was supported by JSPS KAKENHI Grant Number 26450034.

References

Bai S, Saito T, Sakamoto D, Fujii H, Moriguchi T (2013) Transcriptome analysis of Japanese pear (*Pyrus pyrifolia* Nakai) flower buds transitioning through endodormancy. Plant Cell Physiol 54:1132–1151

Bassett CL, Wisniewski ME, Artlip TS, Norelli JL, Renaut J, Farrel RE Jr (2006) Global analysis of genes regulated by low temperature and photoperiod in peach bark. J Am Soc Hortic Sci 131:551–563

Ben Mohamed H, Vadel AM, Geuns JMC, Khemira H (2012) Effects of hydrogen cyanamide on antioxidant enzymes' activity, proline and polyamine contents during bud dormancy release in Superior Seedless grapevine buds. Acta Physiol Plant 34:429–437

Bohnert HJ, Jensen RG (1996) Strategies for engineering water-stress tolerancein plants. Trends Biotechnol 14:89–97

Campoy J, Ruiz D, Egea J (2011) Dormancy in temperate fruit trees in a global warming context: a review. Sci Hortic 130:357–372

Chen SY (1991) Injury of membrane lipid peroxidation to plant cell. Plant Physiol 27:84–90

Díaz-Riquelme J, Grimplet J, Martínez-Zapater JM, Carmona MJ (2012) Transcriptome variation along bud development in grapevine (*Vitis vinifera* L.). BMC Plant Biol 12:181

Doğramacı M, Horvath DP, Chao WS, Foley ME, Christoffers MJ, Anderson JV (2010) Low temperatures impact dormancy status, flowering competence, and transcript profiles in crown buds of leafy spurge. Plant Mol Biol 73:207–226

Doğramacı M, Horvath DP, Christoffers MJ, Anderson JV (2011) Dehydration and vernalization treatments identify overlapping molecular networks impacting endodormancy maintenance in leafy spurge crown buds. Funct Integr Genomics 11:611–626

Faust M, Erez A, Rowland LJ, Wang SY, Norman HA (1997) Bud dormancy in perennial fruit trees: physiological basis for dormancy induction, maintenance, and release. HortSci 32:623–629

Fowler SG, Cook D, Thomashow MF (2005) Low temperature induction of Arabidopsis CBF1, 2, and 3 is gated by the circadian clock. Plant Physiol 137:961–968

Habu T, Yamane H, Sasaki R, Yano K, Fujii H, Shimizu T, Yamamoto T, Tao R (2014) Custom microarray analysis for transcript profiling of dormant vegetative buds of Japanese apricot during prolonged chilling exposure. J Jpn Soc Hortic Sci 83:1–16

Halaly T, Pang X, Batikoff T, Crane O, Keren A, Venkateswari J, Ogrodovitch A, Sadka A, Lavee S, Or E (2008) Similar mechanisms might be triggered by alternative external stimuli that induce dormancy release in grape buds. Planta 228:79–88

Juhnke H, Krems B, Kotter P, Entian KD (1996) Mutants that show increased sensitivity to hydrogen peroxide reveal an important role for the pentose phosphate pathway in protection of yeast against oxidative stress. Mol Gen Genet 252:456–464

Kamata T, Uemura M (2004) Solute accumulation in wheat seedlings during cold acclimation: contribution to increased freezing tolerance. CryoLetters 25:311–322

Keilin T, Pang X, Venkateswari J, Halaly T, Crane O, Keren A, Ogrodovitch A, Ophir R, Volpin H, Galbraith D, Or E (2007) Digital expression profiling of grape EST collection leads to new insight into molecular events during grape-bud dormancy release. Plant Sci 173:446–457

Koster KL, Lynch DV (1992) Solute accumulation and compartmentation during the cold acclimation of puma rye. Plant Physiol 98:108–113

Kotak S, Larkindale J, Lee U, von Koskull-Doring P, Vierling E, Scharf KD (2007) Complexity of the heat stress response in plants. Curr Opin Plant Biol 10:310–316

Kozlowski TT (2002) Acclimation and adaptive responses of woody plants to environmental stresses. Bot Rev 68:270–334

Kuroda H, Sugiura T, Ito D (2002) Changes in hydrogen peroxide content in flower buds of Japanese pear (*Pyrus pyrifolia* Nakai) in relation to breaking of endodormancy. J Jpn Soc Hortic Sci 71:610–616

Lang GA (1987) Dormancy: a new universal terminology. HortSci 22:817–820

Lee GJ, Vierling E (2000) A small heat shock protein cooperates with heat shock protein 70 systems to reactivate a heat-denatured protein. Plant Physiol 122:189–197

Li M, Ji L, Yang X, Meng Q, Guo S (2012) The protective mechanisms of CaHSP26 in transgenic tobacco to alleviate photoinhibition of PSII during chilling stress. Plant Cell Rep 31:1969–1979

Liu W, Zheng P, Xu T, Chen L, Liu D, Hussain S, Teng Y (2012) Transcriptomic analysis of 'Suli' pear (*Pyrus pyrifolia* white pear group) buds during the dormancy by RNA-Seq. BMC Genomics 13:700

Mittler R (2002) Oxidative stress, antioxidants, and stress tolerance. Trends Plant Sci 7:405–410

Mittler R, Zilinskas BA (1992) Molecular cloning and characterization of a geneencoding pea cytosolic ascorbate peroxidase. J Biol Chem 267:21802–21807

Neill SJ, Desikan R, Clarke A, Hurst RD, Hancock JT (2002) Hydrogen peroxide and nitric oxide as signalling molecules in plants. J Exp Bot 53:1237–1247

Nir G, Shulman Y, Fanberstein L, Lavee S (1986) Changes in the activity of catalase in relation to the dormancy of grapevine (*Vitis vinifera* L.) buds. Plant Physiol 81:1140–1142

Nishitani C, Saito T, Ubi BE, Shimizu T, Itai A, Saito T, Yamamoto T, Moriguchi T (2012) Transcriptome analysis of *Pyrus pyrifolia* leaf buds during transition from endodormancy to ecodormancy. Sci Hortic 147:49–55

Or E, Vilozny I, Eyal Y, Ogrodovitch A (2000) The transduction of the signal for grape bud dormancy breaking induced by hydrogen cyanamide may involve the SNF-like protein kinase GDBRPK. Plant Mol Biol 43:483–494

Or E, Vilozny I, Fennell A, Eyal Y, Ogrodovitch A (2002) Dormancy in grape buds: isolation and characterisation of catalase cDNA and analysis of its expression following chemical induction of bud dormancy release. Plant Sci 162:121–130

Panchuk II, Volkov RA, Schoffl F (2002) Heat stress- and heat shock transcription factor-dependent expression and activity of ascorbate peroxidase in Arabidopsis. Plant Physiol 129:838–853

Pandolfi PP, Sonati F, Rivi R, Mason P, Grosveld F, Luzzatto L (1995) Targeted disruption of the housekeeping gene encoding glucose 6-phosphate dehydrogenase (G6PD): G6PD is dispensable for pentose synthesis but essential for defense against oxidative stress. EMBO J 14:5209–5215

Pang X, Halaly T, Crane O, Keilin T, Keren-Keiserman A, Ogrodovitch A, Galbraith D, Or E (2007) Involvement of calcium signalling in dormancy release of grape buds. J Exp Bot 58:3249–3262

Pérez FJ, Burgos B (2004) Alterations in the pattern of peroxidase isoenzymes and transient increases in its activity and in H_2O_2 levels take place during the dormancy cycle of grapevines buds: the effect of hydrogen cyanamide. Plant Growth Regul 43:213–220

Pérez FJ, Lira W (2005) Possible role of catalase in post-dormancy bud-break in grapevines. J Plant Physiol 162:301–308

Pérez FJ, Vergara R, Rubio S (2008) H_2O_2 is involved in the dormancy-breaking effect of hydrogen cyanamide in grapevine buds. Plant Growth Regul 55:149–155

Renaut J, Lutts S, Hoffmann L, Hausman JF (2004) Responses of poplar to chilling temperatures: proteomic and physiological aspects. Plant Biol 6:81–90

Renaut J, Hausman J, Bassett C, Artlip T, Cauchie H, Witters E, Wisniewski M (2008) Quantitative proteomic analysis of short photoperiod and low-temperature responses in bark tissues of peach (*Prunus persica* L. Batsch). Tree Genet Genomes 4:589–600

Ruttink T, Arend M, Morreel K, Storme V, Rombauts S, Fromm J, Bhalerao RP, Boerjan W, Rohde A (2007) A molecular timetable for apical bud formation and dormancy induction in poplar. Plant Cell 19:2370–2390

Saure MC (1985) Dormancy release in delicious fruit trees. Hortic Rev 7:239–300

Seeley SD, Powell LE (1981) Seasonal changes of free and hydrolysable abscisic acid in vegetative apple buds. J Am Soc Hortic Sci 106:405–409

Sharp RE, Hsiao TC, Silk WK (1990) Growth of the maize primary root at low water potentials: II. Role of growth and deposition of hexose and potassium in osmotic adjustment. Plant Physiol 93:1337–1346

Simmonds JA, Simpson GM (1972) Regulation of Krebs cycle and pentose phosphate pathway activities in the control of dormancy of Avenafatua. Can J Bot 50:1041–1048

Souza ELD, Souza ALKD, Tiecher A, Girardi CL, Nora L, Silva JAD, Argenta LC, Rombaldi CV (2011) Changes in enzymatic activity, accumulation of pro-teins and softening of persimmon (*Diospyros kaki* Thunb.) flesh as a function of pre-cooling acclimatization. Sci Hortic 127:242–248

Sugiura T, Kuroda H, Sugiura H (2007) Influence of the current state of global warming on fruit tree growth in Japan. Hortic Res (Japan) 6:257–263 (In Japanese with English abstract)

Sung DY, Kaplan F, Guy CL (2001) Plant Hsp70 molecular chaperones: protein structure, gene family, expression and function. Physiol Plant 113:443–451

Swindell WR, Huebner M, Weber AP (2007) Transcriptional profiling of Arabidopsis heat shock proteins and transcription factors reveals extensive overlap between heat and non-heat stress response pathways. BMC Genomics 8:125

Takemura Y, Kuroki K, Jiang M, Matsumoto K, Tamura F (2015a) Identification of the expressed protein and the impact of change in ascorbate peroxidase activity related to endodormancy breaking in *Pyrus pyrifolia*. Plant Physiol Biochem 86:121–129

Takemura Y, Kuroki K, Shida Y, Araki S, Takeuchi Y, Tanaka K, Ishige T, Yajima S, Tamura F (2015b) Comparative transcriptome analysis of the less-dormant Taiwanese pear and the dormant Japanese pear during winter season. PLoS One 10:e0139595

Tamura F, Tanabe K, Banno K (1992) Effect of chilling treatment on intensity of bud dormancy, respiration and endogenous growth regulators in Japanese pear 'Nijisseiki'. J Jpn Soc Hortic Sci 60:763–769 (in Japanese with English abstract)

Tamura F, Tanabe K, Itai A, Tanaka H (1998) Protein changes in the flower buds of Japanese pear during breaking of dormancy by chilling or high-temperature treatment. J Am Soc Hortic Sci 123:532–536

Tohbe M, Mochioka R, Horiuchi S, Ogata T, Shiozaki S, Kurooka H (1998) Role of ACC and glutathione during breaking of dormancy in grapevine buds by high temperature treatment. J Jap Soc Hortic Sci 67:912–916 (In Japanese with English summary)

Ubi BE, Sakamoto D, Ban Y, Shimada T, Ito A, Nakajima I, Takemura Y, Tamura F, Saito T, Moriguchi T (2010) Molecular cloning of dormancy-associated MADS-box gene homologs and their characterization during seasonal endodormancy transitional phases of Japanese Pear. J Am Soc Hortic Sci 135:174–182

Uemura M, Steponkus PL (1994) A contrast of the plasma membrane lipid composition of oat and rye leaves in relation to freezing tolerance. Plant Physiol 104:479–496

Uemura M, Joseph RA, Steponkus PL (1995) Cold acclimation of Arabidopsis thaliana: effect on plasma membrane lipid composition and freeze-induced lesions. Plant Physiol 109:15–30

Ueno S, Klopp C, Leplé JC, Derory J, Noirot C, Léger V, Prince E, Kremer A, Plomion C, Le Provost G (2013) Transcriptional profiling of bud dormancy induction and release in oak by next-generation sequencing. BMC Genomics 14:236–250

Vierling E (1991) The role of heat shock proteins in plants. Annu Rev Plant Physiol 42:579–620

Wang SY, Faust M (1994) Changes in the antioxidant system associated with budbreak in 'anna' apple (*Malus domestica* Borkh.) buds. J Am Soc Hortic Sci 119:735–741

Wanner LA, Junttila O (1999) Cold-induced freezing tolerance in Arabidopsis. Plant Physiol 120:391–400

Welling A, Palva ET (2006) Molecular control of cold acclimation in trees. Physiol Plant 127:167–181

Westwood MN (1978) Temperate-zone pomology. W. H. Freeman and Co, New York, pp 199–303

Wisniewski M, Close TJ, Artlip T, Arora R (1996) Seasonal patterns of dehydrins and 70-kDa heat-shock proteins in bark tissues of eight species of woody plants. Physiol Plant 96:496–505

Xue Y, Peng R, Xiong A, Li X, Zha D, Yao Q (2009) Yeast heat-shock protein gene HSP26 enhances freezing tolerance in Arabidopsis. J Plant Physiol 166:844–850

Yooyongwech S, Sugaya S, Sekozawa Y, Gemma H (2009) Differential adaptation of high- and low-chill dormant peaches in winter through aquaporin gene expression and soluble sugar content. Plant Cell Rep 28:1709–1715

Zhang J, Huang W, Pan Q, Liu Y (2005) Improvement of chilling tolerance and accumulation of heat shock proteins in grape berries (*Vitis vinifera* cv. Jingxiu) by heat pretreatment. Postharvest Biol Technol 38:80–90

Zhang JH, Wang LJ, Pan QH, Wang YZ, Zhan JC, Huang WD (2008) Accumulation and subcellular localization of heat shock proteins in young grape leaves during cross-adaptation to temperature stresses. Sci Hortic 117:231–240

Zhong W, Gao Z, Zhuang W, Shi T, Zhang Z, Ni Z (2013) Genome-wide expression profiles of seasonal bud dormancy at four critical stages in Japanese apricot. Plant Mol Biol 83:247–264
Zhuang WB, Shi T, Gao ZH, Zhang Z, Zhang JY (2013) Differential expression of proteins associated with seasonal bud dormancy at four critical stages in Japanese apricot. Plant Biol 15:233–242
Zou J, Liu C, Liu A, Zou D, Chen X (2012) Overexpression of OsHsp17.0 and OsHsp23.7 enhances drought and salt tolerance in rice. J Plant Physiol 169:628–635

Chapter 5
Heat Shock Proteins in Wild Barley at "Evolution Canyon", Mount Carmel, Israel

Zujun Yang and Eviatar Nevo

Abstract The analysis of stress-responsiveness in cereal plants is an important route to the discovery of genes conferring stress tolerance and their use in breeding programs. High temperature is one of the environmental stress factors that can affect the growth and quality characteristics of barley (*Hordeum vulgare*). Almost all stresses induce the production of a group of proteins called heat-shock protein (HSPs) or stress-induced proteins. The induction of transcription of these different types of heat shock proteins reflects an adaptation to tolerate the heat stress. The "Evolution Canyon" I at lower Nahal Oren, Mount Carmel, Israel (EC I), reveals evolution in action across life at a microsite caused by interslope microclimatic divergence. The adaptation, speciation, domestication and rich genetic diversity of wild barley, *H. spontaneum*, was a good model to study the evolution and adaptation at both *macro*- and *micro*-scale levels. The genetic divergence and haplotype diversity of heat shock protein genes were significantly different among the populations at EC I. The diversity was also correlated with microclimatic divergence interslopes. We briefly review the remarkable interslope incipient adaptive sympatric speciation of wild barley at "Evolution Canyon", focusing on HSPs which highlight barley improvement for stress tolerances.

Keywords Heat shock protein genes • "Evolution Canyon" • *Hordeum spontaneum* • Molecular adaptive evolution

---

Z. Yang
School of Life Science and Technology, University of Electronic Science and Technology of China, Chengdu 610054, Sichuan, China

E. Nevo (✉)
International Graduate Center of Evolution, Institute of Evolution, University of Haifa, 199 Aba Khoushy Ave., Mount Carmel, Haifa 31905, Israel
e-mail: nevo@research.haifa.ac.il

A.A.A. Asea et al. (eds.), *Heat Shock Proteins and Plants*,
Heat Shock Proteins 10, DOI 10.1007/978-3-319-46340-7_5

Abbreviations

ABA	abscisic acid
ABRE	abscisic acid responsive cis-elements
ACD	alpha-crystallin domain
CRISPR	clustered regularly interspaced short palindromic repeats
EC	Evolution Canyon
EcoTILLING	ecotype targeting induced local lesions in genomes
Hd	haplotype diversity
Hsf	heat shock factor
HSP	heat shock proteins
NCBI	National Center for Biotechnology Information
NFS	north-facing slopes
OGCs	orthologous gene clusters
QTL	quantitative trait loci
SNP	single nucleotide polymorphisms
SFS	south-facing slopes
sHSPs	small heat shock proteins
SSR	simple sequence repeats
TE	transposable element

5.1 Introduction

Cereal crops constitute over 50 % of total food production for humans and the crop species including wheat, barley and rye are essentially components for human and animal nutrition (http://www.fao.org). Barley ranks the fifth in the world production and is widely used as animal feeds and food industry. As the first domestication crop species in Fertile Crescent about 10,000 years ago, the progenitor of cultivated barley, wild barley (*Hordeum spontaneum*) significantly contributed to the beginning of agriculture and the breakthrough of human civilization (Morrell and Clegg 2007). The natural distribution of wild barley ranged from the Middle East to Southwest Asia, and the abundant diversity is vital for population genetics and genomics studies, which is also used in barley pre-breeding researches for improving quality, enhancing biotic and abiotic stress tolerances (Nevo and Chen 2010). However, the vast genome size of the barley and the complexity of genome sequences lagged our understanding of barley studies behind rice and other plants. Recently, the situation has changed dramatically, since the next generation sequences technology leads to the completion of draft genome sequences and genome-wide transcriptional studies (Mayer et al. 2012; Dai et al. 2014). The achievement of population genomics and the representative genetic resources will be helpful in the better understanding of the wild barley species with respect to dealing with the environmental stress during its evolutionary process. Heat shock proteins (HSP) consists of large protein families, and they are highly conserved for all organisms including cereal species with large genome. The expression and function of the HSP gene families responsible for stress responses to heat and drought were clearly demonstrated in wild barley resources. However, the genomic evolution and adaptation of the HSP gene families in cereal genomes, including the barley species, need further investigation. This chapter focuses on the advances of genomic and evolutionary studies of HSP protein genes in wild barley, particularly the researches on unique wild barley populations at "Evolution Canyon", Mount Carmel, Israel.

5.2 "Evolution Canyon" Model

The "Evolution Canyon" I (EC I, Fig. 5.1) microsite has ecologically divergent environments (biomes) on a microscale level, mirroring abutting continents, which reinforces evolutionary biological studies exploring the interaction between the organisms and their environment (Nevo list of "Evolution Canyons" studies at http://evolution.haifa.ac.il, reviewed in Nevo 1995, 1997, 2001, 2006, 2009, 2012, 2015). The "Evolution Canyon" model consists of four representative microsites in four regions of Israel, including Mount Carmel (EC I), the western Upper Galilee Mountains (EC II), Negev Desert (EC III), and Golan Heights (EC IV) (Fig. 5.2). The "Evolution Canyon" model (Fig. 5.3) reveals *evolution in action* at a microscale

Fig. 5.1 "Evolution Canyon" I (EC I) at lower Nahal Oren, Mount Carmel, Israel. The opposite slopes, separated on average by 250 m, represent sharply divergent microclimatic and ecological biomes: On the right, the hot and dry savannoid south facing slope (SFS), dubbed "African" slope (AS). On the left, the cool and humid forested, north facing slope (NFS), or "European" slope. The ECI has been a microsite model for studying biodiversity evolution, adaptation, and incipient sympatric speciation across life from bacteria to mammals

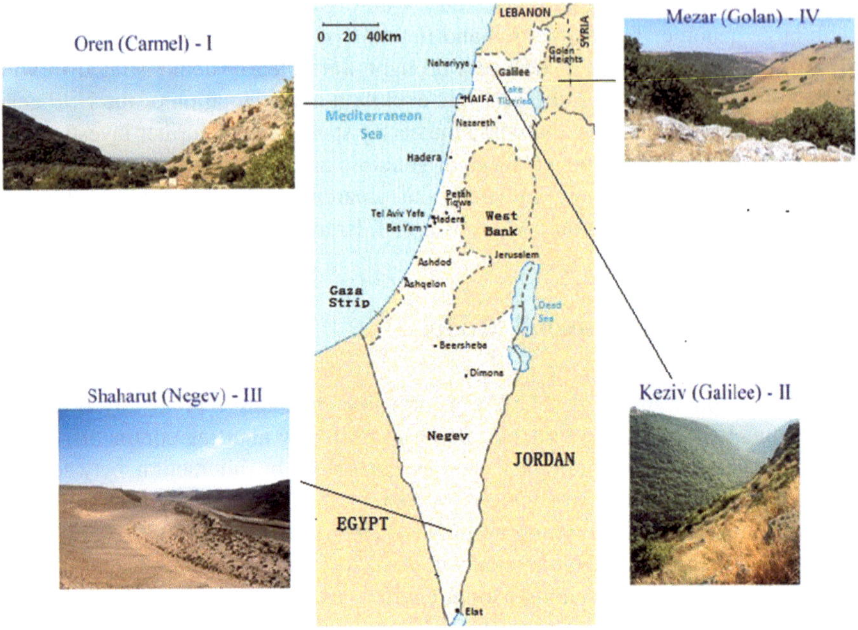

Fig. 5.2 Location of the four "Evolution Canyons" in Israel (EC I–IV). Note the savannoid south facing slope (SFS) on the *right hand side*, and the forested north facing slope (NFS) on the *left hand* in EC I, II, and IV. The orientation at EC III is opposite to the orientation in EC I, II, and IV. The SFS is on the *left hand side* covered primarily with cyanobacteria, whereas the NFS is on the *right hand side*, covered with lichens, and the scarce flowering annual and perennial plant species grow only in the valley

involving biodiversity evolution, adaptive ecological radiation, and incipient sympatric ecological speciation across life from viruses, bacteria and fungi to plants and animals (Nevo 2006, 2009, 2012; Nevo et al. 2012). Most studies have been conducted to date in "Evolution Canyon" I (ECI) at Lower Nahal Oren, Mount Carmel (Fig. 5.3) including 100 bacterial species, 500 fungi, 320 flowering plants, 1500 insects, and 80 vertebrates. This "Evolution Canyon" I consists of two abutting slopes separated, on average, by 250 m. The opposing "African" (AS) or south-facing (SFS) and "European" (ES) or north-facing (NFS) slopes are 100 m apart at the valley bottom, 400 m apart at the valley top, and 250 m at mid-slope (Nevo et al. 2005, 2006). The SFS at ECI in Mount Carmel receives 200–800 % higher solar radiation than the ES. The higher solar radiation on the AS leads to a hot and dry slope. By contrast, on the ES, the 17 % higher relative humidity on the ES, and less solar radiation leads to the humid and cooler ES (Pavlicek et al. 2003, Fig. 5.3a). The slopes share an identical Plio-Pleistocene evolutionary history, presumably 3–5 million years ago, sharing geology (upper Cenomanian Limestone), terra rossa soils (Nevo 1998), and regional Mediterranean climate and vegetation (Nevo et al. 1999). This dramatic microclimate divergence causes the drastic interslope divergence

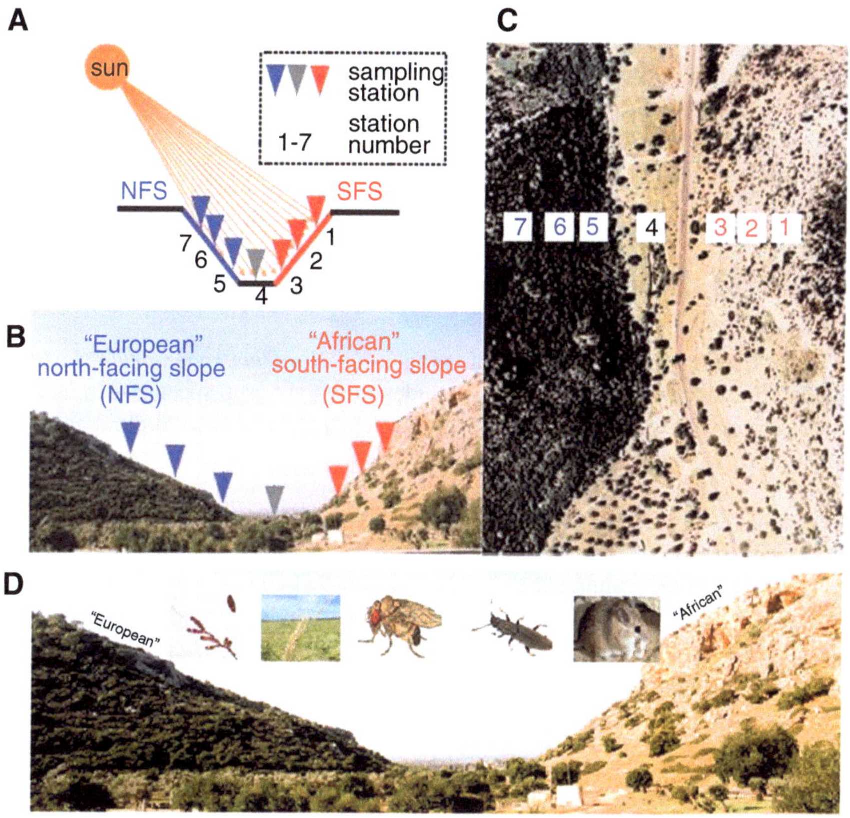

Fig. 5.3 "Evolution Canyon" I model in Israel. (**a**) represents a *microclimatic model*. (**b**) shows the cross section of "Evolution Canyon" (EC), with *arrows* pointing to collecting stations at AS = SFS in *red*, and to the ES = NFS in blue. Station 4 in the valley is in *gray*. (**c**) represents an air view of EC, with the seven collection stations (populations): 1–3 on AS = SFS, and 5–6 on ES = NFS. Station 4 is in the creek. The sharp divergence of savanna and forest habitats are seen in both the cross section of "Evolution Canyon" (EC) in B, and its air view in C. Collecting stations nos. 1–3 on the "African", tropical, savannoid south – facing slope and stations nos. 5–7 on the abutting "European", temperate, forested north- facing slope are seen in both A, B and C. (**d**) Cross section of EC I with the five model organisms that incipiently speciate in it. From *left* to *right*: soil bacterium, *Bacillus simplex*, wild barley, *Hordeum spontaneum*, fruit fly, *Drosophila melanogaster*, beetle, *Oryzaephilus surinamennsis*, and spiny mouse, *Acomys cahirinus* (From Nevo 2014)

across life from bacteria to mammals. A drastic example of the interslope genomic divergence is demonstrated by whole genome comparison of fruit fly, *Drosophila melanogaster* at EC I. Hubner et al. (2013) identified chromosomal differentiation "islands" which accumulate interslope genetic changes. Moreover, Kim et al. (2014), referring to the repeatome paper of fruit flies, found that almost half of the slope unique mobile element insertions in *Drosophila melanogaster* at EC I, disrupt coding sequences of genes critical for cognition, olfaction, and thermotolerance, which may be adaptive to the divergent interslope ecologies. Consequently, local biodiversity divergence across a few 100 m displays globally divergent patterns (Nevo 1995, 1997, 2001, 2009, 2011, 2012, 2015). The overall studies on organisms of the ECs model are from phenotypes and genotypes to genomic and epigenetic patterns. Recently, the EC microclimatic model was also successfully used to track global warming at a microscale across life from viruses and bacteria to mammals in Israel (Nevo 2012).

5.3 Wild Barley, *Hordeum spontaneum*

Barley (*Hordeum vulgare* L.) and its progenitor, *Hordeum spontaneum* (Fig. 5.4), belong to the tribe Triticeae, which includes important crop species such as wheat (*Triticum aestivum* L.) and rye (*Secale cereale* L.). As an annual diploid grass species, barley has a large haploid genome of 5.1-Gb and a high abundance of repetitive elements (Mayer et al. 2012). The genomic composition of transposable element (TEs) makes up at least 80 % of the overall Triticeae genome content. Although more than 200 TE families were identified in each species, approximately 50 % of the overall genome comprised 12–15 TE families (Middleton et al. 2013). As shown in Fig. 5.5, the barley genome is composed of long-tandem repeat sequences in subtelomeric or telomere-associated regions and the tandem repeated satellite DNA in centromeric and related regions (Cuadrado and Schwarzacher 1998; Heslop-Harrison and Schwarzacher 2011). Remarkably, the retrotransposon BARE1 of the

Fig. 5.4 A lush field of wild barley, *Hordeum spontaneum*, the progenitor of cultivated barley, growing on a Natufian terrace at the entrance to the Paleolithic Oren cave, at Lower Nahal Oren, "Evolution Canyon" I, in Mount Carmel, Israel. The rocks belong to upper Cenomanian limestone, covered with a gray patina involving some 50 species of cyanobacteria. The picture was taken in the spring, after spiking and before shattering. This field suggests the domestication of barley at the cave by Natufian and Neolithic human populations

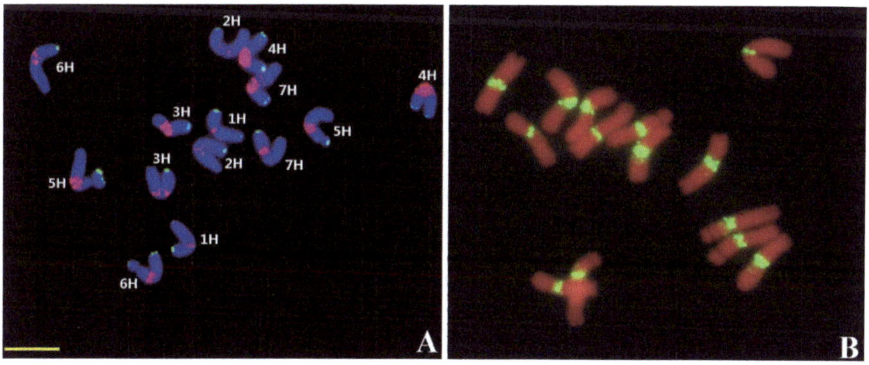

Fig. 5.5 Metaphase chromosomes of wild barley, *Hordeum spontaneum*, from ECI enriched in tandem repeated sequences revealed by fluorescence *in situ* hybridization (FISH). (**a**) FISH (*red*) and tandemly repeated DNA sequence probe Oligo-pSc119.2 (*green*; Tang et al. 2014) and simple sequence repeats probe Oligo- $(GAA)_7$ (*red*) by DAPI staining (*blue*); (**b**) FISH with simple sequence repeats probe Oligo- $(CAA)_7$ (*green*) by PI stain (*red*). Bar shows 10 μm (Data from Zujun Yang, unpublished)

TE clade in barley form approximately 10 % of the genome, and the study showed that BARE-1 retrotransposon dynamics was adaptive to the sharp microclimatic divergence on the opposite slopes of EC I, forming two divergent clusters on the AS and the ES, respectively (Kalendar et al. 2000). Thus, the TEs have been implicated in shaping the long evolutionary history of adaptation, speciation, and domestication at ECI. The TEs adaptive function in wild barley and *Drosophila* fruit flies was possibly through direct remobilization after stress, causing mutations and a source of new genetic variation (González et al. 2010; Kim et al. 2014).

As a founder crop in this process, and the most important steps in its domestication, barley was essential for human civilization's transition to agriculture-based societies. The domestication of barley displayed unique evolution processes and a molecular system that was distinct from other cereal crops, such as rice and maize (Sakuma et al. 2011; Zohary et al. 2012; Pourkheirandish et al. 2015). As one of the earliest crops domesticated in the Near East, barley is well adapted to semi-arid conditions (Nevo 2014), and the natural distribution of wild barley was dispersed widely from the Near East to Central Asia and the Tibetan Plateau (Poukheirandish and Komatsuda 2007; Dai et al. 2012, 2014). The Fertile Crescent is the center of its distribution and, most likely, the region is a primary center of barley origin, diversity, and domestication, where cultivated barley was domesticated from wild barley (Badr et al. 2000; Zohary et al. 2012). Moreover, a large number of wild barleys have been collected from the Tibetan Plateau, which is characterized by an extreme environment. Dai et al. (2012) used genome-wide diversity array technology markers to analyze the genotypic division between wild barley from the Near East and Tibet, and they suggested a concept of polyphyletic domestication of cultivated barley and indicated that the Tibetan Plateau and its vicinity is also one of the centers of domestication of cultivated barley. Moreover, Pourkheirandish et al. (2015) found the genes Btr1 and Btr2, which prevent grains from shattering in cereals, and revealed that present-day cultivars derived from two ancient domestication centers. They confirmed that domestication of wild barley occurred at about 12,000 years ago in the Near East.

Wild barley occupies a variety of local habitats that differ by soil type, temperature, precipitation, and altitude. This ecological diversity is reflected in the genetic population structure in the natural species range (Nevo et al. 1979; Morrell et al. 2003; Morrell and Clegg 2007). On a smaller geographic scale, both phenotypic and genetic variations of wild barley are highly variable along an environmental gradient in Israel, which comprises a transition from cool and moist Mediterranean climate in the North to the arid desert climate in the Negev (Nevo et al. 1979; Volis et al. 2002a,b; Vanhala et al. 2004). Because of its occurrence in a wide range of different habitats, the wild progenitor of barley was therefore an important resource to study the evolution, adaptation, and domestication of cultivated barley (Nevo 1992). Recently, Dai et al. (2014) compared transcriptome profiling of cultivated and wild barley genotypes, and revealed that the domestication of barley may have occurred over time in geographically distinct regions. Zeng et al. (2015) sequenced the Tibetan hulless barley lines, and the expansion of gene families related to stress responses was found in Tibetan hulless barley. Both studies revealed high levels of genetic variation in Tibetan wild barley, and they suggested adaptive correlations of Tibet barley genes under selection with extensive environmental stresses.

Climate change is a major environmental stress threatening biodiversity and human civilization. Nevo et al. (2012) used ten wild barley populations in Israel, sampling them in 1980 and again in 2008, and performed phenotypic and genotypic analyses on the collected samples. Profound adaptive changes of these wild cereals were witnessed in Israel over the last 28 years in both flowering time and simple sequence repeat allelic turnover. Since global warming does affect phenotypic and genotypic characteristics (Nevo et al. 2012), it is important to follow its current evolution under global warming and preserve its rich genetic resources for crop improvement (Nevo 2015).

The rich genetic diversity of wild barley with its success in adapting to different environments has made it an excellent model to reveal the important genes influencing stress and evolution. Ma et al. (2012) observed that drought stress and adaptive natural selection might have been important determinants in the sequence variation of HvABCG31 promoter. Dehydrins are water-soluble, lipid vesicle-associating proteins involved in the adaptive responses of plants to drought, low temperature, and salinity (Close et al. 2000). Cronin et al. (2007) found highly significant correlations between diversity at the *Isa* locus, which has a putative role in plant defense and key water variables, e.g., evaporation, rainfall, humidity, and latitude. Natural microclimatic selection was apparently the major evolutionary driving force causing interslope divergence and adaptive evolution of these genes. Liu et al. (2014) reported that the genetic variation in amy1-2 and amy2-2 was at least partly ecologically determined in these populations, representing adaptive patterns generated by natural selection. The only likely explanatory model for divergent interslope polymorphisms at EC I is adaptive natural selection as the major evolutionary divergent driving force (Yang et al. 2009; Zhang et al. 2014).

5.4 Heat Shock Proteins (HSP) in Grasses

In plants, HSP genes are accumulated in response to a large number of stress factors such as heavy metals, water stress, light, hormones, abscisic acid, wounding, excess NaCl, chilling, heat, cold, and anoxic conditions (Süle et al. 2004). The plant's HSPs play a crucial role in protecting plants against stress and in the reestablishment of cellular homeostasis (Wang et al. 2004). Heat stress is common in most cereal-growing areas of the world. Most of the world crops are exposed to heat stress during some stages of their life cycle (Stone 2001). No single 'thermo-tolerant' gene has been found in cereals to control heat tolerance. The heat stress response and expression of heat shock proteins (HSPs) in thermo-tolerance is important for protecting the yield and quality of cereal. In particular, cereals consist of a number of polyploid species; the molecular diversity of HSPs within each plant is increased due to the contribution of the sub-genomes. Furthermore, HSP gene families have more members in even diploid cereals than in dicots contributing to the molecular diversity (Maestri et al. 2002). Heat shock proteins are commonly grouped in plants into five classes according to their approximate molecular weight: (1) Hsp100, (2) Hsp90, (3) Hsp70, (4) Hsp60, and (5) small heat-shock proteins (sHsps). Higher plants have at least 20 sHsps, and there might be 40 kinds of these sHsps in one plant species.

5.4.1 Small Heat Shock Proteins in Barley

A majority of these small proteins weighing 12–42 kDa are small heat shock proteins (sHSPs) and act as chaperones that play an important role in various developmental, and biotic and abiotic stresses. The sHSP family possesses a conserved domain of approximately 80–100 amino acids, called alpha-crystalline domain (ACD), and are flanked by N- and C-terminal regions. Although sHSPs were first discovered as induced by heat stress, the patterns of sHSP expression are complex and vary between organisms and developmental stages (Santhanagopalan et al. 2015). The sHSPs are extremely diverse and variable in plants. Some plants have over 30 individual sHSPs. Land plants possess distinct sHSP subfamilies (Waters 2013). The *Arabidopsis* sHsp gene family revealed that there are 19 coding genes (Scharf et al. 2001; Basha et al. 2013). Phylogenetic analysis showed that 23 out of these 40 alpha-crystallin domain (ACD) genes constitute sHsps of rice; 17 of 23 rice sHsp genes were noted to be intronless (Sarkar et al. 2009; Wang et al. 2014). A genome-wide sequence survey identified 20 non-redundant small heat shock proteins (sHsp) and 22 heat shock factor (Hsf) genes in barley (Reddy et al. 2014).

They found that all three major classes (A, B, and C) of Hsfs are localized in the nucleus, and the 20 sHsp gene family members are localized in different cell organelles like cytoplasm, mitochondria, plastids, and peroxisomes (Reddy et al. 2014). Pandey et al. (2015) report genome-wide identification and characterization of 27 new *TaHsp20* candidate genes in wheat and 13 *Hvhsp20* in barley. The structural analysis highlights that this gene family possesses a conserved ACD region at the C-terminal. Based on the wheat Hsp17, Hsp20, and Hsp90 EST sequence (Xue et al. 2014), a genome-wide search was made at the National Center for Biotechnology Information (NCBI) GenBank database and IPK barley BLAST server of Morex assembled contigs. A total of 51 contigs homologous to these Hsp were predicted, and the chromosomal location was shown in Fig. 5.6. The results showed that these Hsp families clearly showed duplications among the different chromosomes, in which the chromosome 6H was enriched in these Hsps genes, with a few Hsps located in chromosome 1H.

The expression of HSPs is regulated by multiple mechanisms, mainly on a transcriptional level. Heat shock transcription factor (Hsf) is the master regulator in this process, playing critical roles in high-temperature stress responses and thermal tolerance. It is believed that diversification of these Hsf proteins is essential for cereal stress responses. Among A, B, and C classes of Hsf genes, the class A and C Hsf proteins possess an inserted sequence of 21 and 7 amino acid residues, respectively, which is absent from class B Hsfs (Scharf et al. 2012). Yang et al. (2014) performed a genome-wide survey of *Hsf* genes in five grass species, including rice, maize, sorghum, setaria, and *Brachypodium*, by describing their phylogenetic relationships, adaptive evolution, and expression patterns under abiotic stresses. As shown in Table 5.1, the Hsf genes in grasses were divided into over 20 orthologous gene clusters (OGCs) based on phylogenetic relationships and synteny, suggesting that 24 Hsf genes were present in the ancestral grass genome. The duplication and gene-loss events were identified in the tested genomes. Based on the genomic distribution of the Hsf genes, it was clearly noted that the polyploid wheat consists of more members because of its polyploidization of three different sub-genomes. Sarkar et al. (2009) determined that positive selection acted on only one of the duplicated genes in 8 of 9 paralogous pairs, suggesting that neo-functionalization contributed to the evolution of these duplicated pairs. They also investigated the expression patterns of rice and maize Hsf genes under heat, salt, drought, and cold stresses. The results revealed divergent expression patterns between the duplicated genes among the various plant Hsps (i.e., Hsp100, Hsp90, and Hsp70), Hsp20, or sHsps (Sarkar et al. 2009).

Most of the sHSPs are highly up-regulated in response to heat and other stressors. Others are selectively expressed in seeds and pollen, and a few are constitutively expressed. The evolutionary patterns for most of the sHSP subfamilies reflect selection to maintain consistent function across species (Waters 2013). Hsf and sHsp members are differentially regulated during drought and at different seed developmental stages, suggesting the importance of the chaperone role in plant development and stress tolerance (Waters et al. 1996; Waters 2013). Reddy et al. (2014) found that Hsf and sHsp members are differentially regulated during drought and at different

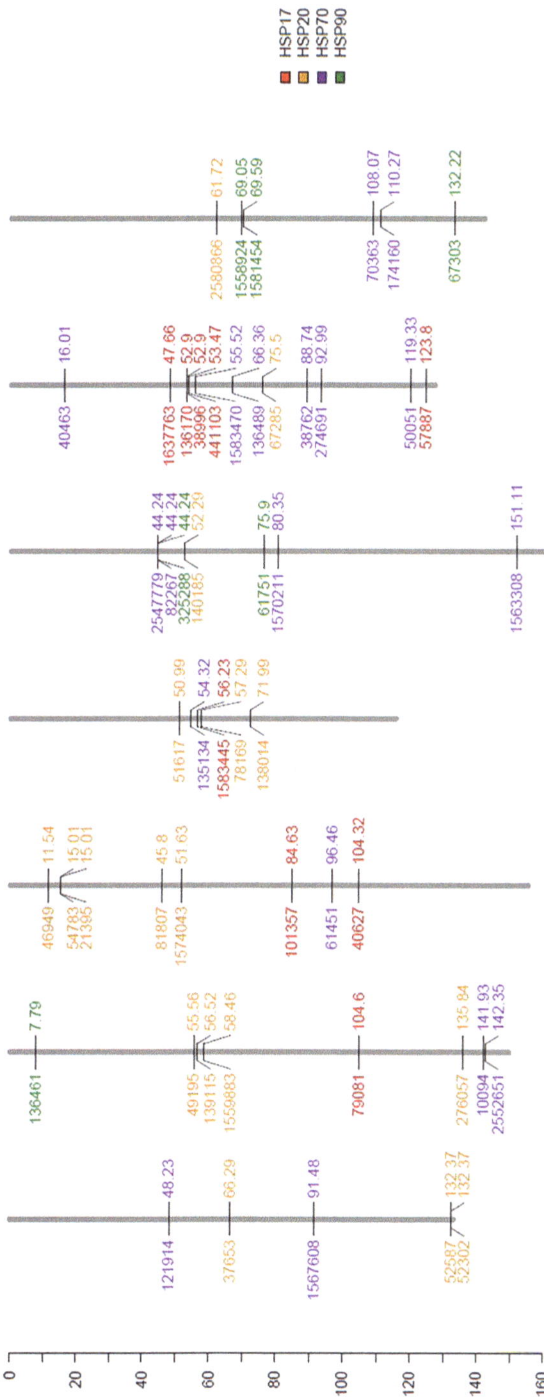

Fig. 5.6 Chromosome distribution of putative HSPgene families in barley. The *left* is the contig number of barley cv. Morex (Data from Zujun Yang, unpublished)

Table 5.1 Plant heat shock factors

Cereals	*Arabidopsis*	Rice	Maize	*Brachypodium*	Barley	Wheat
HsfA1	a, b, d, e	a	a, c	a	A	a, b
HsfA2	a	a, b, e	a, b, e	a, b, e	a, b, c, d, e, f	a, b, c, h, i
HsfA3	a	a	a	a	A	a
HsfA4	a, c	a, d	a, b	a, d	b, d	a, b, c, d, e
HsfA5	a	a	a	a	A	a, b,
HsfA6	a, b	a	a, b	a		a, e
HsfA7	a, b	a, b	a, b	a, b		a, b
HsfA8	a	a	a, b	a		a,
HsfA9	a		a, b		A	
HsfB1	a	a	a, b	a	A	a, b, c,
HsfB2	a, b	a, b, c	a, b, c, d	a, b, c	a, b, c	a, b, c, d, e
HsfB3	a					
HsfB4	a	a, b, c, d	a, b, d	b, c, d		a
HsfB5						
HsfC1		a, b	a, b, c,	a, b	a, b	a, b, c, d, e,
HsfC2		a, b	a, b	a, b	a, b	a, b, d, e, f, g
Total	21	25	30	23	20	40

seed developmental stages suggesting the importance of the chaperone role under drought as well as seed development. They isolated the barley Hsp 17.5 gene and found that it was predominantly up-regulated under drought stress treatments and also preferentially expressed in developing seeds. *In silico* cis-regulatory motif analysis of Hsf promoters showed enrichment with abscisic acid responsive cis-elements (ABRE), implying a regulatory role of ABA in mediating transcriptional responses of HvsHsf genes (Reddy et al. 2014; Pandey et al. 2015).

5.5 HSP and Stress and Evolution

The evolutionary processes that generate complex gene families include gene duplication, recombination, and gene loss (Ohnu 1970; Taylor and Raes 2004). It is likely that all these evolutionary forces, i.e., gene duplication, gene conversion, neo-functionalization, and genome duplication, have played a role in the evolution of the sHSPs (Waters 1995; Waters et al. 2008). The clear differences in the evolutionary patterns among the plant sHSP subfamilies suggest undiscovered functional differences among subfamilies (Waters 2013). The sHSPs may be differentially expressed in different accessions within a species which were exposed to the stress conditions. The ability of different genotypes to adapt to various stress conditions may be related to their genetic diversity of allelic variations in natural populations (Sun et al. 2002; Zhang et al. 2014). Several studies suggested that the analysis of gene sequences within species could also elucidate important evolutionary patterns among the *sHsps* (Wu et al. 2007; Waters et al. 2008; Liao et al. 2010). Recently, Ashoub et al. (2015) used leaf proteomics of wild barley and revealed that the response was distinguished from the response to transient heat stress, which included the up-regulation of a broad range of HSP products. Xia et al. (2013) investigated single nucleotide polymorphisms (SNPs) of the HSP17.8 gene across 210 barley accessions collected from 30 countries using EcoTILLING (Ecotype targeting induced local lesions in genomes) technology. They found that 4 SNPs in *Hsp17.8* were associated with at least one of the agronomic traits evaluated except for spike length, namely, the number of grains per spike, kernel weight, plant height, flag leaf area and leaf color. The association between SNP and these agronomic traits may provide new insight for the study of the gene's potential contribution to drought tolerance of barley. Taken together, all of these studies indicate that further studies of sHSP sequence diversity within and between closely related species under different stress environments would provide important insights into sHSP evolution.

5.6 Barley HSP17 at Evolution Canyon

Gene duplication is a major evolutionary driving force for establishing new gene functions. However, very little is known about the involvement of the structural divergence of recent duplicated genes in local adaptation. Small heat shock protein 17 genes (*Hsp17*) represented a model to identify the role of genetic polymorphism in phenotypic variations in response to environmental stresses by the powerful approach of association analysis (Yu and Buckler 2006; Zhu et al. 2008). The HSP17 is produced abundantly in plant cells under heat and other stress conditions and may play an important role in plant tolerance to stressful environments. Montfort et al. (2001) found and isolated the crystal structure of the HSP17 from wheat and wild barley genotypes, but not in drought-sensitive genotypes under drought stress. Grigorova et al. (2011) found that the *Hsp17* transcripts were present at a higher level in the drought-tolerant wheat cultivar than in the non-tolerant wheat cultivar. The changes in gene expression of hsp17 genes at the transcriptional level in barley leaves during the reproductive stage under drought conditions were observed. Guo et al. (2009) used the 22K Affymetrix Barley 1 microarray to screen two drought-tolerant barley genotypes; the *Hsp17* showed expression exclusively in drought-tolerant barley. Based on the comparative sequence analysis of the *Hsp17*s the extensive gene duplication and conversion in the grass genome was indicated (Waters et al. 2008). The physiological and molecular population genetic studies of wild barley (Nevo et al. 1997; Nevo 2015; Cronin et al. 2007) confirmed that wild barley at EC displays dramatic interslope adaptive genomic divergence, slope-specific fitness components, and incipient sympatric speciation on the opposite slopes (Nevo 2006, 2009; Yang et al. 2009). The evolutionary driving force of the

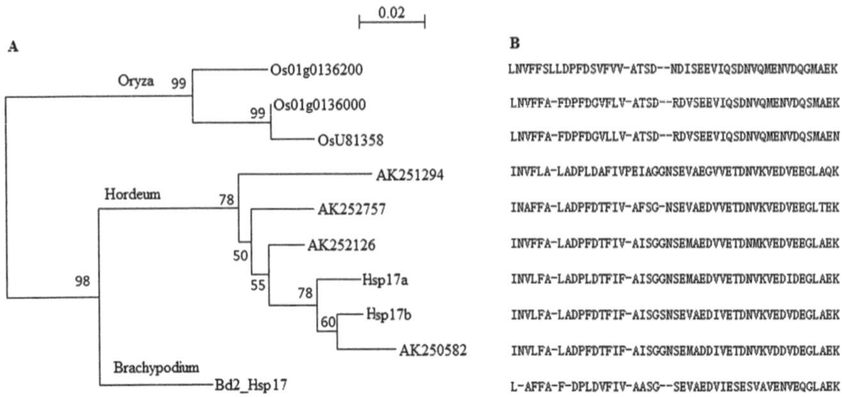

Fig. 5.7 (**a**) A maximum parsimony tree of the *Hsp17* homologous families based on nucleotide sequences of coding regions. Bootstrap values based on 1000 replicates are indicated above the branches. The lengths of tree branches are proportional to the number of mutations. (**b**) Amino acid sequence alignment of the *Hsp17* homologous families. Only variable sites are shown and N-terminal and C-terminals are indicated above the amino acid alignments. The scale bar is 10-nt substitutions (From Zhang et al. 2014)

sHSP can be further investigated both across the species and within the stressed population on the microevolution level using the EC model.

A total of 9 putative *Hsp17* sequences were located on 4 chromosomes of barley (Fig. 5.6). Two *Hsp17* copies, named *Hsp17a* and *Hsp17b* of wild barley, were very likely orthologous to the locus on chromosome 3H, collinear with rice chromosome 1 (Bossolini et al. 2007; Sato et al. 2009). A phylogenetic tree, based on the amino acid sequences of these *Hsp17* homologs, was generated (Fig. 5.7) and indicated that the duplication of *Hsp17a* and *Hsp17b* sequences likely occurred about 2–3 Mya. We evaluated the nucleotide variation of recent duplicated pair copies of the Hsp17 locus for small heat shock proteins, namely, the older copy *Hsp17a* and the younger copy *Hsp17b* (Fig. 5.7). Forty wild barley genotypes from seven climatically divergent sites of "Evolution Canyon" I, Mount Carmel, Israel, were investigated for sequence diversity of their *Hsp17* copies (Zhang et al. 2014). Populations of *H. spontaneum* were sampled in ECI at 7 collection sites (populations) (see Fig. 5.3), and at 3 elevations that were 30 m apart (60, 90, and 120 m above sea level) on each slope: AS = SFS: 1, 2 and 3; and ES = NFS: 5, 6 and 7, and 1 site (station 4) at the valley bottom (VB). Seeds from 40 genotypes, including 4–9 genotypes from each population, were used for the study.

We sequenced the 456 bp coding sequence and 397 bp 5′-UTR. The *Hsp17a* contained a 3′-UTR of 99 bp, while the *Hsp17b* had a 3′-UTR of 492 bp. The single nucleotide polymorphisms (SNPs) of the *Hsp17* locus were investigated, and 150 SNPs of *Hsp17a* and 135 SNPs of *Hsp17b* in the coding region were identified. Total amino acid substitutions were present at 43 positions of Hsp17a and 35 positions of *Hsp17b* among the seven populations. The total number of nucleotide changes in the populations from the 2 slopes (36 in AS vs 16 in the ES) is significant

($p < 0.05$). The variation in the *Hsp17b* 3′-UTR insertion in the AS was significantly higher than variations in the ES, indicating a nonrandom distribution. The genetic divergence and haplotype diversity were significantly different among the populations. For *Hsp17b*, the site 21 L was detected as positively selected; for *Hsp17a*, the two amino acid sites 30G and 60D were positively selected.

In order to represent the relationships between genotypes of the coding region of *Hsp17a* and *Hsp17b*, cluster analysis (NJ) was used to generate a dendrogram based on the calculated nucleotide diversity (π) values (Nei 1987). Two main clusters of *Hsp17a* and *Hsp17b* were clearly evident (Fig. 5.8). Most of the genotypes from either the AS or ES were clustered in different groups, reflecting their high similarity within the slopes and the large divergence among the slopes. An analysis of Pearson correlation coefficients was performed between climatic parameters of the EC with haplotype diversity (Hd) and nucleotide diversity (π). Remarkable correlation values were obtained between Hudd (daily relative humidity difference), Tdd (daily temperature difference), Tm (average ambient temperature) with Hd of *Hsp17a*, and negative correlation of Hsp17a Hd to Pc (general plant cover), which support the hypothesis that the *Hsp17a* gene was subjected to adaptive natural microclimatic selection.

The expression pattern of the *Hsp17a* and Hsp17b genes was also conducted after the dehydration treatment of the seedlings was carried out at 0, 3, 6, 9, 12, and 24 h. The results indicated that the *Hsp17* of wild barley was clearly upregulated by heat shock. The Hsp17a was immediately expressed by heat treatment and was maintained continuously. We have not observed the differences of *Hsp17a* expression, possibly because of the upregulation occurring so fast. However, the expression of Hsp17b was different among the examined genotypes. The interslope divergence of the *Hsp17* expression patterns observed between AS and ES stations was remarkably related to the changes of 3UTR insertion.

The *Hsp17a* gene sequences showed a considerably higher evolutionary rate of amino acid substitutions than that of the *Hsp17b* sequences. The alternative models of functional diversification following gene duplication (Lynch and Conery 2000; Zhang 2003) might explain the divergence of Hsp17b from Hsp17a. This model is supported by evidence that nucleotide substitution rates in Hsp17a differ significantly from those in Hsp17b. Moreover, one copy might evolve a new function via positive selection. During this process, gene duplication and the resulting functional divergence enable lineages to have better fitness in rather different environments (Hanada et al. 2008; Zou et al. 2009). The *Hsp17* diversity was also correlated with microclimatic interslope divergence factors in "Evolution Canyon" I. The natural microclimatic selection appears to adaptively diversify the duplicated copies of *Hsp17* of wild barley between the opposite warmer (AS) and cooler (ES) slopes, suggesting that the recent duplication of stress-responsive genes were subjected to adaptive evolution. Gene duplication and the resulting functional divergence enable plants to have better fitness in local climate change. Comparing the recent studies of the stress-induced drought resistant dehydrin genes *Dhn1* (Yang et al. 2009) and *Dhn6* (Yang et al. 2011), Hsp17 has higher interslope divergences, while the disease resistance-related gene *Isa* had fewer polymorphisms (Cronin et al. 2007). The ECI

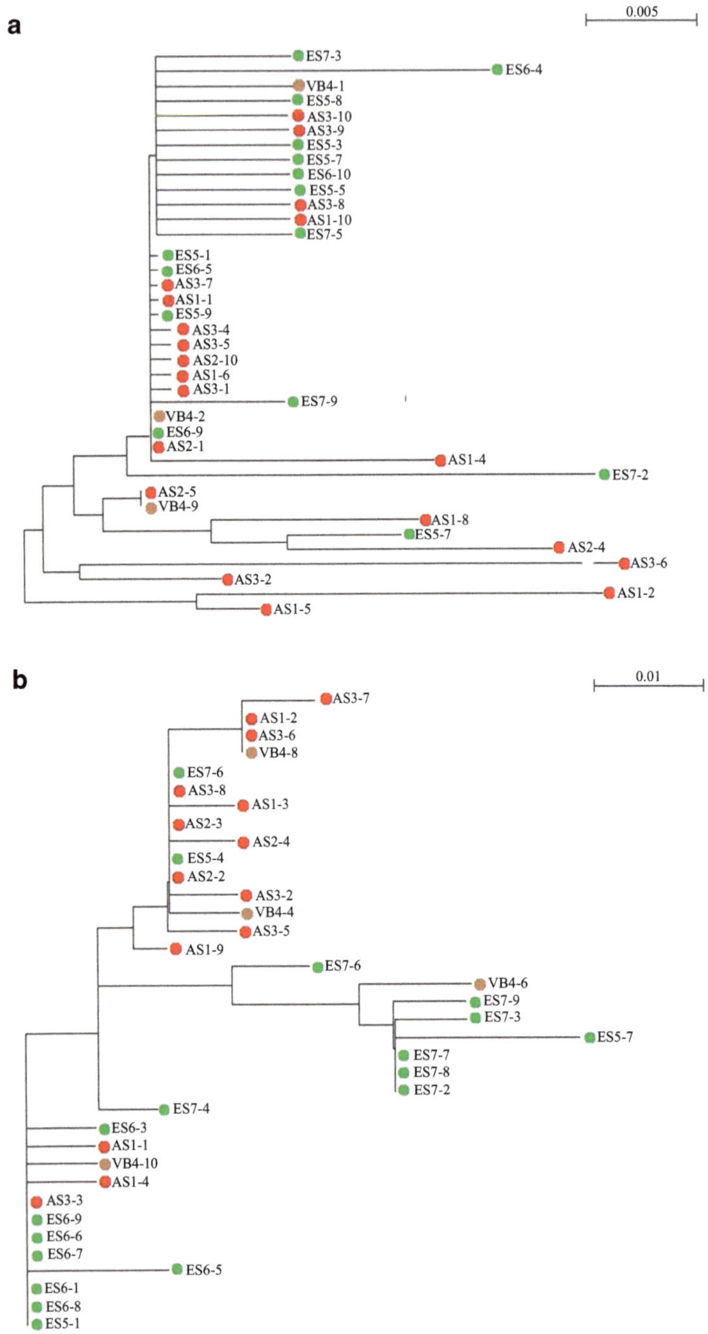

Fig. 5.8 Dendrogram based on genetic relationships of the genotypes of wild barley from seven populations, representing the Hsp17a (**a**) and Hsp17b (**b**) in "Evolution Canyon" I, Mount Carmel, Israel. These genotypes were obtained from the Hsp17 alignment sequence, based on Nei's (1987) calculated nucleotide diversity (p-distance) values (see scale), using the NJ method. Numbers on branches are percentage values from bootstrap analysis (1000 replicates) (From Zhang et al. 2014)

model provides clues for future evolutionary studies on both adaptive stress genes and speciation genes. The study of sHSP evolution in microsite populations may aid in understanding how structural variation was significantly related to environmental stresses. The nucleotide polymorphism of *Drosophila* hsp23 and hsp26 shows a latitudinal cline in Australia, suggesting they are under selection in natural populations (Frydenberg et al. 2003). Carmel et al. (2011) carried out similar studies on fruit flies (*Drosophila melanogaster*) of "Evolution Canyon" I (EC I). They found differential expression of small heat shock protein genes Hsp23 and Hsp40 in fruit flies on a microclimatic gradient of the NFS and SFS, and revealed that a positive correlation between the expression of Hsp40 gene after the applied mild heat treatment and the flies' thermo-tolerance. They also found that a significant difference in Hsp40 expression level occurred between flies originating from the opposite EC I slopes.

5.7 Future Studies on HSP at the "Evolution Canyon" Model

The evaluation of a specific heat shock proteins' contribution to barley drought resistance required the combination of genetic, genomic, ecological, and physiological studies. Quantitative trait loci for salinity tolerance identified under-drained and waterlogged conditions and their association with flowering time in barley, and the agronomic trait involving yield formation (Tondelli et al. 2006; Islamovic et al. 2013; Ma et al. 2015). Through the relationships with markers common between the consensus and other linkage maps of barley, the position of cold and drought tolerance QTLs have been added to the consensus function map (Cattivelli et al. 2002; Nevo et al. 2012; Islamovic et al. 2013; Ma et al. 2015). Moreover, the candidate gene strategy shows promise in bridging quantitative genetic approaches with molecular and genetic approaches in studying complex traits, the transcriptional regulators are better candidates to give a measurable trait at the QTL level than effector genes (Cattivelli et al. 2002; Tondelli et al. 2006). The genome-wide search of the heat shock protein-related genes in a barley map will also shed light on the utilization for barley improvement.

Bedada et al. (2014a) sequenced 34 short genomic regions in 54 accessions of wild barley collected throughout Israel and from the opposing slopes of two "Evolution Canyons" (EC I and EC II). There is a high level of genetic clustering throughout Israel and within EC, which roughly differentiates the slopes. Accessions from the hot and dry south-facing slope (AS = SFS) have significantly reduced genetic diversity and are genetically more distinct from accessions from the north-facing slope (ES = NFS). Bedada et al. (2014b) found a strong physiological and genomic differentiation between the desert and Mediterranean wild barley ecotypes and a closer relationship of the Mediterranean to cultivated barley. Nevo et al. (2012) reported that the existing global climate warming during 28 years (from 1980 to 2008) significantly caused phenotypic (earliness in flowering) and genotypic (SSR turnover, involving genetic diversity reduction and appearance of new possibly adaptive alleles) differences, affecting similarly wild barley and wild emmer wheat, the progenitors of barley and wheat, across Israel. Since wild cereals are the best hope for crop improvement, the wild progenitors need to be preserved *in situ* and *ex situ* (Nevo 1998; Nevo and Chen 2010). Greater responses to climate change in the xeric populations compared with mesic populations are expected and empirically confirmed. A high level of population structure at large geographic scale shows isolation-by-distance and is also consistent with ongoing natural selection contributed to genetic differentiation on a small geographic scale (Bedada et al. 2014a).

The adaptation of wild barley to drought and salinity environments has accumulated rich adaptive genetic diversities, which is an excellent genetic resource for crop improvement (Nevo and Chen 2010). Advances are still needed to efficiently explore the extensive reservoir of novel agronomic important traits such as drought and salt-tolerant alleles from wild barley germplasm: (1) developing wild barley high throughput phenotypically, genotypically, and epigenetically to screen heat, drought, or salt tolerances related to a regulation network. (2) By utilizing next generation sequencing, increasing availability of sequence information, enhancing the understanding of the potential regulatory relationship between coding and non-coding regions. (3) Attention will be on the dynamic chromatin remodeling genome-wide level particularly for understanding the possible mechanism for barley to fit the environmental changes, such as with fast global warming. (4) To consider the extensive international collaborative projects in ensuring the use of wild barley germplasm for global cereal crop improvement. The whole genome sequence of wild barley, *H. spontaneum* is currently under preparation and could be an important resource for stress genes for barley improvement.

Recently, Pourkheirandish et al. (2015) cloned non-brittle rachis (Btr) genes during the domestication of barley. They found that characteristics evolved by duplication and neofunctionalization of two genes, *Btr1* and *Btr2*. *Btr* and *Btr*-like genes were diverged functionally after a duplication event that occurred specifically in the Pooideae lineage. It is likely that the heat shock proteins endure the significant duplication in cereal genomes (Moore and Purugganan 2005). It is important to view the duplication and neo-functionalization of the Hsp genes and the regulated pathways.

There has been a breakthrough in the technology of genome editing, the CRISPR-Cas9 (clustered regulatory interspaced short palindromic repeats/CRISPR-associated protein 9) technology, in the last decade (Mali et al. 2013). The CRISPR-Cas 9 technology has been successfully applied in model plants such as *Arabidopsis* and tobacco and important crops like wheat, maize, rice, sorghum, tomato, and barley (Jiang et al. 2013; Brooks et al. 2014; Feng et al. 2014; Shan et al. 2014; Gao et al. 2015; Lawrenson et al. 2015). Basak and Nithin (2015) applied the CRISPR-Cas9 technology in editing ncRNAs in plants. With respect to modifying the target heat shock proteins, there is a possibility to regulate the network for barley or cereals to adapt the changing stress environment with the aim of revealing the genetic and epi-genetic mechanism of Hsp evolution and modifying the gene expression for agricultural improvement. It will be of substantial importance to follow the evolutionary dynamics of duplicated Hsps in a planet undergoing global warming.

5.8 Conclusion

Wild barley, *Hordeum spontaneum*, the progenitor of cultivated barley, is an excellent theoretical and empirical evolutionary model for studying adaptive and speciational patterns in both *macro*evolution and *micro*evolution. Wild barley is a hardy generalist species, growing in divergent climatic and edaphic environments, and displays abundant adaptive genetic and genomic diversity exploitable for barley improvement. The "Evolution Canyon" I (EC I) at lower Nahal Oren, Mount Carmel, Israel, reveals *evolution in action* of biodiversity, adaptation, and sympatric speciation, across life at a microsite caused by interslope microclimatic divergence. The studies on genomic divergence of Heat Shock Proteins (HSP) gene family in wild barley at EC I provide a basic understanding of the genetic diversity of HSP, its causation and adaptive expression. HSP diversity is associated with heat and drought environmental stresses on the savannoid hot and dry African slope, separated by 250 m from the forested shaded, cool and humid European slope. Further investigations on the genome-wide level is underway to reveal the structure of HSP genes and the adaptive functional and regulatory divergence in wild barley, and other related cereal species in ECI. The candidate gene variation of HSP genes and the unique wild barley genotypes will be very useful in the genetic improvement of cultivated barley.

Acknowledgements This work was supported by the National Natural Science Foundation of China (31171542) and the Ancell-Teicher Research Foundation for Genetics and Molecular Evolution. We are thankful for Dr. Avigdor Beiles of University of Haifa, Israel for reading and commenting on the manuscript.

References

Ashoub A, Baeumlisberger M, Neupaertl M, Karas M, Brüggemann W (2015) Characterization of common and distinctive adjustments of wild barley leaf proteome under drought acclimation, heat stress and their combination. Plant Mol Biol 87:459–471

Badr A, Muller K, Schafer-Pregl R, El Rabey H, Effgen S, Ibrahim HH, Pozzi C, Rhode W, Salamini F (2000) On the origin and domestication history of barley (*Hordeum vulgare*). Mol Biol Evol 17:499–510

Basak J, Nithin C (2015) Targeting non-coding RNAs in plants with the CRISPR-Cas technology is a challenge yet worth accepting. Front Plant Sci 6:1001

Basha E, Jones C, Blackwell AE, Cheng G, Waters ER, Samsel KA, Siddique M, Pett V, Wysocki V, Vierling E (2013) An unusual dimeric small heat shock protein provides insight into the mechanism of this class of chaperones. J Mol Biol 425:1683–1696

Bedada G, Westerbergh A, Nevo E, Korol A, Schmid KJ (2014a) DNA sequence variation of wild barley *Hordeum spontaneum* (L.) across environmental gradients in Israel. Heredity (Edinb) 112:646–655

Bedada G, Westerbergh A, Müller T, Galkin E, Bdolach E, Moshelion M, Fridman E, Schmid KJ (2014b) Transcriptome sequencing of two wild barley (*Hordeum spontaneum* L.) ecotypes differentially adapted to drought stress reveals ecotype-specific transcripts. BMC Genomics 15:995

Bossolini E, Wicker T, Knobel PA, Keller B (2007) Comparison of orthologous loci from small grass genomes *Brachypodium* and rice, implications for wheat genomics and grass genome annotation. Plant J 49:704–717

Brooks C, Nekrasov V, Lippman ZB, Van Eck J (2014) Efficient gene editing in tomato in the first generation using the clustered regularly interspaced short palindromic repeats/CRISPR-associated 9 system. Plant Physiol 166:1292–1297

Carmel J, Rashkovetsky E, Nevo E, Korol A (2011) Differential expression of small heat shock protein genes Hsp23 and Hsp40, and heat shock gene Hsr-omega in fruit flies (*Drosophila melanogaster*) along a microclimatic gradient. J Hered 102:593–603

Cattivelli L, Baldi P, Crosatti C, Di Fonzo N, Faccioli P, Grossi M, Mastrangelo AM, Pecchioni N, Stanca AM (2002) Chromosome regions and stress-related sequences involved in resistance to abiotic stress in Triticeae. Plant Mol Biol 48:649–665

Close TJ, Choi DW, Venegas M, Salvi S, Tuberosa R, Ryabushkina N, Turuspekov Y, Nevo E (2000) Allelic variation in wild and cultivated barley at the Dhn4 locus, which encodes a major drought-induced and seed protein, DHN 4. In: 8th international barley genetics symposium, 22–27 October 2000, Adelaide, SA, Australia

Cronin JK, Bundock PC, Henry RJ, Nevo E (2007) Adaptive climatic molecular evolution in wild barley at the *Isa* defense locus. Proc Natl Acad Sci U S A 104:2773–2778

Cuadrado A, Schwarzacher T (1998) The chromosomal organization of simple sequence repeats in wheat and rye genomes. Chromosoma 107:587–594

Dai F, Nevo E, Wu D, Comadran J, Zhou M, Qiu L, Chen Z, Beiles A, Chen G, Zhang G (2012) Tibet is one of the centers of domestication of cultivated barley. Proc Natl Acad Sci U S A 109:16969–16973

Dai F, Chen ZH, Wang X, Li Z, Jin G, Wu D, Cai S, Wang N, Wu F, Nevo E, Zhang G (2014) Transcriptome profiling reveals mosaic genomic origins of modern cultivated barley. Proc Natl Acad Sci U S A 111:13403–13408

Feng Z, Mao Y, Xu N, Zhang B, Wei P, Yang DL, Wang Z, Zhang Z, Zheng R, Yang L, Zeng L, Liu X, Zhu JK (2014) Multigeneration analysis reveals the inheritance, specificity, and patterns of CRISPR/Cas-induced gene modifications in *Arabidopsis*. Proc Natl Acad Sci U S A 111:4632–4637

Frydenberg J, Hoffmann AA, Loeschcke V (2003) DNA sequence variation and latitudinal associations in *hsp23*, *hsp26* and *hsp27* from natural populations of *Drosophila melanogaster*. Mol Ecol 12:2025–2032

Gao J, Wang G, Ma S, Xie X, Wu X, Zhang X, Wu Y, Zhao P, Xia Q (2015) CRISPR/Cas9-mediated targeted mutagenesis in *Nicotiana tabacum*. Plant Mol Biol 87:99–110

González J, Karasov TL, Messer PW, Petrov DA (2010) Genome-wide patterns of adaptation to temperate environments associated with transposable elements in *Drosophila*. PLoS Genet 6:e1000905

Grigorova B, Vaseva II, Demirevska K, Feller U (2011) Expression of selected heat shock proteins after individually applied and combined drought and heat stress. Acta Physiol Plant 33:2041–2049

Guo P, Baum M, Grando S, Ceccarelli S, Bai G, Li R, von Korff M, Varshney RK, Graner A, Valkoun J (2009) Differentially expressed genes between drought-tolerant and drought-sensitive barley genotypes in response to drought stress during the reproductive stage. J Exp Bot 12:3531–3544

Hanada K, Zou C, Lehti-Shiu MD, Shinozaki K, Shiu SH (2008) Importance of lineage-specific expansion of plant tandem duplicates in the adaptive response to environmental stimuli. Plant Physiol 148:993–1003

Heslop-Harrison JS, Schwarzacher T (2011) Organisation of the plant genome in chromosomes. Plant J 66:18–33

Hubner S, Rashkovetsky E, Kim YB, Oh JH, Michalak K, Weiner D, Michalak P (2013) Genome differentiation of *Drosophila melanogaster* from a microclimate contrast in "Evolution Canyon," Israel. Proc Natl Acad Sci U S A 110:21059–21064

Islamovic E, Obert DE, Oliver RE, Marshall JM, Miclaus KJ, Hang A, Chao S, Lazo GR, Harrison SA, Ibrahim A, Jellen EN, Maughane PJ, Brown RH, Jackson EW (2013) A new genetic linkage map of barley (*Hordeum vulgare* L.) facilitates genetic dissection of height and spike length and angle. Field Crops Res 154:91–99

Jiang W, Zhou H, Bi H, Fromm M, Yang B, Weeks DP (2013) Demonstration of CRISPR/Cas9/sgRNA-mediated targeted gene modification in *Arabidopsis*, tobacco, sorghum and rice. Nucleic Acids Res 41:e188

Kalendar R, Tanskanen J, Immonen S, Nevo E, Schulman AH (2000) Genome evolution of wild barley (*Hordeum spontaneum*) by BARE-1 retrotransposon dynamics in response to sharp microclimatic divergence. Proc Natl Acad Sci U S A 97:6603–6607

Kim YB, Oh JH, McIver LJ, Rashkovetsky E, Michalak K, Garner HR, Kang L, Nevo E, Korol AB, Michalak P (2014) Divergence of *Drosophila melanogaster* repeatomes in response to a sharp microclimate contrast in Evolution Canyon, Israel. Proc Natl Acad Sci U S A 111:10630–10635

Lawrenson T, Shorinola O, Stacey N, Li C, Østergaard L, Patron N, Uauy C, Harwood W (2015) Induction of targeted, heritable mutations in barley and *Brassica oleracea* using RNA-guided Cas9 nuclease. Genome Biol 16:258

Liao PC, Lin TP, Lan WC, Chung JD, Hwang SY (2010) Duplication of the class I cytosolic small heat shock protein gene and potential functional divergence revealed by sequence variations flanking the alpha-crystallin domain in the genus *Rhododendron* Ericaceae. Ann Bot 105:57–69

Liu Y, Wang JR, Liu ZH, Wang Y, Chen X, Henry RJ, Wei YM, Nevo E, Zheng YL (2014) Adaptive evolution of a-amylase genes in wild barley (*Hordeum spontaneum*) on micro and macro scales. J Syst Evol 52:765–778

Lynch M, Conery JS (2000) The evolutionary fate and consequences of duplicate genes. Science 290:1151–1154

Ma X, Sela H, Jiao G, Li C, Wang A, Pourkheirandish M, Weiner D, Sakuma S, Krugman T, Nevo E, Komatsuda T, Korol A, Chen G (2012) Population-genetic analysis of HvABCG31 promoter sequence in wild barley (*Hordeum vulgare* ssp. *spontaneum*). BMC Evol Biol 12:188

Ma Y, Shabala S, Li C, Liu C, Zhang W, Zhou M (2015) Quantitative trait loci for salinity tolerance identified under drained and waterlogged conditions and their association with flowering time in barley (*Hordeum vulgare*. L). PLoS One 10:e0134822

Maestri E, Klueva N, Perrotta C, Gulli M, Nguyen HT, Marmiroli N (2002) Molecular genetics of heat tolerance and heat shock proteins in cereals. Plant Mol Biol 48:667–681

Mali P, Yang L, Esvelt KM, Aach J, Guell M, Dicarlo JE, Norville JE, Church GM (2013) RNA-guided human genome engineering via Cas9. Science 339:823–826

Mayer KF, Waugh R, Brown JW, Schulman A, Langridge P, Platzer M, Fincher GB, Muehlbauer GJ, Sato K, Close TJ, Wise RP, Stein N, International Barley Genome Sequencing Consortium (2012) A physical, genetic and functional sequence assembly of the barley genome. Nature 491:711–716

Middleton CP, Stein N, Keller B, Kilian B, Wicker T (2013) Comparative analysis of genome composition in Triticeae reveals strong variation in transposable element dynamics and nucleotide diversity. Plant J 73:347–356

Montfort RL, Basha E, Friedrich KL, Slingsby C, Vierling E (2001) Crystal structure and assembly of a eukaryotic small heat shock protein. Nat Struct Biol 8:1025–1030

Moore C, Purugganan MD (2005) The evolutionary dynamics of plant duplicate genes. Curr Opin Plant Biol 8:122–128

Morrell PL, Clegg MT (2007) Genetic evidence for a second domestication of barley (*Hordeum vulgare*) east of the Fertile Crescent. Proc Natl Acad Sci U S A 104:3289–3294

Morrell PL, Lundy KE, Clegg MT (2003) Distinct geographic patterns of genetic diversity are maintained in wild barley (*Hordeum vulgare* ssp. *spontaneum*) despite migration. Proc Natl Acad Sci U S A 100:10812–10817

Nei M (1987) Molecular evolutionary genetics. Columbia University Press, New York

Nevo E (1992) Origin, evolution, population genetics and resources for breeding of wild barley, *Hordeum spontaneum*, in the Fertile Crescent. In: Shewry P (ed) Barley, genetics, molecular biology and biotechnology. CAB International, Wallingford, pp 19–43
Nevo E (1995) Asian, African and European biota meet at 'Evolution Canyon', Israel, local tests of global biodiversity and genetic diversity patterns. Proc R Soc B Biol Sci 262:149–155
Nevo E (1997) Evolution in action across phylogeny caused by microclimatic stresses at "Evolution Canyon". Theor Popul Biol 52:231–243
Nevo E (1998) Genetic diversity in wild cereals, regional and local studies and their bearing on conservation *ex situ* and *in situ*. Genet Resour Crop Evol 45:355–370
Nevo E (2001) Evolution of genome-phenome diversity under environmental stress. Proc Natl Acad Sci U S A 98:6233–6240
Nevo E (2006) 'Evolution Canyon', a microcosm of life's evolution. Isr J Ecol Evol 52:183–184
Nevo E (2009) Evolution in action across life at "Evolution Canyons", Israel. Trends Evol Biol 1:e-3
Nevo E (2011) Evolution under environmental stress at macro-and microscales. Genome Biol Evol 2:1039–1052
Nevo E (2012) "Evolution Canyon," a potential microscale monitor of global warming across life. Proc Natl Acad Sci U S A 109:2960–2965
Nevo E (2014) Evolution in action, adaptation and incipient sympatric speciation with gene flow across life at "Evolution Canyon", Israel. Isr J Ecol Evol 60:85–98
Nevo E (2015) Evolution of wild barley at "Evolution Canyon", adaptation, speciation, pre-agricultural collection, and barley improvement. Isr J Plant Sci 62:22–32
Nevo E, Chen G (2010) Drought and salt tolerances in wild relatives for wheat and barley improvement. Plant Cell Environ 33:670–685
Nevo E, Zohary D, Brown A, Haber M (1979) Genetic diversity and environmental associations of wild barley, *Hordeum spontaneum* in Israel. Evolution 33:815–833
Nevo E, Apelbaum-Elkaher I, Garty J, Beiles A (1997) Natural selection causes microscale allozyme diversity in wild barley and a lichen at 'Evolution Canyon' Mt Carmel, Israel. Heredity 78:373–382
Nevo E, Fragman O, Dafni A, Beiles A (1999) Biodiversity and interslope divergence of vascular plants caused by microclimatic differences at 'Evolution Canyon', Lower Nahal Oren, Mount Carmel, Israel. Isr J Plant Sci 47:49–59
Nevo E, Beharav A, Meyer RC, Hackett CA, Forster BP, Russell JR, Powell W (2005) Genomic microsatellite adaptive divergence of wild barley by microclimatic stress in 'Evolution Canyon', *Israel*. Biol J Linnean Soc 84:205–224
Nevo E, Lu Z, Pavlicek T (2006) Global evolutionary strategies across life caused by shared ecological stress, fact or fancy? Isr J Plant Sci 54:1–8
Nevo E, Fu YB, Pavlicek T, Khalifa S, Tavasi M, Beiles A (2012) Evolution of wild cereals during 28 years of global warming in Israel. Proc Natl Acad Sci U S A 109:3412–3415
Ohnu S (1970) Evolution by gene duplication. Springer, Berlin
Pandey B, Kaur A, Gupta OP, Sharma I, Sharma P (2015) Identification of HSP20 gene family in wheat and barley and their differential expression profiling under heat stress. Appl Biochem Biotechnol 175:2427–2446
Pavlicek T, Sharon D, Kravchenko V, Saaroni H, Nevo E (2003) Microclimatic interslope differences underlying biodiversity contrasts in 'Evolution Canyon', Mt Carmel, Israel. Isr J Earth Sci 52:1–9
Poukheirandish M, Komatsuda T (2007) The importance of barley genetics and domestication in a global perspective. Ann Bot 100:999–1008
Pourkheirandish M, Hensel G, Kilian B, Senthil N, Chen G, Sameri M, Azhaguvel P, Sakuma S, Dhanagond S, Sharma R, Mascher M, Himmelbach A, Gottwald S, Nair SK, Tagiri A, Yukuhiro F, Nagamura Y, Kanamori H, Matsumoto T, Willcox G, Middleton CP, Wicker T, Walther A, Waugh R, Fincher GB, Stein N, Kumlehn J, Sato K, Komatsuda T (2015) Evolution of the grain dispersal system in barley. Cell 162:527–539

Reddy PS, Kavi Kishor PB, Seiler C, Kuhlmann M, Eschen-Lippold L, Lee J, Reddy MK, Sreenivasulu N (2014) Unraveling regulation of the small heat shock proteins by the heat shock factor HvHsfB2c in barley, its implications in drought stress response and seed development. PLoS One 9:e89125

Rodin SN, Riggs AD (2003) Epigenetic silencing may aid evolution by gene duplication. J Mol Evol 56:718–729

Sakuma S, Salomon B, Komatsuda T (2011) The domestication syndrome genes responsible for the major changes in plant form in the Triticeae crops. Plant Cell Physiol 52:738–749

Santhanagopalan I, Basha E, Ballard KN, Bopp NE, Vierling E (2015) Model chaperones, small heat shock proteins from plants. In: Tanguay RM, Hightower LE (eds) The big book on small heat shock proteins, vol 8, Heat shock proteins. Springer International Publishing, Cham

Sarkar NK, Kim YK, Grover A (2009) Rice sHsp genes, genomic organization and expression profiling under stress and development. BMC Genomics 10:339

Sato K, Shin-I T, Seki M, Shinozaki K, Yoshida H, Takeda K, Yamazaki Y, Conte M, Kohara Y (2009) Development of 5006 full-length cDNAs in barley, a tool for accessing cereal genomics resources. DNA Res 16:81–89

Scharf KD, Siddique M, Vierling E (2001) The expanding family of *Arabidopsis thaliana* small heat stress proteins and a new family of proteins containing alpha-crystallin domains (Acd proteins). Cell Stress Chaperones 6:225–237

Scharf KD, Berberich T, Ebersberger I, Nover L (2012) The plant heat stress transcription factor (Hsf) family, structure, function and evolution. Biochim Biophys Acta 1819:104–119

Shan Q, Wang Y, Li J, Gao C (2014) Genome editing in rice and wheat using the CRISPR/Cas system. Nat Protoc 9:2395–2410

Stone P (2001) The effects of heat stress on cereal yield and quality. In: Basra AS (ed) Crop responses and adaptations to temperature stress. Food Products Press, Binghamton, pp 243–291

Süle A, Vanrobaeys F, Hajós G, Van Beeumen J, Devreese B (2004) Proteomic analysis of small heat shock protein isoforms in barley shoots. Phytochemistry 65:1853–1863

Sun W, Van Montagu M, Verbruggen N (2002) Small heat shock proteins and stress tolerance in plants. Biochimica Biophysica Acta 1577:1–9

Tang Z, Yang Z, Fu S (2014) Oligonucleotides replacing the roles of repetitive sequences pAs1, pSc119.2, pTa-535, pTa71, CCS1, and pAWRC.1 for FISH analysis. J Appl Genet 55:313–318

Taylor JS, Raes J (2004) Duplication and divergence, the evolution of new genes and old ideas. Annu Rev Genet 38:615–643

Tondelli A, Francia E, Barabaschi D, Aprile A, Skinner JS, Stockinger EJ, Stanca AM, Pecchioni N (2006) Mapping regulatory genes as candidates for cold and drought stress tolerance in barley. Theor Appl Genet 112:445–454

Vanhala TK, van Rijn CPE, Buntjer J, Stam P, Nevo E, Poorter H, Eeuwijk FA (2004) Environmental, phenotypic and genetic variation of wild barley (*Hordeum spontaneum*) from Israel. Euphytica 137:297–309

Volis S, Mendlinger S, Ward D (2002a) Adaptive traits of wild barley plants of Mediterranean and desert origin. Oecologia 133:131–138

Volis S, Mendlinger S, Ward D (2002b) Differentiation in populations of *Hordeum spontaneum* Koch along a gradient of environmental productivity and predictability, plasticity in response to water and nutrient stress. Biol J Linn Soc 75:301–312

Wang W, Vinocur B, Shoseyov O, Altman A (2004) Role of plant heat-shock proteins and molecular chaperones in the abiotic stress response. Trends Plant Sci 9:244–252

Wang Y, Lin S, Song Q, Li K, Tao H, Huang J, Chen X, Que S, He H (2014) Genome-wide identification of heat shock proteins (Hsps) and Hsp interactors in rice, Hsp70s as a case study. BMC Genomics 15:344

Waters ER (1995) The molecular evolution of the small heat-shock proteins in plants. Genetics 141:785–795

Waters ER (2013) The evolution, function, structure, and expression of the plant sHSPs. J Exp Bot 64:391–403

Waters ER, Lee GJ, Vierling E (1996) Evolution, structure and function of the small heat shock proteins in plants. J Exp Bot 47:325–338
Waters ER, Aevermann BD, Sanders-Reed Z (2008) Comparative analysis of the small heat shock proteins in three angiosperm genomes identifies new subfamilies and reveals diverse evolutionary patterns. Cell Stress Chaperones 13:127–142
Wu ML, Lin TP, Lin MY, Cheng YP, Hwang SY (2007) Divergent evolution of the chloroplast small heat shock protein gene in the genera *Rhododendron* (Ericaceae) and *Machilus* (Lauraceae). Ann Bot 99:461–475
Xia Y, Li R, Ning Z, Bai G, Siddique KH, Yan G, Baum M, Varshney RK, Guo P (2013) Single nucleotide polymorphisms in HSP17.8 and their association with agronomic traits in barley. PLoS One 8:e56816
Xue GP, Sadat S, Drenth J, McIntyre CL (2014) The heat shock factor family from *Triticum aestivum* in response to heat and other major abiotic stresses and their role in regulation of heat shock protein genes. J Exp Bot 65:539–557
Yang Z, Zhang T, Bolshov A, Beharav A, Nevo E (2009) Adaptive microclimatic structural and expressional dehydrin 1 evolution in wild barley, *Hordeum spontaneum*, at 'Evolution Canyon', Mount Carmel, Israel. Mol Ecol 18:2063–2075
Yang Z, Zhang T, Li G, Nevo E (2011) Adaptive microclimatic evolution of the dehydrin 6 gene on wild barely at "Evolution Canyon", Israel. Genetica 139:1429–1438
Yang Z, Wang Y, Gao Y, Zhou Y, Zhang E, Hu Y, Yuan Y, Liang G, Xu C (2014) Adaptive evolution and divergent expression of heat stress transcription factors in grasses. BMC Evol Biol 14:147
Yu J, Buckler ES (2006) Genetic association mapping and genome organization of maize. Curr Opin Biotechnol 17:155–160
Zeng X, Long H, Wang Z, Zhao S, Tang Y, Huang Z, Wang Y, Xu Q, Mao L, Deng G, Yao X, Li X, Bai L, Yuan H, Pan Z, Liu R, Chen X, WangMu Q, Chen M, Yu L, Liang J, DunZhu D, Zheng Y, Yu S, LuoBu Z, Guang X, Li J, Deng C, Hu W, Chen C, TaBa X, Gao L, Lv X, Abu YB, Fang X, Nevo E, Yu M, Wang J, Tashi N (2015) The draft genome of Tibetan hulless barley reveals adaptive patterns to the high stressful Tibetan Plateau. Proc Natl Acad Sci U S A 112:1095–1100
Zhang J (2003) Evolution by gene duplication, an update. Trends Ecol Evol 18:292–298
Zhang T, Li GR, Yang Z, Nevo E (2014) Adaptive evolution of duplicated hsp17 genes in wild barley from microclimatically divergent sites of Israel. Genet Mol Res 13:1220–1232
Zhu C, Gore M, Buckler ES, Yu J (2008) Status and prospects of association mapping in plants. Plant Genome 1:5–20
Zohary D, Hopf M, Weiss E (2012) Domestication of plants in the Old World. Oxford University Press, Oxford
Zou C, Lehti-Shiu MD, Thomashow M, Shiu S (2009) Evolution of stress-regulated gene expression in duplicate genes of *Arabidopsis thaliana*. PLoS Genet 5:e1000581

Chapter 6
Insights into the Mechanism of Heat Shock Mitigation Through Protein Repair, Recycling and Degradation

Rashid Mehmood Rana, Muhammad Ammar Khan, Muhammad Kausar Nawaz Shah, Zulfiqar Ali, and Hongsheng Zhang

Abstract Exposed to elevated temperature, plants exhibit a characteristic set of conserved responses. These signature responses include an accelerated transcription and translation of stress specific proteins such as heat shock factors (HSFs), heat shock proteins (HSPs), molecular chaperones and co-chaperones, to repair misfolded/unfolded proteins that occur as a consequence of stress factor. The formation of all these stress specific proteins is regulated through different proteins including heat shock factor binding protein (HSBP), HSP70, and BAG (Bcl2 associated athanogene). The review focuses on stress response mechanism, and discussed information about the gene families that are directly involved in protein repair, recycling and degradation. The recent advances made in this field have also been critically discussed to provide readers with logical sense of the mechanism of protein repair, recycling and degradation.

Keywords Autophagy • HSP • HSF • HSR • HSE

R.M. Rana (✉) • M.K.N. Shah
Department of Plant Breeding and Genetics, PMAS-Arid Agriculture University, Rawalpindi, Pakistan
e-mail: rashid_cabb@hotmail.com

M.A. Khan
University College of Agriculture & Environmental Sciences, The Islamia University of Bahawalpur, Bahawalpur, Pakistan

Z. Ali
Department of Plant Breeding and Genetics, University of Agriculture Faisalabad, Faisalabad, Pakistan

Department of Plant Breeding and Genetics, Muhammad Nawaz Shareef University of Agriculture (MNSUA), Multan, Pakistan

H. Zhang
State Key Laboratory of Crop Genetics and Germplasm Enhancement, Nanjing Agricultural University, Nanjing 210095, China

A.A.A. Asea et al. (eds.), *Heat Shock Proteins and Plants*,
Heat Shock Proteins 10, DOI 10.1007/978-3-319-46340-7_6

Abbreviations

ABA	abscisic acid
ATG6	Autophagy related protein 6
BAG	Bcl-2 associated athanogenes
CAT	catalase
cDNA	complementary DNA
DnaJ	bacterial homologue of Hsp40
DnaK	bacterial homologue of Hsp70
HS	heat shock
HSBP	heat shock factor binding protein
HSC	heat shock cognate protein
HSE	heat shock element
HSF	heat shock transcription factor
HSP	heat shock protein
HSR	heat shock response
POD	peroxidase
ROS	reactive oxygen species
SOD	superoxide dismutase(s)

6.1 Introduction

Plants, being sessile, are prone to environmental stresses including high temperature. Therefore, it is essential for plants to have higher levels of tolerance to abiotic and biotic stress factors. High temperature, a potentially destructive environmental factor, affects numerous biological processes of all organisms. An evolutionarily conserved mechanism called the heat shock response (HSR) is therefore induced. Severe heat stress (HS) leads to cellular damage and cell death (Lindquist 1986; Schöffl et al. 1998). In the consequence of HSR, cells and organisms are protected from severe damage by resumption of normal cellular and physiological activities, resulting in higher level of thermotolerance. HSR is a transient reprogramming of cellular activities by rapid and specific induction of heat shock proteins (HSPs) that act as molecular chaperones.

The expression of molecular chaperones targeting the unstable or inactive proteins increases dramatically during heat stress. Temperature perception and multiple signal transduction pathways lead to the activation of heat shock factors (HSFs) that bind to the conserved heat shock element (HSE) and induce the expression of heat shock genes (Sung et al. 2003). The first evidence for the induction of HSP gene expression through HSF, after binding to HSEs as a binding site, was demonstrated through DNA-protein interaction studies on nuclei isolated from *Drosophila melanogaster* cells (Parker and Topol 1984; Wu 1984). Afterwards a number of research groups have independently confirmed this mechanism of HSF gene expression

(Rieping and Schöffl 1992; Schöffl et al. 1998; Miller and Mittler 2006) and hence is universally accepted. HSPs and HSFs are regulated in a stoichiometric manner through HSBP (heat shock factor binding protein), HSP70 and BAG (Bcl-2-associated athanogene).

When the mechanisms described above are not enough to repair the unfolded proteins, cell launches another defense mechanism called autophagy. Autophagy is a specialized lysosomal pathway, intended to recycle the cellular constituents including deshaped/unfolded proteins in eukaryotes, when they come across the unfavorable conditions (Harrison-Lowe and Olsen 2008). The mechanism of cell death and survival is summarized in the Fig. 6.1.

Here, we summarized the molecular mechanism of protein repair and degradation. A brief detail about some important gene families is also given, that will provide a constitutive understanding of these fundamental pathways.

6.2 Heat Shock Proteins as Molecular Chaperones

The HSPs or other proteins highly homologous to HSPs are produced under stress conditions, as well as in some cells either constitutively, under cell cycle or developmental control. The evolutionary conservation of HSPs and their correlation with cellular resistance to high temperature, suggest that accumulation of HSPs lead to increased thermotolerance. Heat shock proteins are "molecular chaperones", to keep other proteins from 'inappropriate' aggregations. They are also involved in transport, folding, unfolding, assembly and disassembly of multi-structured protein units and degradation of misfolded or aggregated proteins (Sørensen et al. 2003). Chaperone activities are provoked during four different stages in the life cycle of proteins (Gething and Sambrook 1992; Gething 1997), where they perform the following functions: (i) keep nascent polypeptides in proper folding until the whole sequence information is available; (ii) retain or create a partially unfolded form of proteins intended for the import into mitochondria or plastids; (iii) repair/stabilize damaged proteins; (iv) assist/control assembly and disassembly of multi-protein complexes (Baniwal et al. 2004).

In all eukaryotes, HSPs are grouped into five conserved classes: small HSPs, (sHSPs), HSP60, HSP70, HSP90 and HSP100 (Waters et al. 1996). Carboxylic terminal found in all HSPs, is a characteristic feature of HSPs and called as heat-shock domain (Helm et al. 1993). HSPs with molecular weights ranging from 10 to 200 KD are characterized as chaperones, and participate in the signal induction during heat stress (Schöffl et al. 1999).

Small heat shock proteins consist of a family of stress inducible molecular chaperones that range in size from ~16 to 42 kDa, containing α-crystalline domain (80–100 amino acid residues) located in the C-terminal region (Seo et al. 2006). sHSPs form large oligomeric complexes, ranging from 200–800 kDa in size, both in prokaryotic and eukaryotic cells. These proteins prevent the aggregation of denatured proteins by forming complexes with them, which are subsequently refolded by

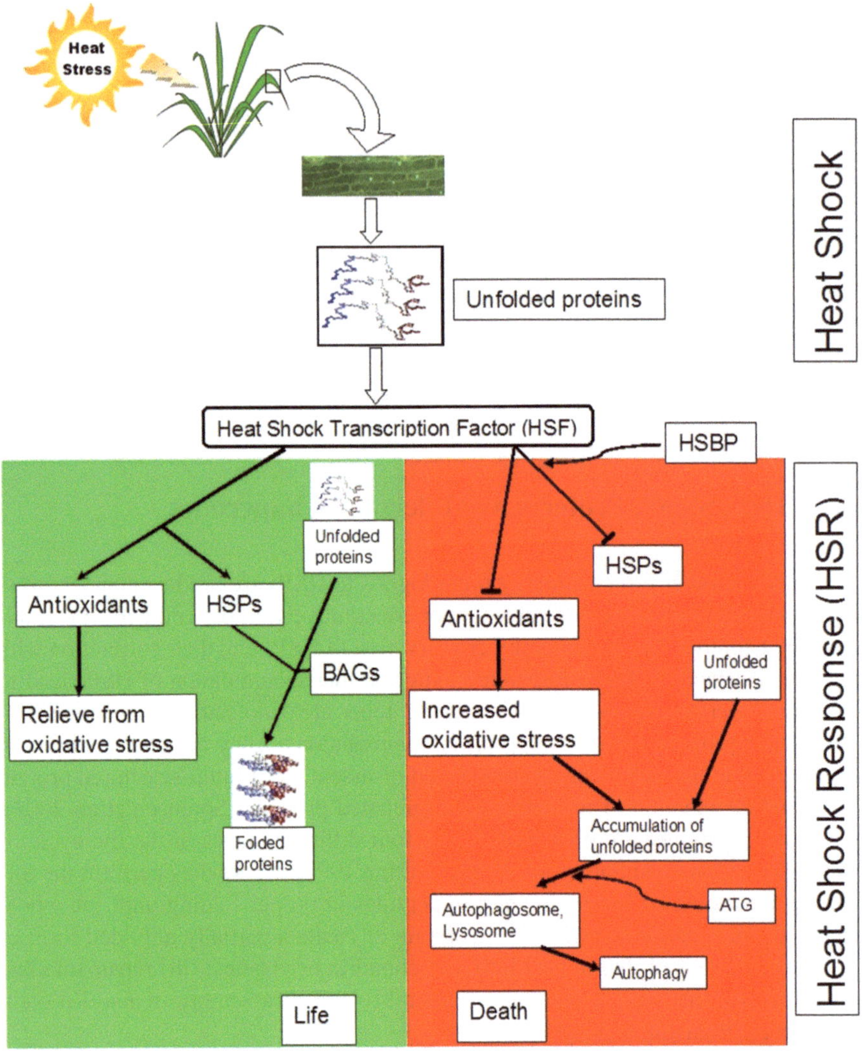

Fig. 6.1 Mechanism of cell death and survival. Heat stress degrades the cell proteins by unfolding the polypeptide chain which otherwise are folded properly to perform specific functions. The accumulation of misfolded/unfolded proteins to a threshold level induces the expression of HSF which in turn triggers the expression of HSPs and antioxidants. Antioxidants protect cells from oxidative damage while HSPs in combination with BASs refold the misfolded/unfolded proteins, resulting in survival of the cell. Increased levels of HSBP decrease the activity of HSF and ultimately considerably reduce the production of HSPs and antioxidants. The decreased levels of antioxidants reduces the cells ability to cope with oxidative stress, while decreased HSPs causes the accumulation of misfolded/unfolded proteins in the cell which provokes autophagy and ultimately cell dies

Hsp100/Hsp70 and co-chaperones in ATP-dependent pathway during the recovery phase (Cashikar et al. 2005).

In *Arabidopsis*, 19 genes of sHSP family are placed into seven nucleo-cytoplasmic (I, II, III, IV, V, VI and VII) and two mitochondrial (MI and MII) subfamilies (Siddique et al. 2008). Whereas, sHSP gene family in bacteria and lower eukaryotes is not very complex, as *E. coli* and *S. cerevisiae* have 2 sHSPs each. In human and zebra-fish, 10 and 13sHsps have been reported respectively (Elicker and Hutson 2007). Higher diversification of sHsps in plant may reflect an adaptation to stress conditions that is unique to plants. sHSPs localizing in chloroplast and mitochondria are considered important for their role during heat tolerance. Rice (*Oryza sativa* L.) genome contained 23 sHSPs, with nine subfamilies, which is more complex than *Arabidopsis*. Expression analysis of rice sHSPs showed that 19 sHSP genes were upregulated by high temperature stress (Sarkar et al. 2009).

The HSP60, also called as chaperonins, found in bacteria (GroEL), mitochondria, and chloroplasts, is generally agreed to be important in assisting plastid proteins such as Rubisco (Wang et al. 2003). Some studies suggested that this class might contribute in folding of proteins that are transported to chloroplasts and mitochondria (Lubben et al. 1989). Functionally, plant chaperonins are limited and are involved in attaining functional conformation of newly imported proteins to the chloroplast (Jackson-Constan et al. 2001).

The HSP70 proteins act as molecular chaperones in almost all organisms. They prevent the accumulation/aggregation of newly synthesized proteins and fold them in a proper way during their transfer to final destination (Su and Li 2008). HSP70 are also reported to contribute during protein transport by binding to the precursor of a protein to be transferred through the membranes into the organelles (e.g. chloroplast) (Jackson-Constan et al. 2001). HSP70B was observed in stroma of chloroplasts and was found involved in photo protection and repairing of the photo-system II, during and after photo-inhibition. In *A. thaliana*, HSP70 was located in stomata and found necessary for differentiation of germinating seeds and its tolerance of heat (Su and Li 2008). In rice, 26 HSP70 genes has been reported and found involved in different abiotic stresses including heat stress (Hu et al. 2009).

The HSP90 facilitates protein folding within the cell by interacting with substrate proteins (client proteins). Conformational changes in the chaperone occur as a result of ATP binding and hydrolysis in an ATP-dependent manner, that are required for client protein activation. HSP90 often works together with HSP70 or HSC70 (heat shock cognate protein 70), the two chaperones considered as parts of a single multi-chaperone machinery (Young et al. 2004). In *A. thaliana*, cytoplasmic HSP90 found involved in the inhibition of HSF in the absence of heat stress, but not under heat stress (Yamada et al. 2007). In rice, 9 members of HSP90 has been reported (Hu et al. 2009), while some found essential for innate-immune response and pathogenic resistance (Thao et al. 2007).

The HSP100 proteins belonging to ClpB group, contain two ATP-binding domains which are critical for their hexamerization and chaperone activity (Schirmer et al. 1996). ClpB/HSP100 proteins also contain a spacer region that separates the two NBDs (nucleotide binding domains), and found essential for chaperone activity

(Cashikar et al. 2002). Sometimes, HSP100 requires alliance with HSP70/HSP110 chaperone machinery for disaggregation of misfolded proteins (Doyle et al. 2007). In yeast, ClpB/HSP100 family comprises of only one member, and this protein has a notable capacity to rescue aggregated proteins (Parsell et al. 1994). Yeast ClpB/HSP100 was found involved in temperature stress as defective mutants showed a clear thermo-sensitive phenotype (Singh and Grover 2010). In rice 10 members of HSP100 has been reported (Hu et al. 2009), mostly induced by heat stress. Indica rice cultivar over-expressing AtHSP101 showed increased thermotolerance (Katiyar-Agarwal et al. 2003).

In conclusion, HSPs are molecular chaperone that play important role in refolding of denatured proteins, partake in finishing of the *de novo* synthesized proteins and reduce the protein aggregation.

6.3 Heat Shock Factors

Stress-induced transcription of HSPs requires activation of specific transcription factors called as heat shock factor (HSF) that binds to the heat shock promoter element (HSE) (Wu 1995). HSE consists of alternating units of pentameric nucleotides (5′-nGAAn-3′) and at least three alternating units (5′-nGAAnnTTCnnGAAn-3′) are required for efficient HSF binding (Schöffl et al. 1998). Unlike yeast and *Drosophila*, regulation of the heat shock response in plants and larger animals is complex due to the presence of multiple HSFs, providing specialized transcriptional control of stress signal transduction pathway (Morimoto 1998; Chen et al. 2006).

The HSF family contains several highly conserved domains with conserved modular structure, including DNA binding domain (DBD, also known as the HSF domain), the oligomerization domain and the nuclear localization signal domain (NLS); however, some HSFs also have the C-terminal activation domains (CTAD) and nuclear export signal (NES) (Harrison et al. 1994; Green et al. 1995; Nover et al. 1996). Hydrophobic core of DBD forms a helix-turn-helix (H2-T-H3), which is required for the specific binding to conserved motif of HSE (Littlefield and Nelson 1999). HSFs utilize their oligomerization domains (connected through a linker to the C-terminal of DBD) with heptad hydrophobic repeat (HR-A/B), which is characterized by a coiled-coil structure to form trimer and function as sequence-specific trimeric DNA binding proteins (Schöffl et al. 1998).

Based on the difference of the flexible linkers between the A and B parts of heptad hydrophobic repeat (HR-A/B), as well as difference in the sequence region between the DBD and HR-A/B, plant HSFs are classified into three major classes, viz. A, B, and C (Nover et al. 2001). The HR-A/B regions of class B HSFs have no insert sequences, whereas, HSFs of class A and C have an extended HR-A/B region with an insertion of 21 and 7 amino acid residues respectively. The region between the DBD and HR-A/B, spans from 9 to 39 amino acid residues for class A, 50–78 amino acid residues for class B, and 14–49 amino acid residues for class C HSFs (Nover et al. 2001). An arginine and lysine rich nuclear localization signal (NLS) determines nuclear localization of HSFs. While leucine-rich nuclear export signal determines cytoplasmic distribution of HSF proteins (Lyck et al. 1997). AHA motif (aromatic and large hydrophobic amino acid residues embedded in an acidic surrounding), embedded in C-terminal activation domain (CTAD), is an additional characteristic of class A HSFs that binds to some basic transcription protein complexes and activates the transcription of HSPs; whereas class B HSFs and C HSFs lack this function due to the absence of AHA motif (Nover et al. 2001).

Environmental stresses and dysfunctions in the ubiquitin-proteasome pathway induce accumulation of aberrant proteins, which in turn leads to activation of the HSFs and ultimately leads to synthesis of HSPs. In heat-induced pathway, HSF1 and HSF3 are activated and work collaboratively under severe stress. Beyond the heat shock response, HSF3 binds to c-Myb and sufficiently enhances the supply of HSPs; however, this interaction is disrupted when p53 binds to HSF3 leading to ubiquitin-mediated degradation of c-Myb, and consequently down-regulation of HSPs expression. HSF2 is also an activator of heat shock genes. Moreover, both HSF1 and HSF2 are likely to have additional target genes (Pirkkala et al. 2001).

The HSF gene family has been thoroughly characterized in *Arabidopsis*, rice and maize (Nover et al. 2001; Guo et al. 2008; Mittal et al. 2009; Lin et al. 2011), and studied comprehensively in tomato (Scharf et al. 1998). Table 6.1 shows HSFs and their isoforms in rice, *Arabidopsis* and maize. In tomato, HSFA1 has been reported as a master regulator of thermotolerance (Mishra et al. 2002). In *Arabidopsis*, *AtHSFA1a*, *AtHSFA1b* and *AtHSFA2* play important role in the induction and maintenance of HSP expression (Nover et al. 2001; Lohmann et al. 2004; Charng et al. 2007).

6.4 Regulation of Heat Shock Response

Studies on *E. coli*, yeast, and other higher eukaryotes suggested an auto-regulation of the HSR (Schöffl et al. 1998). Phosphorylation of Ser residues has been proved to play a role in repression of HSFs under normal conditions, as well as upon the stimulation a mitogen-activated protein kinase pathway (Chu et al. 1996; Kline and Morimoto 1997). Previous researches depicted that, in a negative feedback response, increased levels of HSP70 inactivate HSF by the disassembly of trimeric HSFs, and consequently, shut off the heat-shock response (Boorstein and Craig 1990; Baler et al. 1996; Schöffl et al. 1998). Negative regulation of HSF through HSP70 was also observed during the analysis of HSP70 knocked-down *Arabidopsis* plants (Lee and Schoffl 1996). Besides HSPs and other factors, regulating HSFs activity, heat shock factor binding protein-1 (HSBP1) also binds to hydrophobic heptad-repeats of active-trimerized form of HSF; and negatively regulates HSR by inactivating it (Satyal et al. 1998; Hsu et al. 2010). HSBP has also been reported for their role in other developmental process as their absence causes abnormal development or premature abortion of embryo, both in plants and other eukaryotes (Fu et al. 2002; Su et al. 2006; Hsu and Jinn 2010). Despite of the regulation through HSF, HSR is also

Table 6.1 HSFs and their isoforms in Rice, *Arabidopsis* and Maize

	Rice[1]	*Arabidopsis*[2]	Maize[3]
1.	*OsHsfA1a*	*AtHsfA1a*	*ZmHsf-01*
2.	*OsHsfA2a.1*	*AtHsfA1b*	*ZmHsf-02*
3.	*OsHsfA2a.2*	*AtHsfA1c*	*ZmHsf-03*
4.	*OsHsfA2a.3*	*AtHsfA1d*	*ZmHsf-04*
5.	*OsHsfA2a.4*	*AtHsfA1e*	*ZmHsf-05*
6.	*OsHsfA2a.5*	*AtHsfA2/2a*	*ZmHsf-06*
7.	*OsHsfA2b*	*AtHsfA2b*	*ZmHsf-07*
8.	*OsHsfA2c.1*	*AtHsfA2c*	*ZmHsf-08*
9.	*OsHsfA2c.2*	*AtHsfA2d*	*ZmHsf-09*
10.	*OsHsfA2c.3*	*AtHsfA2e*	*ZmHsf-10*
11.	*OsHsfA2c.4*	*AtHsfA3*	*ZmHsf-11*
12.	*OsHsfA2d.1*	*AtHsfA4a*	*ZmHsf-12*
13.	*OsHsfA2d.2*	*AtHsfA4b*	*ZmHsf-13*
14.	*OsHsfA2e.1*	*AtHsfA4c*	*ZmHsf-14*
15.	*OsHsfA2e.2*	*AtHsfA4d*	*ZmHsf-15*
16.	*OsHsfA3.1*	*AtHsfA5*	*ZmHsf-16*
17.	*OsHsfA3.2*	*AtHsfA6a*	*ZmHsf-17*
18.	*OsHsfA4b*	*AtHsfA6b*	*ZmHsf-18*
19.	*OsHsfA4d*	*AtHsfA7a*	*ZmHsf-19*
20.	*OsHsfA5*	*AtHsfA7b*	*ZmHsf-20*
21.	*OsHsfA7*	*AtHsfA8*	*ZmHsf-21*
22.	*OsHsfA9.1*	*AtHsfA9*	*ZmHsf-22*
23.	*OsHsfA9.2*	*AtHsfB1*	*ZmHsf-23*
24.	*OsHsfA9.3*	*AtHsfB2a*	*ZmHsf-24*
25.	*OsHsfB1*	*AtHsfB2b*	*ZmHsf-25*
26.	*OsHsfB2a*	*AtHsfB2c*	*ZmHsf-01*
27.	*OsHsfB2b.1*	*AtHsfB3/B3a*	
28.	*OsHsfB2b.2*	*AtHsfB3b*	
29.	*OsHsfB2c.1*	*AtHsfB4/4a*	
30.	*OsHsfB2c.2*	*AtHsfB4b*	
31.	*OsHsfB2c.3*	*AtHsfB4c*	
32.	*OsHsfB4a*	*AtHsfB4d*	
33.	*OsHsfB4b*	*AtAtHsfC1/1a*	
34.	*OsHsfB4c*	*AtHsfC1b*	
35.	*OsHsfB4d*	*AtHsfC2a*	
36.	*OsHsfC1a.1*	*AtHsfC2b*	
37.	*OsHsfC1a.2*		
38.	*OsHsfC1b*		
39.	*OsHsfC2a*		
40.	*OsHsfC2b*		

[1]Mittal et al. (2009),[2]Baniwal et al. (2004),[3]Lin et al. (2011)

affected by some other negative regulators of chaperones including BAG1 (Bcl-2-associated athanogene 1). BAG1 binds to ATPase domain of HSP70 and therefore, negatively regulate the chaperon activity of HSP70 (Bimston et al. 1998).

6.4.1 *HSBP as HSF Regulator*

Heat shock Factor Binding Protein 1 (HsHSBP1) was first identified by yeast two-hybrid screening for interacting proteins of HSF1. Further investigation revealed that HsHSBP1 localizes in nucleus and negatively regulates the DNA binding affinity of HSF1 during HSR (Satyal et al. 1998). Structure analysis of revealed that HSBP family proteins are small (less than 10 kD) and contain two arrays of highly conserved (over 60 %) hydrophobic heptad-repeats (designated as HR-N and HR-C), suggesting that, it most likely regulates the function of other proteins through coiled-coil interactions (Tai et al. 2002).

Heat shock factor 1 (HSF1) is activated in response to the accumulation of unfolded proteins and the requirement for molecular chaperones (Hsp90, Hsp70, and Hdj1) that refold the misfolded proteins. HSF1 exists in the control state in cytoplasm or nucleus as an inert monomer, through transient interactions with chaperones such as Hsp90 and Hsp70. Stress perception induces the activation of HSF1, which consequently re-localized into nucleus. Phosphorelation of DNA binding site and oligomerization of HSF1 occurs. The HSF1 trimer binds heat shock element and induces the expression of heat shock genes. During attenuation of the heat shock response, HSP70 and Hdj-1 directly bind to HSF and repress their transcriptional activity. The trimers of HSF1 are negatively regulated by HSF binding protein 1 (HSBP1), which binds to the hydrophobic heptad-repeat of HSF1 and HSF1 trimers is dissociation into inert monomers.

A comparison of HSBP homologs revealed that animals and dicots such as *Arabidopsis* contain only a single copy of HSBP, whereas, monocots such as maize and rice contain two HSBP isoforms (Satyal et al. 1998; Fu et al. 2002; Hsu and Jinn 2010). In plants, first HSBP-like protein was reported in maize as EMPTY PERICARP2 (EMP2), showing aborted kernels at coleoptiles stage/stage 1, followed by necrosis and re-absorption of kernel contents, showing greatly increased *HSP* expression and un-attenuated HSR (Fu et al. 2002; Fu and Scanlon 2004). In addition to containing *EMP2*, maize also contains another *HSBP* paralog, *ZmHSBP2*, whereas, both interact non-redundantly with specific HSFs, suggesting their distinct functions during plant development and HSR; however, the functions of ZmHSBP2 needs further clarification (Fu et al. 2006). *Arabidopsis* HSBP1 protein is reported to be localized in cytoplasm and moves to nucleus in response to HSR, whereas, knock-out mutants showed premature seed abortion, therefore considered essential for seed development (Hsu and Jinn 2010; Hsu et al. 2010). Rice HSBP1 and HSBP2 were found as negative regulators of thermotolerance as well as essential for seed development (Rana et al. 2012c).

6.4.2 BAG as Co-chaperones

The Bcl-2-associated athanogene (BAG) family proteins, sometimes regarded as co-chaperones. An evolutionarily conserved region located at the C-terminus of the BAG have been described (Sondermann et al. 2001), and proven in a variety of organisms including human (Sondermann et al. 2001), Drosophila (Coulson et al. 2005), *Bombyx mori* (silk worm) (Moribe et al. 2001) and *Arabidopsis thaliana* (Doukhanina et al. 2006). The over-expression of BAG1 resulted in inhibited or delayed cell death caused by growth factor deprivation, heat shock, and p53 (Takayama et al. 2003). All BAG proteins share a common HSP70/HSC70 interaction domain, but generally differ in the N-terminal region, which imparts specificity to particular proteins and pathways. Ubiquitin-like domain at the N terminus of human BAG proteins (BAG1 and BAG6) is probably functionally relevant and conserved in yeast, plants, and worms. BAG proteins regulate diverse physiological processes in animals, including apoptosis, tumorigenesis, neuronal differentiation, stress responses, and the cell cycle (Song et al. 2001; Lau and Chan 2003; Ueda et al. 2004; Suzuki et al. 2011). Recent studies showed that BAG3 is involved macro-autophagy in cooperation with small heat-shock protein HSPB8 (McCollum et al. 2010).

BAG proteins are being extensively studied in animals. Seven homologs of the BAG family have been reported in *A. thaliana*. Domain organization of four *AtBAGs* was found similar to their animal homologs. While three members contain a calmodulin-binding domain near binding domain (BD), a novel feature associated with plant BAG proteins reflecting possible divergent mechanisms involved with plant-specific PCD (Doukhanina et al. 2006). AtBAG7 has been reported essential for the proper maintenance of the unfolded protein response (Williams et al. 2010). Six BAG homologs were also found involved in different stress responses as their expression was considerable affected by stresses (Rana et al. 2012b).

6.5 Unfolded Protein Response and Autophagy

Cells have evolved multifaceted mechanisms to ensure the accuracy, as only properly folded proteins are allowed to leave the endoplasmic reticulum (ER). The accumulation of misfolded/unfolded proteins in the ER is collectively called as ER-stress, provoking self-protective mechanisms termed unfolded protein response (UPR) (Schroder and Kaufman 2005). Under severe ER-stress, UPR triggers autophagy. Autophagy is an evolutionarily conserved lysosomal pathway involved in the recycling of misfolded and unfolded proteins (Harrison-Lowe and Olsen 2008). Therefore, autophagy is considered as a protective mechanism attempting to reduce cellular damages. The process of autophagy emerges through the engulfment of cytoplasmic components into vesicles, which are subsequently degraded into vacuole/lysosome (Fisher and Williams 2008). Chaperone-mediated autophagy (CMA) is a proteolytic pathway that plays an important role in protein quality control, and involves the chaperone molecules (HSP70 and HSP90) and co-chaperones like BAG3 (Fink 1999; Massey et al. 2006; McCollum et al. 2010).

6.5.1 ATG6/Beclin1 Proteins

Among other proteins involved in autophagy, Autophagy-related proteins (ATG) are critical for autophagosome formation. In yeast, 15 ATG proteins are involved in autophagosome formation (Suzuki et al. 2001). Yeast autophagy protein 6 (ATG6/Vps30) has been reported to be essential for autophagosome formation during starvation (Kametaka et al. 1998). ATG6 has been found in many eukaryotes, including mammals and plants etc., and has been assigned as a conserved domain (PF04111). Mammalian homolog of yeast ATG6 is recognized as Beclin-1, which has been reported to interact with Bcl-2 family, and is involved in regulation of autophagy and anti-apoptotic pathways (Sinha and Levine 2008). The plant ortholog of yeast ATG6 was first identified in *Nicotiana tobacum* during screening of the genes affecting tobacco mosaic virus-induced hypersensitive response programmed cell death (HR PCD), suggesting ATG6 as a negative regulator of PCD, as well as responsible for controlling virus replication (Liu et al. 2005). No physiological defects were observed in tobacco ATG6 knock-down plants; however, *Arabidopsis* ATG6 (*AtATG6*) T-DNA insertion mutants exhibited defects in pollen germination (Fujiki et al. 2007; Qin et al. 2007). Three ATG6 homologs in rice also showed their involvement in abiotic stresses response (Rana et al. 2012a).

6.6 Recent Research Advances on HSR

Plants, being sessile organisms, have evolved a variety of defense mechanisms to respond adverse conditions; of which, increased synthesis of heat shock proteins (HSP) is most important. These proteins function as molecular chaperones and involved in repairing damaged protein (Sørensen et al. 2003). Expression of HSPs is controlled by specific transcription factor called heat shock factors (HSFs) by recognition of heat shock element (HSE) present in the promoter region of HSPs (Wu 1995). Differential patterns for rice HSFs binding affinities with HSEs were observed that may lead to cellular functioning under a range of different physiological and environmental conditions (Mittal et al. 2011). Several reports have been published on identification and functional elucidation of rice HSFs and HSPs. To date, up to 25 rice HSFs have been identified and divided into three classes, HSFA, HSFB and HSFC, according to their protein structure (Von Koskull-Döring et al. 2007; Wang et al. 2009). Microarray analysis of rice HSFs and HSPs under different abiotic stresses (heat, cold, drought and salinity) suggested an overlapping behavior of most of these proteins, while some were significantly specific to distinct stress

(Hu et al. 2009). Rice spotted leaf (lesion-mimic) gene, Spl7 (Os01g54550, *HSFA4b*), found involved in heat stress responses and their mutants showed lesion-mimic phenotype in rice leaf under high temperature (Yamanouchi et al. 2002). Rice HSF, OsHSFA2e showed enhanced tolerance to environmental stresses in transgenic *Arabidopsis* (Yokotani et al. 2008). Over-expression of *OsHSF7* showed higher basal thermotolerance in transgenic *Arabidopsis*. Besides modulating the expression of *HSPs*, *Arabidopsis HSFA1* is reported to control *BAG6*, which has been involved in apoptosis under HS (Kang et al. 2006; Liu et al. 2011).

In a study, conducted to investigate rice leaf proteome in response to heat stress, 18 HSPs were identified including seven HSP70s, three HSP100s, one HSP60, and seven sHSPs (Lee et al. 2007). Transgenic rice over-expressing *HSP* genes showed improved heat tolerance, as rice plant over-expressing an *Arabidopsis* HSP101 gene showed significant heat tolerance (Katiyar-Agarwal et al. 2003). Over-expression of *HSP17.7* in rice plant also conferred heat, drought and UV-B tolerance (Murakami et al. 2004; Sato and Yokoya 2008). Over-expression of mitochondrial *HSP70* showed suppression of heat induced programmed cell death in rice protoplasts (Qi et al. 2011). A rice chloroplast sHSP (*OsHSP26*) gene, when over-expressed in *E. coli* showed heat, as well as oxidative stress tolerance (Lee et al. 2000).

Recently, modulators of HSFs are in focus, and among these, HSF binding protein (HSBP) showed interaction with *Arabidopsis HSFA1a*, *HSFA1b* and *HSFA2* and negatively regulate HSR (Hsu et al. 2010). Previously, two maize homologs of HSBP have been reported, showing non-redundant interaction with specific HSFs, whereas, *ZmHSBP1* reported as negative regulator of HSR (Fu et al. 2002, 2006). Protein families, other than HSPs and HSFs, such as Bcl-2-associated athanogene (BAG) and autophagy-related proteins (ATG; especially ATG6) has been reported to play important role in *Arabidopsis*, as well as in mammals and yeast in defense mechanism (Fu et al. 2002; Williams et al. 2010; Suzuki et al. 2011).

6.7 Conclusion

Conclusively, HSBPs modulate the expression of HSPs by binding to HSFs, while BAG proteins modulate HSPs negatively or positively during stresses. If damage caused by stress is beyond the limits of protein repair mechanism, autophagy related proteins (ATG) are activated, ATG6 take part in the formation of autophagosome, where protein degradation takes place.

Acknowledgements The authors would like to thank Dr Huang Ji (Nanjing Agricultural University, Nanjing, PR China) and Dr M Ishaq A Rehmani (Ghazi University, D.G. Khan, Pakistan) for revising an early draft of this manuscript, and Dr Yongmei Bao and Hongxia Lan (Nanjing Agricultural University, Nanjing, PR China) for useful discussion.

References

Baler R, Zou J, Voellmy R (1996) Evidence for a role of Hsp70 in the regulation of the heat shock response in mammalian cells. Cell Stress Chaperon 1:33–39

Baniwal SK, Bharti K, Chan KY, Fauth M, Ganguli A, Kotak S, Mishra SK, Nover L, Port M, Scharf KD (2004) Heat stress response in plants: a complex game with chaperones and more than twenty heat stress transcription factors. J Biosci 29:471–487

Bimston D, Song J, Winchester D, Takayama S, Reed JC, Morimoto RI (1998) BAG-1, a negative regulator of Hsp70 chaperone activity, uncouples nucleotide hydrolysis from substrate release. EMBO J 17:6871–6878

Boorstein WR, Craig EA (1990) Transcriptional regulation of SSA3, an HSP70 gene from Saccharomyces cerevisiae. Mol Cell Biol 10:3262–3267

Cashikar AG, Schirmer EC, Hattendorf DA, Glover JR, Ramakrishnan MS, Ware DM, Lindquist SL (2002) Defining a pathway of communication from the C-terminal peptide binding domain to the N-terminal ATPase domain in a AAA protein. Mol Cell 9:751–760

Cashikar AG, Duennwald M, Lindquist SL (2005) A chaperone pathway in protein disaggregation. Hsp 26 alters the nature of protein aggregates to facilitate reactivation by Hsp 104. J Biol Chem 280:23869–23875

Charng YY, Liu HC, Liu NY, Chi WT, Wang CN, Chang SH, Wang TT (2007) A heat-inducible transcription factor, HsfA2, is required for extension of acquired thermotolerance in *Arabidopsis*. Plant Physiol 143:251–262

Chen XJ, Ye CJ, Lu HY, Xu MX, Li W, Zhang LM, Wang C, Luo SP, Zhu BG (2006) Cloning of *GmHSFA1* gene and its overexpression leading to enhancement of heat tolerance in transgenic soybean. Yi Chuan 28:1411–1420

Chu B, Soncin F, Price BD, Stevenson MA, Calderwood SK (1996) Sequential phosphorylation by mitogen-activated protein kinase and glycogen synthase kinase 3 represses transcriptional activation by heat shock factor-1. J Biol Chem 271:30847–30857

Coulson M, Robert S, Saint R (2005) *Drosophila* starvin encodes a tissue-specific BAG-domain protein required for larval food uptake. Genetics 171:1799

Doukhanina EV, Chen S, van der Zalm E, Godzik A, Reed J, Dickman MB (2006) Identification and functional characterization of the BAG protein family in *Arabidopsis thaliana*. J Biol Chem 281:18793–18801

Doyle SM, Hoskins JR, Wickner S (2007) Collaboration between the ClpB AAA+ remodeling protein and the DnaK chaperone system. Proc Natl Acad Sci U S A 104:11138–11144

Elicker KS, Hutson LD (2007) Genome-wide analysis and expression profiling of the small heat shock proteins in zebrafish. Gene 403:60–69

Fink AL (1999) Chaperone-mediated protein folding. Physiol Rev 79:425–449

Fisher EA, Williams KJ (2008) Autophagy of an oxidized, aggregated protein beyond the ER: a pathway for remarkably late-stage quality control. Autophagy 4:721–723

Fu S, Meeley R, Scanlon MJ (2002) Empty pericarp2 encodes a negative regulator of the heat shock response and is required for maize embryogenesis. Plant Cell 14:3119–3132

Fu S, Rogowsky P, Nover L, Scanlon MJ (2006) The maize heat shock factor-binding protein paralogs EMP2 and HSBP2 interact non-redundantly with specific heat shock factors. Planta 224:42–52

Fu S, Scanlon MJ (2004) Clonal mosaic analysis of EMPTY PERICARP2 reveals nonredundant functions of the duplicated HEAT SHOCK FACTOR BINDING PROTEINs during maize shoot development. Genetics 167:1381–1394

Fujiki Y, Yoshimoto K, Ohsumi Y (2007) An *Arabidopsis* homolog of yeast ATG6/VPS30 is essential for pollen germination. Plant Physiol 143:1132–1139

Gething MJ (1997) Guidebook to molecular chaperones and protein-folding catalysts. A Sambrook and Tooze Publication: Oxford University Press, Oxford

Gething MJ, Sambrook J (1992) Protein folding in the cell. Nature 355:33–45

Green M, Schuetz TJ, Sullivan EK, Kingston RE (1995) A heat shock-responsive domain of human HSF1 that regulates transcription activation domain function. Mol Cell Biol 15:3354–3362
Guo J, Wu J, Ji Q, Wang C, Luo L, Yuan Y, Wang Y, Wang J (2008) Genome-wide analysis of heat shock transcription factor families in rice and *Arabidopsis*. J Genet Genomics 35:105–118
Harrison CJ, Bohm AA, Nelson H (1994) Crystal structure of the DNA binding domain of the heat shock transcription factor. Science 263:224–227
Harrison-Lowe NJ, Olsen LJ (2008) Autophagy protein 6 (ATG6) is required for pollen germination in *Arabidopsis thaliana*. Autophagy 4:339–348
Helm KW, LaFayette P, Nagao R, Key J, Vierling E (1993) Localization of small heat shock proteins to the higher plant endomembrane system. Mol Cell Biol 13:238–247
Hsu SF, Jinn TL (2010) AtHSBP functions in seed development and the motif is required for subcellular localization and interaction with AtHSFs. Plant Signal Behav 5:1042–1044
Hsu SF, Lai HC, Jinn TL (2010) Cytosol-localized heat shock factor-binding protein, AtHSBP, functions as a negative regulator of heat shock response by translocation to the nucleus and is required for seed development in *Arabidopsis*. Plant Physiol 153:773–784
Hu W, Hu G, Han B (2009) Genome-wide survey and expression profiling of heat shock proteins and heat shock factors revealed overlapped and stress specific response under abiotic stresses in rice. Plant Sci 176:583–590
Jackson-Constan D, Akita M, Keegstra K (2001) Molecular chaperones involved in chloroplast protein import. Biochim Biophys Acta 1541:102–113
Kametaka S, Okano T, Ohsumi M, Ohsumi Y (1998) Apg14p and Apg6/Vps30p form a protein complex essential for autophagy in the yeast, *Saccharomyces cerevisiae*. J Biol Chem 273:22284–22291
Kang CH, Jung WY, Kang YH, Kim JY, Kim DG, Jeong JC, Baek DW, Jin JB, Lee JY, Kim MO et al (2006) AtBAG6, a novel calmodulin-binding protein, induces programmed cell death in yeast and plants. Cell Death Differ 13:84–95
Katiyar-Agarwal S, Agarwal M, Grover A (2003) Heat-tolerant basmati rice engineered by overexpression of hsp101. Plant Mol Biol 51:677–686
Kline MP, Morimoto RI (1997) Repression of the heat shock factor 1 transcriptional activation domain is modulated by constitutive phosphorylation. Mol Cell Biol 17:2107–2115
Lau PP, Chan L (2003) Involvement of a chaperone regulator, Bcl2-associated athanogene-4, in apolipoprotein B mRNA editing. J Biol Chem 278:52988–52996
Lee JH, Schoffl F (1996) An Hsp70 antisense gene affects the expression of HSP70/HSC70, the regulation of HSF, and the acquisition of thermotolerance in transgenic *Arabidopsis thaliana*. Mol Gen Genet 252:11–19
Lee BH, Won SH, Lee HS, Miyao M, Chung WI, Kim IJ, Jo J (2000) Expression of the chloroplast-localized small heat shock protein by oxidative stress in rice. Gene 245:283–290
Lee DG, Ahsan N, Lee SH, Kang KY, Bahk JD, Lee IJ, Lee BH (2007) A proteomic approach in analyzing heat-responsive proteins in rice leaves. Proteomics 7:3369–3383
Lin YX, Jiang HY, Chu ZX, Tang XL, Zhu SW, Cheng BJ (2011) Genome-wide identification, classification and analysis of heat shock transcription factor family in maize. BMC Genomics 12:76
Lindquist S (1986) The heat-shock response. Annu Rev Biochem 55:1151–1191
Littlefield O, Nelson HC (1999) A new use for the 'wing' of the 'winged' helix-turn-helix motif in the HSF–DNA cocrystal. Nat Struct Biol 6:464–470
Liu Y, Schiff M, Czymmek K, Talloczy Z, Levine B, Dinesh-Kumar SP (2005) Autophagy regulates programmed cell death during the plant innate immune response. Cell 121:567–577
Liu HC, Liao HT, Charng YY (2011) The role of class A1 heat shock factors (HSFA1s) in response to heat and other stresses in *Arabidopsis*. Plant Cell Environ 34:738–751
Lohmann C, Eggers-Schumacher G, Wunderlich M, Schoffl F (2004) Two different heat shock transcription factors regulate immediate early expression of stress genes in *Arabidopsis*. Mol Genet Genomics 271:11–21

Lubben TH, Donaldson GK, Viitanen PV, Gatenby AA (1989) Several proteins imported into chloroplasts form stable complexes with the GroEL-related chloroplast molecular chaperone. Plant Cell 1:1223–1230

Lyck R, Harmening U, HoÈhfeld I, Treuter E, Scharf KD, Nover L (1997) Intracellular distribution and identification of the nuclear localization signals of two plant heat-stress transcription factors. Planta 202:117–125

Massey AC, Zhang C, Cuervo AM (2006) Chaperone-mediated autophagy in aging and disease. Curr Top Dev Biol 73:205–235

McCollum AK, Casagrande G, Kohn EC (2010) Caught in the middle: the role of Bag3 in disease. Biochem J 425:e1-3

Miller G, Mittler R (2006) Could heat shock transcription factors function as hydrogen peroxide sensors in plants? Ann Bot 98:279–288

Mishra SK, Tripp J, Winkelhaus S, Tschiersch B, Theres K, Nover L, Scharf KD (2002) In the complex family of heat stress transcription factors, HsfA1 has a unique role as master regulator of thermotolerance in tomato. Genes Dev 16:1555–1567

Mittal D, Chakrabarti S, Sarkar A, Singh A, Grover A (2009) Heat shock factor gene family in rice: genomic organization and transcript expression profiling in response to high temperature, low temperature and oxidative stresses. Plant Physiol Biochem 47:785–795

Mittal D, Enoki Y, Lavania D, Singh A, Sakurai H, Grover A (2011) Binding affinities and interactions among different heat shock element types and heat shock factors in rice (*Oryza sativa* L.). FEBS J 278:3076–3085

Moribe Y, Niimi T, Yamashita O, Yaginuma T (2001) Samui, a novel cold-inducible gene, encoding a protein with a BAG domain similar to silencer of death domains (SODD/BAG-4), isolated from Bombyx diapause eggs. Eur J Biochem 268:3432–3442

Morimoto RI (1998) Regulation of the heat shock transcriptional response: cross talk between a family of heat shock factors, molecular chaperones, and negative regulators. Genes Dev 12:3788–3796

Murakami T, Matsuba S, Funatsuki H, Kawaguchi K, Saruyama H, Tanida M, Sato Y (2004) Over-expression of a small heat shock protein, sHSP17. 7, confers both heat tolerance and UV-B resistance to rice plants. Mol Breed 13:165–175

Nover L, Scharf KD, Gagliardi D, Vergne P, Czarnecka-Verner E, Gurley WB (1996) The Hsf world: classification and properties of plant heat stress transcription factors. Cell Stress Chaperones 1:215–223

Nover L, Bharti K, Doring P, Mishra SK, Ganguli A, Scharf KD (2001) *Arabidopsis* and the heat stress transcription factor world: how many heat stress transcription factors do we need? Cell Stress Chaperones 6:177–189

Parker CS, Topol J (1984) A drosophila RNA polymerase-II transcription factor binds to the regulatory site of an HSP70 gene. Cell 37:273–283

Parsell DA, Kowal AS, Singer MA, Lindquist S (1994) Protein disaggregation mediated by heat-shock protein HSP104. Nature 372:475–478

Pirkkala L, Nykanen P, Sistonen L (2001) Roles of the heat shock transcription factors in regulation of the heat shock response and beyond. FASEB J 15:1118–1131

Qi Y, Wang H, Zou Y, Liu C, Liu Y, Wang Y, Zhang W (2011) Over-expression of mitochondrial heat shock protein 70 suppresses programmed cell death in rice. FEBS Lett 585:231–239

Qin G, Ma Z, Zhang L, Xing S, Hou X, Deng J, Liu J, Chen Z, Qu LJ, Gu H (2007) *Arabidopsis* AtBECLIN 1/AtAtg6/AtVps30 is essential for pollen germination and plant development. Cell Res 17:249–263

Rana RM, Dong S, Ali Z, Huang J, Zhang HS (2012a) Regulation of ATG6/Beclin-1 homologs by abiotic stresses and hormones in rice (*Oryza sativa* L.). Genet Mol Res 11:3676–3687

Rana RM, Dong S, Ali Z, Khan AI, Zhang HS (2012b) Identification and characterization of the Bcl-2-associated athanogene (BAG) protein family in rice. Afr J Biotechnol 11:88–99

Rana RM, Dong S, Tang H, Ahmad F, Zhang HS (2012c) Functional analysis of OsHSBP1 and OsHSBP2 revealed their involvement in the heat shock response in rice (*Oryza sativa* L.). J Exp Bot 63:6003–6016
Rieping M, Schöffl F (1992) Synergistic effect of upstream sequences, CCAAT box elements, and HSE sequences for enhanced expression of chimaeric heat shock genes in transgenic tobacco. Mol Gen Genet 231:226–232
Sarkar NK, Kim YK, Grover A (2009) Rice sHSP genes: genomic organization and expression profiling under stress and development. BMC Genomics 10:393
Sato Y, Yokoya S (2008) Enhanced tolerance to drought stress in transgenic rice plants overexpressing a small heat-shock protein, sHSP17.7. Plant Cell Rep 27:329–334
Satyal SH, Chen D, Fox SG, Kramer JM, Morimoto RI (1998) Negative regulation of the heat shock transcriptional response by HSBP1. Gene Dev 12:1962–1974
Scharf K, Heider H, Hohfeld I, Lyck R, Schmidt E, Nover L (1998) The tomato Hsf system: HsfA2 needs interaction with HsfA1 for efficient nuclear import and may be localized in cytoplasmic heat stress granules. Mol Cell Biol 18:2240–2251
Schirmer EC, Glover JR, Singer MA, Lindquist S (1996) HSP100/Clp proteins: a common mechanism explains diverse functions. Trends Biochem Sci 21:289–296
Schöffl F, Prändl R, Reindl A (1998) Regulation of the heat-shock response. Plant Physiol 117:1135–1141
Schöffl F, Prändl R, Reindl A (1999) Molecular responses to heat stress. In: Shinozaki K, Yamaguchi-Shinozaki K (eds) Molecular responses to cold, drought, heat and salt stress in higher plants. R.G. Landes Co., Austin, pp 81–98
Schroder M, Kaufman RJ (2005) The mammalian unfolded protein response. Annu Rev Biochem 74:739–789
Seo JS, Lee YM, Park HG, Lee JS (2006) The intertidal copepod *Tigriopus japonicus* small heat shock protein 20 gene (*Hsp20*) enhances thermotolerance of transformed *Escherichia coli*. Biochem Biophys Res Commun 340:901–908
Siddique M, Gernhard S, von Koskull-Doring P, Vierling E, Scharf KD (2008) The plant sHSP superfamily: five new members in *Arabidopsis thaliana* with unexpected properties. Cell Stress Chaperones 13:183–197
Singh A, Grover A (2010) Plant Hsp100/ClpB-like proteins: poorly-analyzed cousins of yeast ClpB machine. Plant Mol Biol 74:395–404
Sinha S, Levine B (2008) The autophagy effector Beclin 1: a novel BH3-only protein. Oncogene 27:S137–S148
Sondermann H, Scheufler C, Schneider C, Höhfeld J, Hartl FU, Moarefi I (2001) Structure of a Bag/Hsc70 complex: convergent functional evolution of Hsp70 nucleotide exchange factors. Science 291:1553–1557
Song J, Takeda M, Morimoto RI (2001) Bag1-Hsp70 mediates a physiological stress signalling pathway that regulates Raf-1/ERK and cell growth. Nat Cell Biol 3:276–282
Sørensen JG, Kristensen TN, Loeschcke V (2003) The evolutionary and ecological role of heat shock proteins. Ecol Lett 6:1025–1037
Su PH, Li HM (2008) *Arabidopsis* stromal 70-kD heat shock proteins are essential for plant development and important for thermotolerance of germinating seeds. Plant Physiol 146:1231–1241
Su LJ, Chang CF, Han HP, Ma H, Xu CS (2006) Analysis of changes about hsbp1, hsf1, hsf2 AND hsp70's expression levels in rat's regenerating liver. Fen Zi Xi Bao Sheng Wu Xue Bao 39:258–264
Sung DY, Kaplan F, Lee KJ, Guy CL (2003) Acquired tolerance to temperature extremes. Trends Plant Sci 8:179–187
Suzuki K, Kirisako T, Kamada Y, Mizushima N, Noda T, Ohsumi Y (2001) The pre-autophagosomal structure organized by concerted functions of APG genes is essential for autophagosome formation. EMBO J 20:5971–5981

Suzuki M, Iwasaki M, Sugio A, Hishiya A, Tanaka R, Endo T, Takayama S, Saito T (2011) BAG3 (BCL2-associated athanogene 3) interacts with MMP-2 to positively regulate invasion by ovarian carcinoma cells. Cancer Lett 303:65–71

Tai LJ, McFall SM, Huang K, Demeler B, Fox SG, Brubaker K, Radhakrishnan I, Morimoto RI (2002) Structure-function analysis of the heat shock factor-binding protein reveals a protein composed solely of a highly conserved and dynamic coiled-coil trimerization domain. J Biol Chem 277:735–745

Takayama S, Reed JC, Homma S (2003) Heat-shock proteins as regulators of apoptosis. Oncogene 22:9041–9047

Thao NP, Chen L, Nakashima A, Hara S, Umemura K, Takahashi A, Shirasu K, Kawasaki T, Shimamoto K (2007) RAR1 and HSP90 form a complex with Rac/Rop GTPase and function in innate-immune responses in rice. Plant Cell 19:4035–4045

Ueda K, Kosako H, Fukui Y, Hattori S (2004) Proteomic identification of Bcl2-associated athanogene 2 as a novel MAPK-activated protein kinase 2 substrate. J Biol Chem 279:41815–41821

Von Koskull-Döring P, Scharf KD, Nover L (2007) The diversity of plant heat stress transcription factors. Trends Plant Sci 12:452–457

Wang W, Vinocur B, Altman A (2003) Plant responses to drought, salinity and extreme temperatures: towards genetic engineering for stress tolerance. Planta 218:1–14

Wang C, Zhang Q, Shou HX (2009) Identification and expression analysis of *OsHsfs* in rice. J Zhejiang Univ Sci B 10:291–300

Waters ER, Lee GJ, Vierling E (1996) Evolution, structure and function of the small heat shock proteins in plants. J Exp Bot 47:325–338

Williams B, Kabbage M, Britt R, Dickman MB (2010) AtBAG7, an *Arabidopsis* Bcl-2-associated athanogene, resides in the endoplasmic reticulum and is involved in the unfolded protein response. Proc Natl Acad Sci U S A 107:6088–6093

Wu C (1984) Activating protein factor binds in vitro to upstream control sequences in heat shock gene chromatin. Nature 311:81–84

Wu C (1995) Heat shock transcription factors: structure and regulation. Annu Rev Cell Dev Biol 11:441–469

Yamada K, Fukao Y, Hayashi M, Fukazawa M, Suzuki I, Nishimura M (2007) Cytosolic HSP90 regulates the heat shock response that is responsible for heat acclimation in *Arabidopsis thaliana*. J Biol Chem 282:37794–37804

Yamanouchi U, Yano M, Lin H, Ashikari M, Yamada K (2002) A rice spotted leaf gene, Spl7, encodes a heat stress transcription factor protein. Proc Natl Acad Sci U S A 99:7530–7535

Yokotani N, Ichikawa T, Kondou Y, Matsui M, Hirochika H, Iwabuchi M, Oda K (2008) Expression of rice heat stress transcription factor *OsHsfA2e* enhances tolerance to environmental stresses in transgenic *Arabidopsis*. Planta 227:957–967

Young JC, Agashe VR, Siegers K, Hartl FU (2004) Pathways of chaperone-mediated protein folding in the cytosol. Nat Rev Mol Cell Biol 5:781–791

Part II
Large Plant Heat Shock Proteins (HSP)

Chapter 7
Plant Stress Response: Hsp70 in the Spotlight

Doel Ray, Anupama Ghosh, Soumyajit Banerjee Mustafi, and Sanghamitra Raha

Abstract Heat Shock Protein 70 (Hsp70) is an evolutionarily conserved family of proteins which carry out multiple cellular functions such as protein biogenesis, protection during stress, prevention of formation of protein aggregates, assistance in protein translocation and many others. Hsp70, being the major cytoprotective molecular chaperone, plays a crucial role in protecting against a stunning array of stresses and in the re-establishment of cellular homeostasis. This book chapter gives an overview of the multifaceted Hsp70s in plants, with special emphasis on their association with plant response to various stress conditions and eventually, stress acclimation. The contribution of plant stress-responsive proteomics studies towards putting Hsp in the spotlight has also been brought forth. The road ahead is to decipher the underlying mechanisms of Hsp70-mediated multiple cross tolerance, that is likely to lead to new strategies to enhance crop tolerance to environmental stress.

Keywords Acclimation • Biotic and abiotic stress • Molecular chaperones • Hsp70 • Plant • Stress proteomics

D. Ray
School of Computational and Integrative Sciences, Jawaharlal Nehru University, New Mehrauli Road, New Delhi 110067, India

A. Ghosh
Division of Plant Biology, Bose Institute,
Centenary Campus P1/12, C.I.T. Road, Scheme-VIIM, Kolkata 700054, India

S.B. Mustafi
University of Oklahoma Health Sciences Center, Oklahoma City, USA

S. Raha (✉)
Department of Biotechnology and ISERC, Visva-Bharati, Santiniketan 731235, India
e-mail: srr1987@gmail.com

A.A.A. Asea et al. (eds.), *Heat Shock Proteins and Plants*,
Heat Shock Proteins 10, DOI 10.1007/978-3-319-46340-7_7

Abbreviations

ABA	abscisic acid
ABRE	ABA-responsive element
AREB	ABRE binding protein
ABF	ABRE binding factor
BiP	binding protein, Hsp70 homolog
CDPKs	calcium-dependent protein kinases
CHIP	carboxy terminus of Hsc70-interacting protein
CNV	cucumber necrosis virus
DREB	dehydration-responsive element binding
ER	endoplasmic reticulum
GAPDH	glyceraldehyde-3-phosphate dehydrogenase
Hsc	heat shock cognate
HSE	heat shock element
Hsf	heat shock factor
Hsp	heat shock protein(s)
hsp	heat shock protein gene
HSR	heat shock response
H_2O_2	hydrogen peroxide
HM	heavy metal
LEA	late embryogenesis abundant proteins
MAPKs	mitogen activated protein kinases
MgProto or MgProtoMe	Mg-protoporphyrin IX or its monomethyl ester
NO	nitric oxide
PCD	programmed cell death
PSII	photosystem II
ROS	reactive oxygen species
UPR	unfolded protein response

7.1 Introduction

The molecular function of a protein closely correlates to its native structure. Nascent proteins achieve the optimal conformation either spontaneously or more often with the assistance from specialized proteins called molecular chaperones. Molecular chaperones maintain cellular homeostasis both under physiological or pathological conditions through a well-regulated and streamlined series of events; thus preventing either premature folding of nascent proteins or aggregation of the matured proteins enduring stress (Baniwal et al. 2004).

Molecular chaperones are a diversified group of proteins that range in molecular weight (~10 kDa to ~110 kDa); cellular distribution (nuclear, cytosolic, ER) and function (chloroplast development, thermal stress response). Heat shock proteins are ubiquitously expressed stress responsive chaperone proteins found across both plant and animal kingdoms. In line with the name itself, "heat shock proteins" were initially discovered as proteins induced in response to heat shock in *Drosophila melanogaster* (Ritossa 1962). Several instances since then have been noticed involving different stress factors other than heat shock to induce expression of this family of proteins. The host range was also found not to be limited to *D. melanogaster*, but similar behavior of Hsps could be evidenced in other organisms including plants, fungi and bacteria. These stresses mostly include oxidative stress, UV irradiation and cold stress. Most of the structural and functional insights on Hsp70 protein family have been derived from information obtained in *E. coli*, yeast and mammalian cells. Although this group of proteins is highly conserved in evolution, plant Hsp70 shows distinct species specificity and is best characterized in rice, maize and *Arabidopsis*. In the rice genome, the *hsp70* super family represents 24 *hsp70* and 8 *hsp110/sse* subfamily members; Arabidopsis genome contains at least 18 *hsp70*, 14 *dnaK* (yeast mitochondrial *hsp70*) subfamily and 4 *hsp110/sse* subfamily (Sarkar et al. 2013; Wu et al. 1988).

In plants heat shock protein family is categorized into five classes based on the molecular weights of each of the members. Members belonging to Hsp100 family are the ones with highest molecular weights and are involved mostly in thermotolerance of the host plant. Hsp90 on the other hand plays a role in maintaining cellular homeostasis through stabilization of different client proteins by means of its chaperone activities. Besides these, Hsp90 has also been found to participate in the mechanism of disease resistance of certain plants. For instance in *Arabidopsis thaliana* R protein RPM1 gains stability and ability to induce resistance against *Pseudomonas syringae* infection through its interaction with cytosolic AtHsp90 (Hubert et al. 2003). The other three classes include Hsp60, Hsp70 and small Hsps. While Hsp60 family proteins are involved primarily in maintaining organellar protein homeostasis (Lubben et al. 1989), those belonging to Hsp70 family, have been found to have most diverse roles in plants. Like other Hsps, small Hsps also exert their function primarily in the stress response of the host by exhibiting an ATP-independent chaperone activity (Garrido et al. 2012).

Hsp70 group of proteins are highly conserved and are encoded by a multi-gene family whose members are both developmentally regulated as well as can be differentially expressed under biotic and abiotic stress. In a normal physiological environment Hsc70 (Hsp70 cognate) exists as a member of the multiprotein complex with other class of chaperones like Hsp90 and often function in concert with a variety of co-chaperones. The composition of this chaperone/–co- chaperone complex in which Hsp70 plays a central role is dynamic and dictates the functional specificity like maturation of de novo synthesized proteins and transportation of precursor proteins (Wang et al. 2014). Based on the sequence homology that has distinct identifiable motifs at the C-terminus, plant Hsp70 protein families can be divided into 4 major subgroups with different cellular localization: cytosol (EEVD motif), ER (HDEL motif), plastids (PEGDVIDADFTDSK motif) and mitochondria (PEAEYEEAKK) motif (Guy and Li 1998). Apart from this, Hsp70 can also be found at subcellular compartments (glyoxysomes and protein bodies). Unlike yeast and higher eukaryotes, plants contain multiple members belonging to plant Hsp70 family that reside within the ER lumen. For instance tobacco codes at least five *hsp70* genes and the encoded products (designated as BiP) which possess an ER luminal localization. Similar examples can be found in case of soyabean, *Arabidopsis* and maize where four, three and two members respectively from Hsp70 family could be found to participate in protein folding functions inside ER lumen.

7.2 Hsp70 Function and Crosstalk in Plants

7.2.1 *Hsp70 Chaperone Function and Regulation*

Hsp70 proteins are ATP driven molecular chaperones with an N-terminal ATPase domain and a C-terminal peptide binding domain. The substrates gain their native confirmation as it goes through repeated cycles of binding and release as Hsp70 switches between the low-affinity (ATP bound) and high-affinity (ADP bound) states with the ATPase cycle regulated by DnaJ (Hsp40) like co-chaperones (Mayer and Bukau 2005).

Despite the vital functions carried out by Hsp70s and their physiological and agronomic importance, the functional analyses of Hsp70s in plants remain very limited due to (i) gene redundancy which limits genetic analysis (ii) lethality of mutants and (iii) the absence of specific and efficient inhibitors. Therefore, little is known about the specific contributions of Hsp70s to various pathways and their molecular targets in eukaryotes, and particularly in plants. Single Hsp70 mutants were not identified in forward genetic approaches nor did those display any mutant phenotypes in forward genetic screens, probably due to a high level of functional redundancy. It is possible that loss of function of a single Hsp70 can be easily compensated by other co-localized Hsp70s. Furthermore, even when its function is not replaceable, the high level of substrate promiscuity of Hsp70 could cause pleiotropic phenotypes that make it difficult to pinpoint the immediate molecular targets. Moreover, double mutants such as *hsc70-2/hsc70-4* are lethal in seedlings (Cazale et al. 2009; Sung and Guy 2003). Therefore, the physiological characterization of transgenic plants overexpressing Hsp70 members emerged as an important tool to dissect Hsp70 functions *in vivo* in various plants including model plant, Arabidopsis.

7.2.2 *Hsp70 in Growth and Development*

Hsp70s are known to show complex developmental regulation. Several plant cytosolic *hsp70* genes are expressed during seed development, maturation, and/or germination. Hsp70 proteins are noted to be abundant in dry seeds but their levels drastically decline within 72 h after the onset of imbibition's (Sarkar et al. 2013).

Overexpression and knock-down analyses of the *bip1/osbip3* gene demonstrated defects in seed development (Jung et al. 2013). In rice, BiP1 is indicated to be involved in ER quality control during seed maturation (Wakasa et al. 2011). In *A. thaliana*, BiP is involved in the fusion of polar nuclei during female gametophyte development (Maruyama et al. 2010). Hsp70 has also been suggested to play an important role in alfalfa (*Medicago sativa*) nodule development (He et al. 2008). *Arabidopsis* stromal Hsp70s (cpHsc70s) have been reported essential for plastid structure and function as well as plant development. The knockout mutants of *cphsc70-1* exhibited variegated cotyledons, malformed leaves, growth retardation, and impaired root growth under normal conditions while further impairment was observed after heat shock treatment of germinating seeds (Latijnhouwers et al. 2010; Su and Li 2008).

7.2.3 Hsp70 as Translocator

Hsp70s also function as motors driving protein translocation into chloroplast, mitochondria and the ER (Su and Li 2010). Proteins destined for the organelles are targeted from the cytoplasm as unfolded precursors. In the cytosol, unfolded proteins have a high tendency to form cytotoxic, nonspecific aggregates if they accumulate to high levels. Hence, posttranslational targeting to endosymbiotic organelles requires that precursor levels be maintained within limits that do not result in nonspecific aggregate formation while at the same time, the cytosolic regulatory mechanism must not jeopardize the supply of sufficient amounts of proteins to the organelles. Hsp70s play multitude roles in the strategies employed by plant cells for this regulation. For instance, Arabidopsis Hsc70-4 and E3 ligase carboxy terminus of Hsc70- interacting protein (CHIP) mediate plastid-destined precursor degradation to prevent cytosolic precursor accumulation, thereby playing a critical role in embryogenesis (Lee et al. 2009).

7.2.4 Regulation of Induction of HSP70 – Heat Shock Response (HSR)

Hsp70 proteins are transcriptionally regulated by developmental cues such as male gametogenesis and embryogenesis (Neuer et al. 1999), thermal stress (Kotak et al. 2007) and unfolded protein response (UPR) etc. (Wang et al. 2004). Induction of Hsp70 proteins is due to the binding of transactive Heat shock factor 1 (Hsf1) to the evolutionary conserved 5′- aGAAg-3′ sequence in plants, a *cis*-regulatory promoter element (HSE) located in the TATA- box-proximal 5′-flanking regions of heat-shock genes (Barros et al. 1992). Hsf1 from Arabidopsis can bind to the tripartite HSE on *hsp70* promoter post trimerization (Baniwal et al. 2004; Hubel et al. 1995; Hubel and

Schoffl 1994). Plants possess a considerably greater number of genes for *hsf*. Also, the diverse nature of Hsfs is quite unique to the plant kingdom when compared to other eukaryotes such as yeast and mammals (Virdi et al. 2015). Cell survival under high temperature involves the activation of heat stress response (HSR), which in principle is highly conserved among different organisms, but shows remarkable complexity and unique features in plant systems, given their sessile nature and continuous exposure to environmental fluctuations (Bokszczanin and Fragkostefanakis 2013).

In plants, HSR involves multiple pathways, regulatory networks and cellular compartments. The perception of heat is attributed mainly to four putative 'heat sensors' *viz.* a cyclic nucleotide gated calcium channel on the plasma membrane that initiates an inward calcium flux, a histone H2A.Z variant and two unfolded protein sensors in the ER and the cytosol (Lavania et al. 2015; Mittler et al. 2012). The signals generated by these different sensors are probably integrated by a signal transduction network involving various secondary messengers (Ca^{2+}, H_2O_2 and NO), kinases (CDPKs, MAPKs), phosphatases and transcriptional regulators. The activation of different pathways may be tissue specific (Dong et al. 2015; Mittler et al. 2012). Heat stress transcription factors (Hsfs) control core aspects of HSR by regulating the transcriptional activity of a wide range of genes, but they have been mainly implicated with the rapid synthesis and accumulation of Hsps (Scharf et al. 2012). Ca^{2+} entry into the cell after modifications of the plasma membrane induced by heat shock has been documented. Changes in DNA-binding activity of Hsf in plants could be triggered by Ca^{2+}/Calmodulin dependent pathways. Also, hydrogen peroxide and nitric oxide may influence the heat shock response at the Hsf level through modulation of Ca^{2+}/Calmodulin signaling (Saidi et al. 2011). As discussed in the previous sections, Hsp70 functions in multimeric protein complex along with other chaperones such as Hsp90 and Hsp110. Work in tomato plant demonstrates a novel crosstalk between Hsp70 and Hsp90 in regulation of the Hsf1 binding to the HSE. While Hsp70 represses the transcriptional activity by inhibiting DNA binding activity of HsfB1-HsfA2 complex, Hsp90 stimulates DNA binding activity of HsfB1 (Hahn et al. 2011). Hsp70 can also be an important signaling molecule that establishes communication between subcellular compartments performing diverse functions.

Analysis of the light induction of the nuclear Hsp70 in the *Chlamydomonas* provided an early evidence of cross compartmental networking of chaperone proteins in maintaining cellular function. Light within the chloroplast or its envelope can trigger a signaling cascade which leads to the induction of *hsp70* gene in the nucleus. In response to light activation, MgProto and MgProtoMe, are synthesized in the plastid, translocate to the cytosol and nucleus and are identified by transcription factors or signaling molecules that directly or indirectly regulate expression of the nuclear *hsp70* genes (Kropat et al. 2000). As evident from a chloroplast- targeted Hsp70 in *Chlamydomonas*, it plays an important role in protecting photo system II (PSII) against frequent light dependent damages through, de novo synthesis of D1 (a 32 kDa reaction center protein) PSII reassembly and repair (Fig. 7.1) (Schroda et al. 2001; Schroda et al. 1999).

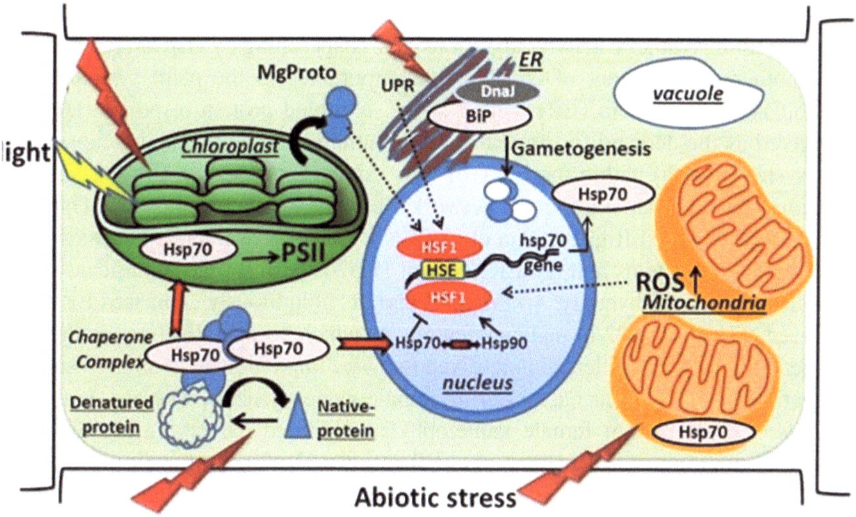

Fig. 7.1 Scheme depicting intracellular cross-talk of Hsp70 within subcellular compartments. Cellular Hsp70 synthesis is induced either due to one of the several abiotic stress or due to normal cellular physiology. ROS generation within the mitochondria regulate Hsp70 transcription. Within the cytosol, Hsp70s participate as multimeric protein complex to chaperone stress induced or nascent unfolded proteins to their native confirmation

The Hsp90- Hsp70 interaction within the nucleus impacts the transcriptional activity of Hsf1. Hsp70 negatively and Hsp90 positively influence Hsf1 binding to the HSE (Heat Shock Element). Hsp70 chaperone activity within the ER (BiP) can be induced by unfolded protein response (UPR). BiPs and its interacting DNA J proteins regulate nuclear fission and gametogenesis. Light induced release of soluble factors from the chloroplast can regulate nuclear transcription of Hsp70. Hsp70 translocated to the chloroplast protects PSII (Photosystem II) from damage and aid in photosynthesis and chlorophyll retention. Excess synthesis of Hsp70 can be auto-regulated through inhibitory feedback loop. Red line arrows indicate translocation; broken line arrows indicate transcriptional regulation. For mechanistic details of the cross-talk, please refer to the text.

Experiments performed by Kim et. al. that involved growing of rice plants at 25 °C versus 40 °C further indicated that chloroplast Hsp70 (OsHsp70CP1) localized in the stroma is essential for normal differentiation of the chloroplasts from the proplastids even under elevated temperatures (Bionda et al. 2016; Kim and An 2013). Augustine et. al elucidated that transgenic sugarcane plants harboring *hsp70* overexpression are more resistant to stress from draught and salinity. Mechanistically, transgenic sugarcane plants showed enhanced chlorophyll retention within the chloroplast along with improved germination ability when compared to non-transgenic counterparts growing under similar saline stress (Augustine et al. 2015a; Augustine et al. 2015b). These studies undertaken by different research groups argue in favor of the existence of a strong intracellular networking coupled to functional cooperativity of photosynthetic activity and molecular chaperoning by Hsp70 (Beck 2005).

Another compartment of considerable importance to the plant system is the Endoplasmic Reticulum (ER). Under stress, unfolded protein response (UPR) is triggered as the demand for protein folding outnumbers the managing capacity of the system. Hsp70 within the ER (BiP1) is critical for inducing ER stress and is functionally operated through a network of other DnaJ proteins like OsP58B, OsERdj3A, and OsERdj3B (Ohta et al. 2013). Analysis of interaction among rice ER-resident J-proteins and the ER-resident Hsp70 (BiP) reveals significant crosstalk and functional diversity. *Arabidopsis* harbors ubiquitously expressed ER stress induced BiP1 and BiP2 in male gametophytes which co-operatively assist in male gametogenesis and pollen competitiveness. *bip1 bip2* double mutant results in defective fusion of polar nuclei during female gametogenesis, whereas, *bip1 bip2 and bip3* triple mutant female gametophytes exhibited defects during the early stages of development. It is likely that ER resident Hsp70 along with its elaborate networks of other J proteins are crucial for initiating signaling cascades leading to gametogenesis. ROS and mitogen activated protein kinase (MAPK) often play a role in gametogenesis and it may be speculated that a crosstalk with the signaling molecules and molecular chaperones are important upstream events within the ER niche to trigger gametogenesis. More recently Maruyama et al. demonstrated that BiP3 is evenly expressed in the pollen tubes independent of the ER stress to meet the high physiological demand of the cellular protein secretion in actively growing cells such as the pollen tube (Maruyama et al. 2014). Growth of pollen tubes is characterized by tip-localized ROS which is required for the normal growth rate of pollen tube (Potocky et al. 2007). So, BiP3 induction can arguably be an indirect effect of the ROS generation.

Arabidopsis responds to ROS generation within the chloroplast by enhanced expression level of various genes encoding heat shock proteins (Hsps) and heat shock transcription factors HsfA2 and HsfA4A that were reported to be regulators of genes involved in stress response of Arabidopsis. So an oxidative stress can trigger a heat stress response within the cellular system. Dehydration priming followed in 2-week-old rice seedlings under continuing dehydration stress demonstrated oxidative stress but did not show any induction of the total Hsp70 level compared to the condition matched controls (Cho and Choi 2009; Goswami et al. 2013). This in contrary indicate that crosstalk between different stress pathways are context dependent, given the induction of Hsp70 expression is positively regulated to acquisition of thermotolerance, salt stress and other abiotic factors in one plant system but not in another.

Hsp70 also play a significant role in rice plant memory imprint to pre-exposure to stress from arsenic and heat, as demonstrated by sustained expression of Hsp70 even after stress withdrawal (Goswami et al. 2010). Such sustained expression of Hsp70 can be attributed to epigenetic changes like methylation and acetylation of histones, and triggering of cellular signaling networks like MAPK pathways.

Functional cooperation of Hsp70 with other chaperones like Hsp90, genetic and epigenetic modifiers and signaling pathways although are well documented in mammalian system is still largely unexplored in the plants.

7.3 Multifaceted Role of HSP70 in Stress Tolerance of Plants

Being sessile organisms plants are exposed to a number of environmental stresses that include both abiotic as well as biotic factors. Under both these situations the response from the plant involves activation of a number of defense mechanisms through initiation of intricate intracellular signaling events. This ultimately results in the differential expression of a number of stress inducible genes. Among these stress inducible genes are included genes coding for functional proteins as well as those coding for regulatory proteins. For instance late embryogenesis abundant (LEA) proteins are enzymes that impart osmoprotection to plants under water stress conditions (Goyal et al. 2005) and are classified under stress inducible functional proteins. They are directly involved in the stress tolerance of the plant. Regulatory proteins on the other hand are mostly transcription factors that target expression of a number of downstream genes involved in stress response. Like for instance the dehydration responsive element binding (DREB) family of transcription factors, DREB2 has been documented to target heat shock transcription factor 3 (AtHsfA3) in *Arabidopsis* leading to thermotolerance (Qin et al. 2011). Both DREB and AtHsfA3 can therefore be grouped under regulatory stress inducible proteins. One of the many downstream genes that are induced by DREB2 family of transcription factors is a heat shock protein gene.

7.3.1 Biotic Stress Tolerance

Information from the recent literature indicates both a protective and an insidious role for Hsp70 in plant defense against pathogens. A direct interaction of pathogen virulence factors with stress tolerance components of the plant host was demonstrated for the *Pseudomonas syringae* type III effector HopI1 that targets Hsp70. HopI1 binds plant Hsp70 and transports it to plant chloroplasts with the formation of a large complex leading to possible suppression of plant defenses (Jelenska et al. 2010). In contrast, Cucumber necrosis virus (CNV) infection of *Nicotiana benthamiana* triggers an increase of *hsp70* transcript and Hsp70 protein level along with higher CNV genomic RNA, coat protein and virion accumulation. In fact, CNV appoints Hsp70 to aid many phases of the infection process (Alam and Rochon 2015). Similarly, Tomato yellow leaf curl virus (TYLCV) DNA decreased in the host plant after inactivation of Hsp70 (Gorovits et al. 2013). Both *hsp70* gene and protein levels underwent changes in three *Solanum* sp. exposed to high and low temperatures and also subjected to infection by powdery mildew. Changes in

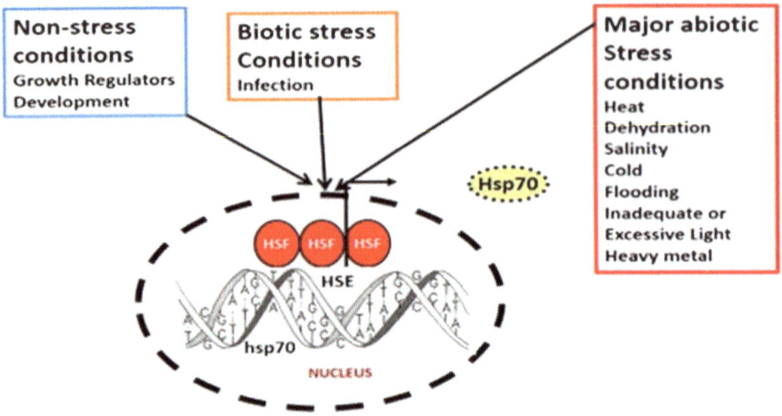

Fig. 7.2 A summary of the different inducers of the heat shock response in plants

temperature induced a rise in mRNA and protein levels of Hsp70. However, fungal infection resulted in the built up of Hsp70 only in susceptible *S. lycopersicum* but not the resistant species. Therefore, the role of Hsp70 may vary from supportive to deleterious depending on the nature of the infectious agent and possibly the host itself (Kubienova et al. 2013).

7.3.2 *Abiotic Stress Tolerance*

Different conditions which induce Hsp70 induction are summarized in Fig. 7.2.

7.3.3 *Heat Stress*

Global warming is increasingly becoming a threat to all living organisms, including plants. Although the reasons behind global warming are still unclear, its adverse consequences are clear and of increasing concern worldwide (Hedhly et al. 2009). The increase in global mean surface temperature is projected to be in the range of 1.5–4 °C by the end of the twenty-first century (Lavania et al. 2015; Liao et al. 2014). A likelihood of more intense, more frequent, and longer lasting heat waves is predicted by the Intergovernmental Panel on Climate Change (Scafaro et al. 2010). Models for major crops in tropical and subtropical regions showed yield losses ranging from 2.5 to 16 % for every 1 °C increase in seasonal temperatures (Lobell et al. 2008; Ahsan et al. 2010). A recent review (Bita and Gerats 2013) has assessed the differential effect of climate change, both in terms of geographic location and the crops that will likely show the most extreme reductions in yield as a result of expected extreme fluctuations in temperature and global warming in general. A temperature rise of 3–4 °C could cause crop yields to fall by 15–35 % in Africa and Asia and by 25–35 % in the Middle East. Sudden rises in temperature to high levels may lead to cell death within short duration as a consequence of extensive protein denaturation and aggregation and loss of membrane integrity. Furthermore, prolonged exposure to moderately high temperatures can lead to reduced cellular function and overall plant fitness (Driedonks et al. 2015). Recurrent heat stress also affects disease resistance in plants by suppressing plant immunity, as plant heat stress and defense responses share important mediators such as calcium ions and Hsps (Barah et al. 2013; Lee et al. 2012).

Elevated temperatures may lead to scorching of leaves and stems, abscission and senescence, shoot and root growth inhibition or fruit damage, which consequently lead to decreased plant productivity (Bita and Gerats 2013). Moreover, the sexual reproductive phases in plants are considered as extremely sensitive and vulnerable to heat stress (Hedhly et al. 2009). Because plants are sessile organisms that cannot escape heat, they are compelled to invest valuable resources in modifying their metabolism to prevent damage caused by heat, in a process generally referred to as acclimation. Alternatively, plants can activate programmed cell death (PCD) in specific cells or tissues, a process that can lead to the shedding of leaves, the abortion of flower or fruit formation, or even death of the entire plant (Mittler et al. 2012).

Heat stress can have a devastating effect on plant metabolism that include misfolding and aggregation of proteins, altered efficiency of enzymatic reactions in the cell, membrane damages, disruption of microtubule organization and accumulation of molecules like reactive oxygen species (ROS) to deleterious levels. Heat stress affects the whole gamut of metabolic processes, acting first on protein complexes, their quaternary structure being lost, one of the side effects is the uncoupling of pathways, disrupting the steady-state flux of metabolites, thereby causing the accumulation of toxic by-products, such as ROS (Fragkostefanakis et al. 2015; Mittler et al. 2012; Timperio et al. 2008). An intimate relationship and multi-level interactions seem to exist between ROS and Hsps, corroborating the hypothesis that during the course of evolution, plants obtained a tight control over ROS levels and were able to use ROS as signals for increasing Hsp levels (Driedonks et al. 2015; Timperio et al. 2008). About 5 % of the plant transcriptome has been found to be upregulated twofold or more in response to heat stress but intriguingly, molecular chaperones including Hsps comprise only a minor fraction of it (Mittler et al. 2012). Nevertheless, the best characterized aspect of acquired thermotolerance is the production of Hsps (Zhou et al. 2012). The homeostasis of cellular proteins (proteostasis) is disturbed when heat stress sets in. Protein structure is of paramount importance for the proper functioning of proteins. When protein structure gets affected by heat stress (in terms of denaturation, misfolding or aggregation), it negatively impacts protein functioning. In order to minimize damage to cellular proteins, cellular levels of chaperone proteins are generally upregulated. Hsps act as molecular chaperones, involved in various aspects of proteostasis such as facilitation of refolding, protecting proteins from terminal aggregation, protein translocation and degradation (Fragkostefanakis et al. 2015; Lavania et al. 2015).

7.3.4 Drought Stress

Drought stress is one of the abiotic stresses considered to be a major cause of crop loss worldwide (Mahajan and Tuteja 2005). Induced expression of host Hsp70 has been evidenced in several instances of drought tolerant plants. For instance proteomic analysis of upland rice variety IRAT 109 at seedling stage during drought stress showed induced expression of *hsp70* genes (Shu et al. 2011). Moreover, examples are also available in the literature that shows induction of drought tolerance in susceptible plants through simple overexpression of an *hsp70* gene. It was shown that heterologous overexpression of Chrysenthemum *hsp70* gene in *Arabidopsis* can induce drought tolerance to the latter (Song et al. 2014). However, the mechanisms through which these increased cellular levels of Hsp70 provide protection against drought tolerance remain an active area of investigation. Nevertheless majority of the studies indicate towards the ability of Hsp70 to stabilize protein structure under different stress condition, to play a major role in stress tolerance. In some cases Hsp70 activity could be linked to induced drought tolerating morphological changes in host plant. For instance in *Erianthus arundinaceus*, a Hsp70 protein has been found to be a key regulator controlling the formation of anisotropic interdigitation in sugarcane in response to drought stress (Augustine et al. 2015a). Accordingly overexpression of the said Hsp70 was demonstrated to impart increased drought and salinity tolerance to sugarcane (Augustine et al. 2015b). Other signaling events and molecular changes associated with drought stress tolerance involve stomatal closure, intracellular accumulation of proline that serves as an osmo-protectant, *abscisic* acid (ABA) dependent and independent drought tolerance signaling pathways. Hsp70 plays an important role in each of these individual events.

7.3.5 Hsp70 in Abscisic Acid (ABA) Dependent Stress Signaling

Water loss due to drought stress leads to several changes within a plant cell. These changes include reduced intracellular volume, disruption of plasma membrane integrity, loss in cell turgor, imbalance of solute concentrations across plasma membrane and finally protein denaturation (Bray 1997). Early perception of these changes by the plant marks the initiation of the signaling events that contribute towards drought tolerance. ABA plays an important role at this point where it acts as a sensor of these cellular changes. In other words ABA synthesis is induced by plant cells in response to drought stress. A subsequent perception of the increased levels of this key phytohormone by its cognate receptors such as pyrabactin resistance 1/PYR1 like/regulatory component of ABA response 1 (PYR/PYL/RCARs) leads to activation of ABA responsive element binding protein/ABRE-binding factor (AREB/ABF) (Umezawa et al. 2010). AREB/ABF then serves as the master regulator transcription factor that regulates a number of downstream drought responsive genes. The encoded proteins of these drought responsive genes then bring about different cellular responses to drought stress including stomatal closure, induction of stress proteins and accumulation of osmoprotective metabolites (Kuromori et al. 2014). Hsp70 is one among the many stress proteins that is induced by ABA dependent stress signaling pathway. Heat-induced enhancements of levels of several Hsps are observed by proteomic studies in diverse plants like rice, wheat grain, a heat and drought-tolerant poplar, and a wild plant *Carissa spinatum* (Kosova et al. 2011).

7.3.6 Salinity Stress

It is predicted that increasing salt content of cultivable land could bring about 50 % loss of land by the year 2050 which will adversely affect global agriculture (Wang et al. 2003; Ngara and Ndimba 2014). Chitteti and Peng (2007) reported the upregulation of Hsp70 in proteome and phosphoproteome differential expression analysis of rice roots subjected to salinity stress. Differential phosphorylation pattern was also observed for putative ribosomal protein S29, dnaK-type chaperone Hsp70, GAPDH, endo-β-1,3-glucosidase and others (Kosova et al. 2011). Hsp70, is upregulated in salt-stressed wheat (Sobhanian et al. 2011). The expression of some defense-related proteins, such as chloroplast Hsp70, is upregulated in *A. lagopoides* as a result of salt stress (Sobhanian et al. 2011). Interesting results have been obtained by Wang et al. (2008) who studied salt stress response in a Chinese common wheat cultivar Jinan 177 and its hybrid with a salt-tolerant *Thinopyrum ponticum*. In both genotypes, significant changes in the expression of several signaling proteins, up-regulation of several transcription and translation factors and Hsp70 and other chaperones was observed upon salt stress. In *Poaceae* species Hsp70 levels are enhanced due to salt stress with variations among species showing differing levels of salt tolerance (Sobhanian et al. 2011).

Mitochondria play a key role in programmed cell death (PCD) in plants. Significant changes in eight PCD-related proteins were observed in the mitochondrial proteome during salt stress- induced PCD in rice root-tip cells. Among the prominently upregulated proteins, mitochondrial Hsp70 was present. Therefore, Hsp70 was indicated as a possible player in PCD regulation (Chen et al. 2009).

7.3.7 Cold Stress

Low temperature has a negative impact on plant productivity, causing significant alterations in plant physiology (such as reduced plant growth, rolled and withered leaves) and plant metabolism, namely decrease in the rate of enzyme-catalyzed reactions. These results in metabolic imbalances associated with an oxidative stress.

Low temperature also increases a potential risk of protein misfolding, resulting in non- functional proteins. Therefore, enhanced accumulation of proteins with chaperone functions (especially several different HSPs) has also been reported (Hashimoto and Komatsu 2007; Kosova et al. 2011).

As a response to chilling stress, Hsps especially Hsp70 were upregulated in roots of rice (Cui et al. 2005), chicory (Degand et al. 2009), and poplar (Ghosh and Xu 2014; Renaut et al. 2004). Hsp70 was also noted to be more abundant in winter wheat upon exposure to cold stress (Kosova et al. 2011). Interestingly, the analysis of changes in the nuclear proteome of *A. thaliana* in response to cold stress (4 °C for 6 h) revealed induction of Hsp70. During exposure to stress, Hsp70 translocates to the nucleus from the cytosol to protect nuclear proteins from aggreagation (Bae et al. 2003). Hsp70 was found to be highly accumulated in mitochondrial proteome of pea subjected to cold stress (4 °C for 36 h) (Taylor et al. 2005). In a study of leaf proteome of three-month-old poplar plants exposed to 4 °C (days 7 and 14), among 30 identifications, about a third represented chaperone-like proteins including Hsp70, found to be more abundant under chilling stress than at optimal growth conditions (Renaut et al. 2004). Interestingly, proteomic analysis of rice seedlings exposed to a progressively low temperature stress treatment from normal temperature to 15, 10, and 5 °C identified Hsp70 as a cold responsive protein. Further, as much as 44 % of the upregulated proteins including Hsp70 were predicted to be chloroplast-localized, thereby implying that the chloroplast is one of the organelles mostly influenced by cold stress (Cui et al. 2005).

7.3.8 Flooding Stress

Flooding is an ever existing environmental constraint that is detrimental to plant performance and yields. Flooding affects about 10 % of the global land area (Mutava et al. 2015). Progressive decreases in soil oxygen concentration and redox potential are major physiological consequences of submergence (Hossain et al. 2009). Another study indicated the involvement of Hsp70 in response to flooding stress and its localization in the plasma membrane. Hsc70 protein was found to be upregulated by more than four times in the soybean plasma membrane proteome upon flooding stress (Komatsu et al. 2009). In flooding-stressed soybean plants, the abundance of 73 and 28 proteins were significantly altered in the root and cotyledon, respectively. The accumulation of only one protein, Hsp70 showed increase in both organs following flooding. The ratio of Hsp70 in the cotyledon was higher than that detected in the root under flooding stress (Komatsu et al. 2013). The investigation of the leaf proteome during flooding stress and the induction of PCD in maize also demonstrated the upregulation of Hsp70 (Chen et al. 2014).

7.3.9 Inadequate or Excessive Light

Plants are photosynthetic organisms and therefore, light plays a crucial role in determining their growth and productivity. However, it is well established that excess light could cause injury to plant tissues, impairing main physiological pathways and finally death. Light stress interferes with oxygenic photosynthesis, a phenomenon known as photoinhibition, which, consists of inhibition of the repair of Photosystem II by the oxidative stress resulting from absorption of the excess energy. When light exceeds photosynthetic capacity, ROS are generated in the chloroplasts and cause oxidative damage (Kosova et al. 2011; Timperio et al. 2008).

Giacomelli et al. studied the effect of excessive irradiance on thylakoid proteome in *A. thaliana* wild-type plants along with vtc2-2 mutants revealing ascorbate deficiency (Giacomelli et al. 2006). After 5 days of plant exposure to an increased irradiance, both wild-type plants and vtc2-2 mutants revealed enhanced accumulation of anthocyanins and ascorbate although mutants exhibited significantly lower levels than wild-type. Proteome analysis detected 45 differentially expressed proteins which included some chloroplast isoforms of Hsp70 (cpHsp70-1 and 2). Chloroplast isoforms of Hsp70 are predominantly found in chloroplast stroma, but they can also associate with thylakoid membranes (Kosova et al. 2011).

7.3.10 Enhanced Concentrations of Heavy Metals

Soil heavy metal (HM) toxicity has become a major environmental concern that restricts plant productivity. Three main reasons of the well-known metal toxicity in plants can be attributed to (i) direct interaction with proteins and their disruption due to the affinities of heavy metals for thioyl-, histidyl- and carboxyl-groups (ii) stimulated generation of ROS that elicit oxidative stress; and (iii) displacement of essential cations from specific binding sites, causing functions to collapse (Hossain et al. 2013; Sharma and Dietz 2009). High levels of cadmium in the environment can produce adverse effects on most living organisms including plants. The main causes of cadmium excess in the soil are anthropogenic activities (Hossain et al. 2012). In a study undertaken to explore the changes in protein profile induced by high and low levels of cadmium toxicity in tomato roots, Hsp70 was identified as differentially expressed (Rodriguez-Celma et al. 2010). Hsp70 was also reported to show altered expression in proteomic analysis of young poplar leaves subjected to cadmium exposure (Kieffer et al. 2008). A putative DnaK-type molecular chaperon Bip, a subfamily of Hsp70 was shown to have increased in response to cadmium stress in germinating rice seedlings exposed to cadmium (Ahsan et al. 2007). Cd exposure of *A. thaliana* cells also caused a marked increase in the expression of several Hsps including Hsp70 proteins (Sarry et al. 2006).

In a comparative proteome analysis of high and low cadmium accumulating soybeans under cadmium stress, more than twofold increased expression of Hsp70 was observed in high cadmium accumulating cultivars Harosoy and RIL CDH-80, while low cadmium accumulating cultivar Fukuyutaka exhibited decreased expression (Hossain et al. 2012). In cadmium-treated flax cell culture, enhanced accumulation of several proteins including Hsp70, heavy-metal binding proteins, fibre annexin etc. was observed while Hsp83 was down-regulated (Kosova et al. 2011).

7.4 Combinatorial Stress and Hsp70

Although most studies focus on plant responses under an ideal condition to a specific stress, plants in nature may have to cope with a variety of such stimuli at the same time. More than one stress may simultaneously affect the plant or be separated in time, thus resulting in a complex network of responses. A primary stress is capable of producing a secondary or even a tertiary stress. In some cases, different primary stresses may produce the same secondary or tertiary stress (Lee et al. 2012; Ngara and Ndimba 2014). Many abiotic stresses are interconnected as they may act through comparable secondary stresses and may cause similar alterations (Wang et al. 2003). Plant acclimation requires a particular response that can counteract the exact environmental situation the plant is exposed to. The intracellular processes initiated by a particular stress condition might not be exactly similar to that launched by a slightly different mixture of environmental stress parameters (Mittler et al. 2012). Consequently, adjustments of signaling pathways are necessary for the plant to deal with various stresses efficiently. In *N. tabacum,* it was reported that Hsp 70 and other Hsps were expressed in individually drought and heat-stressed plants and also in plants subjected to combined stresses. Higher levels of Hsp70 were observed by exposure of plants to a combination of heat and drought stresses when compared to plants subjected to individual stresses (Rizhsky et al. 2002).

7.5 Plant Stress Proteomics and Hsp70

Abiotic or biotic stress factors, either individually or in combination are known to cause several morphological and molecular changes within plants that in turn adversely affect their growth, development and productivity (Wang et al. 2003). Plants are continually challenged to recognize and respond to adverse changes in their environment to avoid their detrimental effects. Environmental stress is a primary cause of crop loss worldwide. Every year more than 70 % of the yield losses of major crops are attributed to various environmental stress factors (Choudhary et al. 2009). The severity of losses also depends on the specific development stage of plant at which the stress occurs. Besides, intensity of the applied stress, duration of stress and combinations of different stress factors constitute other deciding factors to the degree of yield loss (Rodziewicz et al. 2014). An example being the combinatorial effect of drought and heat stresses on reproductive versus vegetative tissues of plants. The reproductive tissues show much higher sensitivity towards the aforementioned stress compared to vegetative tissues (Suzuki et al. 2014). Further, conservative estimates suggest that most crops at best only reach 30 % of the genetic potential for yield as a result of abiotic stress effects in the field (Barkla et al. 2013). The study of plant responses towards various stress have been an active field of research since decades. Initially focused on the model plants, now the area has expanded to various crop plants like rice, wheat, barley, maize, sorghum, legumes and other economically important species.

Exposure to stress leads to a wide range of responses in plants at both the whole plant level as well as at cellular and molecular levels. Plant stress response is an integration of events, ranging from signal perception and transduction to the regulation of gene expression and metabolic changes (Xiao et al. 2009). Understanding the mechanisms that plants employ to tolerate and adapt to stress conditions is of considerable interest for designing genetic engineering strategies to ensure sustainable productivity. Furthermore, alleviating the effects of stress on plants is not only imperative for ensuring agricultural improvement to meet the demand for increased food production but also for improving plant fitness over a wider range of environmental conditions. This in turn would allow for an increased use of degraded or marginal lands for agricultural production (Barkla et al. 2013).

The pathways underlying plants' mechanism of signal perception, transduction and downstream regulatory events to activate adaptive responses have been difficult to study using traditional approaches due to their complexity and the large number of genes and gene products involved. Indeed, the development of current high-throughput "-omics" technologies (e.g. genomics, transcriptomics, proteomics, metabolomics, ionomics) has revolutionized plant science research, providing immense opportunities for dissecting plant stress responses. It has been increasingly clear that the changes in gene expression at transcript level do not often correspond to respective changes at protein level. Under these circumstances, investigation of proteomic changes provides more promising experimental approach towards studying stress response in plants (Kosova et al. 2011). Being one of the fastest growing fields of biological research, proteomics is defined as the systematic analysis of the proteome, the protein complement of genome. In words of Watson and co-workers "as we seek to better understand the gene function and to study the holistic biology of systems, it is inevitable that we study the proteome" (Agrawal et al. 2005). "Proteomics" as a discipline comprises of different avenues including protein profiling, protein quantification, post-translational modifications, and protein/protein interactions. Indeed, proteomics can reveal chemical complexity and biological dynamics, providing functional information on the cellular processes underlying phenotypes that are not accessible by other means (Baginsky 2009; Chen and Harmon 2006).

Plants, as sessile organisms, have to endure varying environmental conditions including adversities, and they rely on proteomic plasticity to remodel themselves and respond appropriately. The plant proteome is highly dynamic and proteome responses may show qualitative (positional shift, present/absent) or quantitative expressional changes following treatment (Ngara and Ndimba 2014) In order to capture the proteome's spatial and temporal expression changes under various stress conditions, proteomic researchers have employed a wide range of experimental designs in their studies. These include the use of different plant species, genotypes, developmental stages, tissues, organelles as well as various types and levels of stress. Further, technical advancements in the last decade have also propelled such comparative stress-responsive proteomic studies, which in turn have allowed us to analyze biochemical pathways involved in the complex plant stress response, thereby revealing interesting modulation of protein candidate(s) that play key roles. Identifying potential protein candidate(s) that play vital roles in plant stress response pathway(s) may prove instrumental in providing opportunities for crop development towards the production of genetically engineered stress-tolerant crop plants (Hossain et al. 2013; Ngara and Ndimba 2014).

A fundamental similarity is noted in the adaptation of plants to stress as plants induce Hsps in response to different stresses (Timperio et al. 2008). Therefore, in this book chapter, special emphasis has been placed on the pivotal role of Hsp70 in stress tolerance by proteomic studies across a wide spectrum of stresses ranging from biotic to abiotic, as well as combination of stresses. Due to the extensiveness of the research area as well as limitation of space and time, we regret not citing all contributions in the field.

7.6 *hsp70* Transgenics: The Road Ahead

For the betterment of agricultural production, development of transgenic crops with a heightened stress tolerance is necessary. The conventional plant breeding methods has not been highly successful against abiotic stresses due to the complex genetics of the underlying biochemical/physiological mechanisms. Production of transgenic plants has arisen as a useful method (Lavania et al. 2015). The numerous studies utilizing *hsp* genes for engineering heat tolerance in plants have been discussed in a recent review (Grover et al. 2013). It is noteworthy that overexpression of Hsp70 has been shown to confer thermotolerance in several studies. Overexpression of DnaK from a halotolerant cyanobacterium *Aphanothece halophytica* enhanced the heat tolerance of tobacco during germination and early growth (Ono et al. 2001). Further, overexpression of *dnak/hsp70* from *A. halophytica* was shown to increase seed yield as well as enhance tolerance for salt and heat stresses in rice and tobacco. Transgenic tobacco exhibited higher activities of ascorbate peroxidase and catalase than wild-type plants. Transgenic rice showed enhanced activities for Calvin-cycle enzymes, faster growth and higher yield of seeds compared to the wild-type rice under normal growth conditions (Uchida et al. 2008). Transgenic tobacco seedlings that constitutively overexpressed NtHSP70-1, a nuclear- localized Hsp70 were shown to exhibit tolerance to drought and heat stress (Cho and Choi 2009). Transgenic *Arabidopsis* expressing the *Trichoderma harzianum hsp70* exhibited enhanced tolerance to heat stress. Furthermore, they did not show any growth inhibition. These transgenic lines when subjected to heat pre-treatment, showed more tolerance towards osmotic, salt and oxidative stresses with respect to the wild-type lines (Montero-Barrientos et al. 2010). Overexpression of a *Chrysanthemum morifolium Hsp70* in *A. thaliana* as well as Chrysanthemum plants enhanced the tolerance against heat, drought and salinity stresses. The transgenic chrysanthemum plants showed increased peroxidase activity, higher proline content and reduced malondialdehyde levels under heat stress, thereby suggesting that *Cghsp70* overexpression may alleviate ROS damage (Song et al. 2014). Notably, the overexpression of *mthsp70* suppressed heat- and H_2O_2-induced PCD in rice protoplasts, most likely by maintaining mitochondrial membrane potential and inhibiting the amplification of ROS (Qin et al. 2011).

As Hsfs constitute the primary regulators of not only *hsp* gene expression but also a number of other defense-related genes, there is great deal of interest in generating transgenics overexpressing *hsfs* (Lavania et al. 2015). To mention a few, *G. max* HsfA1 overexpression in soybean resulted in enhancement of thermotolerance. GmHsfA1 overexpression led to increased activation of Gm*hsp*70 under normal temperature and overexpression of GmHsp70 under high temperature, suggesting that the transgenics showed enhanced heat tolerance through activation of Hsp70 (Zhu et al. 2006). Also, the over-expression of HsfA2 from *Lilium longiflorum* in *A. thaliana* activated the downstream genes including *hsp70, hsp101, hsp25.3* and *APX2*, leading to enhanced heat tolerance in the transgenic plants (Xin et al. 2010).

Members of the J protein family are known to function as molecular chaperones, alone or in association with Hsp70 partners. J proteins, as co-chaperones, work closely with Hsp70 members to execute multiple processes during protein homeostasis. A characteristic of Hsp70 chaperones is a low level basal ATPase activity that can be stimulated many fold by the J proteins (Lavania et al. 2015; Zhou et al. 2012). Using reverse genetic analysis, the *Arabidopsis* J-protein AtDjB1 was shown to facilitate thermotolerance by protecting cells against heat- induced oxidative damage. AtDjB1 was found to localize to mitochondria and directly interact with a mitochondrial Hsp70, stimulating its ATPase activity (Zhou et al. 2012).

7.7 Conclusion

In the present scenario of global climate change, stress adaptation of plants is of prime importance for adequate agricultural yield. In plants, heat shock proteins play a pivotal role as the most important line of defense in plant stress tolerance. Here we have highlighted functions, mechanisms of action of one of the better known heat shock protein-Hsp70 and focused on some modern aspects of stress research such as combinatorial features and commonality of stress responses. Furthermore, in this chapter results from studies employing recent proteomic techniques in plant stress research has been presented. We have also provided insights into modern approaches such as the development of Hsp70 transgenic plants to successfully counteract the adverse environmental factors.

Acknowledgements D.R. acknowledges financial support from CSIR, Govt. of India in the form of Senior Research Associateship (Scientists' Pool Scheme). D.R. would like to thank Dr. Andrew M Lynn, School of Computational and Integrative Sciences, Jawaharlal Nehru University for his kind support and encouragement. A.G. acknowledges the financial support of the Dept. of Science and Technology, Govt. of India for DST inspire faculty fellowship.

References

Agrawal GK, Yonekura M, Iwahashi Y, Iwahashi H, Rakwal R (2005) System, trends and perspectives of proteomics in dicot plants. Part I: technologies in proteome establishment. J Chromatogr B Analyt Technol Biomed Life Sci 815:109–123

Ahsan N, Lee SH, Lee DG, Lee H, Lee SW, Bahk JD, Lee BH (2007) Physiological and protein profiles alternation of germinating rice seedlings exposed to acute cadmium toxicity. C R Biol 330:735–746

Ahsan N, Donnart T, Nouri MZ, Komatsu S (2010) Tissue-specific defense and thermo- adaptive mechanisms of soybean seedlings under heat stress revealed by proteomic approach. J Proteome Res 9:4189–4204

Alam SB, Rochon D (2015) Cucumber necrosis virus recruits cellular heat shock protein 70 homologs at several stages of infection. J Virol 90(7):3302–3317

Augustine SM, Cherian AV, Syamaladevi DP, Subramonian N (2015a) Erianthus arundinaceus HSP70 (EaHSP70) acts as a Key regulator in the formation of anisotropic interdigitation in sugarcane (Saccharum spp. Hybrid) in response to drought stress. Plant Cell Physiol 56:2368–2380

Augustine SM, Narayan JA, Syamaladevi DP, Appunu C, Chakravarthi M, Ravichandran V, Subramonian N (2015b) Erianthus arundinaceus HSP70 (EaHSP70) overexpression increases drought and salinity tolerance in sugarcane (Saccharum spp. hybrid). Plant Sci 232:23–34

Bae MS, Cho EJ, Choi EY, Park OK (2003) Analysis of the Arabidopsis nuclear proteome and its response to cold stress. Plant J 36:652–663

Baginsky S (2009) Plant proteomics: concepts, applications, and novel strategies for data interpretation. Mass Spectrom Rev 28:93–120

Baniwal SK, Bharti K, Chan KY, Fauth M, Ganguli A, Kotak S, Mishra SK, Nover L, Port M, Scharf KD, Tripp J, Weber C, Zielinski D, von Koskull-Doring P (2004) Heat stress response in plants: a complex game with chaperones and more than twenty heat stress transcription factors. J Biosci 29:471–487

Barah P, Jayavelu ND, Mundy J, Bones AM (2013) Genome scale transcriptional response diversity among ten ecotypes of Arabidopsis thaliana during heat stress. Front Plant Sci 4:532

Barkla BJ, Vera-Estrella R, Pantoja O (2013) Progress and challenges for abiotic stress proteomics of crop plants. Proteomics 13:1801–1815

Barros MD, Czarnecka E, Gurley WB (1992) Mutational analysis of a plant heat-shock element. Plant Mol Biol 19:665–675

Beck CF (2005) Signaling pathways from the chloroplast to the nucleus. Planta 222:743–756

Bionda T, Gross LE, Becker T, Papasotiriou DG, Leisegang MS, Karas M, Schleiff E (2016) Eukaryotic Hsp70 chaperones in the intermembrane space of chloroplasts. Planta 243:733–747

Bita CE, Gerats T (2013) Plant tolerance to high temperature in a changing environment: scientific fundamentals and production of heat stress-tolerant crops. Front Plant Sci 4:273

Bokszczanin KL, Fragkostefanakis S (2013) Perspectives on deciphering mechanisms underlying plant heat stress response and thermotolerance. Front Plant Sci 4:315
Bray EA (1997) Plant responses to water deficit. Trends Plant Sci 2:48–54
Cazale AC, Clement M, Chiarenza S, Roncato MA, Pochon N, Creff A, Marin E, Leonhardt N, Noel LD (2009) Altered expression of cytosolic/nuclear HSC70-1 molecular chaperone affects development and abiotic stress tolerance in Arabidopsis thaliana. J Exp Bot 60:2653–2664
Chen S, Harmon AC (2006) Advances in plant proteomics. Proteomics 6:5504–5516
Chen X, Wang Y, Li J, Jiang A, Cheng Y, Zhang W (2009) Mitochondrial proteome during salt stress-induced programmed cell death in rice. Plant Physiol Biochem 47:407–415
Chen Y, Chen X, Wang H, Bao Y, Zhang W (2014) Examination of the leaf proteome during flooding stress and the induction of programmed cell death in maize. Proteome Sci 12:33
Chitteti BR, Peng Z (2007) Proteome and phosphoproteome differential expression under salinity stress in rice (Oryza sativa) roots. J Proteome Res 6:1718–1727
Cho EK, Choi YJ (2009) A nuclear-localized HSP70 confers thermoprotective activity and drought-stress tolerance on plants. Biotechnol Lett 31:597–606
Choudhary MK, Basu D, Datta A, Chakraborty N, Chakraborty S (2009) Dehydration-responsive nuclear proteome of rice (Oryza sativa L.) illustrates protein network, novel regulators of cellular adaptation, and evolutionary perspective. Mol Cell Proteomics 8:1579–1598
Cui S, Huang F, Wang J, Ma X, Cheng Y, Liu J (2005) A proteomic analysis of cold stress responses in rice seedlings. Proteomics 5:3162–3172
Degand H, Faber AM, Dauchot N, Mingeot D, Watillon B, Cutsem PV, Morsomme P, Boutry M (2009) Proteomic analysis of chicory root identifies proteins typically involved in cold acclimation. Proteomics 9:2903–2907
Dong X, Yi H, Lee J, Nou IS, Han CT, Hur Y (2015) Global gene-expression analysis to identify differentially expressed genes critical for the heat stress response in Brassica Rapa. PLoS One 10:e0130451
Driedonks N, Xu J, Peters JL, Park S, Rieu I (2015) Multi-level interactions between heat shock factors, heat shock proteins, and the redox system regulate acclimation to heat. Front Plant Sci 6:999
Fragkostefanakis S, Röth S, Schleiff E, Scharf KD (2015) Prospects of engineering thermotolerance in crops through modulation of heat stress transcription factor and heat shock protein networks. Plant Cell Environ 38:19
Garrido C, Paul C, Seigneuric R, Kampinga HH (2012) The small heat shock proteins family: the long forgotten chaperones. Int J Biochem Cell Biol 44:1588–1592
Ghosh D, Xu J (2014) Abiotic stress responses in plant roots: a proteomics perspective. Front Plant Sci 5:6
Giacomelli L, Rudella A, van Wijk KJ (2006) High light response of the thylakoid proteome in arabidopsis wild type and the ascorbate-deficient mutant vtc2-2. A comparative proteomics study. Plant Physiol 141:685–701
Gorovits R, Moshe A, Ghanim M, Czosnek H (2013) Recruitment of the host plant heat shock protein 70 by Tomato yellow leaf curl virus coat protein is required for virus infection. PLoS One 8:e70280
Goswami A, Banerjee R, Raha S (2010) Mechanisms of plant adaptation/memory in rice seedlings under arsenic and heat stress: expression of heat-shock protein gene HSP70. AoB Plants 2010, plq023
Goswami A, Banerjee R, Raha S (2013) Drought resistance in rice seedlings conferred by seed priming: role of the anti-oxidant defense mechanisms. Protoplasma 250:1115–1129
Goyal K, Walton LJ, Tunnacliffe A (2005) LEA proteins prevent protein aggregation due to water stress. Biochem J 388:151–157
Grover A, Mittal D, Negi M, Lavania D (2013) Generating high temperature tolerant transgenic plants: achievements and challenges. Plant Sci 205–206:38–47
Guy CL, Li QB (1998) The organization and evolution of the spinach stress 70 molecular chaperone gene family. Plant Cell 10:539–556

Hahn A, Bublak D, Schleiff E, Scharf KD (2011) Crosstalk between Hsp90 and Hsp70 chaperones and heat stress transcription factors in tomato. Plant Cell 23:741–755

Hashimoto M, Komatsu S (2007) Proteomic analysis of rice seedlings during cold stress. Proteomics 7:1293–1302

He ZS, Xie R, Wang YZ, Zou HS, Zhu JB, Yu GQ (2008) Cloning and characterization of a heat shock protein 70 gene, MsHSP70-1, in Medicago sativa. Acta Biochimica Et Biophysica Sinica 40:209–216

Hedhly A, Hormaza JI, Herrero M (2009) Global warming and sexual plant reproduction. Trends Plant Sci 14:30–36

Hossain Z, Lopez-Climent MF, Arbona V, Perez-Clemente RM, Gomez-Cadenas A (2009) Modulation of the antioxidant system in citrus under water logging and subsequent drainage. J Plant Physiol 166:1391–1404

Hossain Z, Hajika M, Komatsu S (2012) Comparative proteome analysis of high and low cadmium accumulating soybeans under cadmium stress. Amino Acids 43:2393–2416

Hossain Z, Khatoon A, Komatsu S (2013) Soybean proteomics for unraveling abiotic stress response mechanism. J Proteome Res 12:4670–4684

Hubel A, Schoffl F (1994) Arabidopsis heat-shock factor - isolation and characterization of the gene and the recombinant protein. Plant Mol Biol 26:353–362

Hubel A, Lee JH, Wu C, Schoffl F (1995) Arabidopsis heat-shock factor is constitutively active in drosophila and human-cells. Mol Gen Genet 248:136–141

Hubert DA, Tornero P, Belkhadir Y, Krishna P, Takahashi A, Shirasu K, Dangl JL (2003) Cytosolic HSP90 associates with and modulates the Arabidopsis RPM1 disease resistance protein. EMBO J 22:5679–5689

Jelenska J, van Hal JA, Greenberg JT (2010) Pseudomonas syringae hijacks plant stress chaperone machinery for virulence. Proc Natl Acad Sci U S A 107:13177–13182

Jung KH, Gho HJ, Nguyen MX, Kim SR, An G (2013) Genome-wide expression analysis of HSP70 family genes in rice and identification of a cytosolic HSP70 gene highly induced under heat stress. Funct Integr Genomics 13:391–402

Kieffer P, Dommes J, Hoffmann L, Hausman JF, Renaut J (2008) Quantitative changes in protein expression of cadmium-exposed poplar plants. Proteomics 8:2514–2530

Kim SR, An G (2013) Rice chloroplast-localized heat shock protein 70, OsHsp70CP1, is essential for chloroplast development under high-temperature conditions. J Plant Physiol 170:854–863

Komatsu S, Wada T, Abalea Y, Nouri MZ, Nanjo Y, Nakayama N, Shimamura S, Yamamoto R, Nakamura T, Furukawa K (2009) Analysis of plasma membrane proteome in soybean and application to flooding stress response. J Proteome Res 8:4487–4499

Komatsu S, Makino T, Yasue H (2013) Proteomic and biochemical analyses of the cotyledon and root of flooding-stressed soybean plants. PLoS One 8:e65301

Kosova K, Vitamvas P, Prasil IT, Renaut J (2011) Plant proteome changes under abiotic stress--contribution of proteomics studies to understanding plant stress response. J Proteomics 74:1301–1322

Kotak S, Larkindale J, Lee U, von Koskull-Doring P, Vierling E, Scharf KD (2007) Complexity of the heat stress response in plants. Curr Opin Plant Biol 10:310–316

Kropat J, Oster U, Rudiger W, Beck CF (2000) Chloroplast signalling in the light induction of nuclear HSP70 genes requires the accumulation of chlorophyll precursors and their accessibility to cytoplasm/nucleus. Plant J 24:523–531

Kubienova L, Sedlarova M, Viteckova-Wunschova A, Piterkova J, Luhova L, Mieslerova B, Lebeda A, Navratil M, Petrivalsky M (2013) Effect of extreme temperatures on powdery mildew development and Hsp70 induction in tomato and wild Solanum spp. Plant Prot Sci 49:S41–S54

Kuromori T, Mizoi J, Umezawa T, Yamaguchi-Shinozaki K, Shinozaki K (2014) Drought stress signaling network. In: Howell SH (ed) Molecular biology. Springer, New York, pp 383–409

Latijnhouwers M, Xu XM, Moller SG (2010) Arabidopsis stromal 70-kDa heat shock proteins are essential for chloroplast development. Planta 232:567–578

Lavania D, Dhingra A, Siddiqui MH, Al-Whaibi MH, Grover A (2015) Current status of the production of high temperature tolerant transgenic crops for cultivation in warmer climates. Plant Physiol Biochem 86:100–108

Lee S, Lee DW, Lee Y, Mayer U, Stierhof YD, Jurgens G, Hwang I (2009) Heat shock protein cognate 70–4 and an E3 ubiquitin ligase, CHIP, mediate plastid-destined precursor degradation through the ubiquitin-26S proteasome system in *Arabidopsis*. Plant Cell 21:3984–4001

Lee JH, Yun HS, Kwon C (2012) Molecular communications between plant heat shock responses and disease resistance. Mol Cells 34:109–116

Liao JL, Zhou HW, Zhang HY, Zhong PA, Huang YJ (2014) Comparative proteomic analysis of differentially expressed proteins in the early milky stage of rice grains during high temperature stress. J Exp Bot 65:655–671

Lobell DB, Burke MB, Tebaldi C, Mastrandrea MD, Falcon WP, Naylor RL (2008) Prioritizing climate change adaptation needs for food security in 2030. Science 319:607–610

Lubben TH, Donaldson GK, Viitanen PV, Gatenby AA (1989) Several proteins imported into chloroplasts form stable complexes with the GroEL-related chloroplast molecular chaperone. Plant Cell 1:8

Mahajan S, Tuteja N (2005) Cold, salinity and drought stresses: an overview. Arch Biochem Biophys 444:139–158

Maruyama D, Endo T, Nishikawa S (2010) BiP-mediated polar nuclei fusion is essential for the regulation of endosperm nuclei proliferation in Arabidopsis thaliana. Proc Natl Acad Sci U S A 107:1684–1689

Maruyama D, Sugiyama T, Endo T, Nishikawa S (2014) Multiple BiP genes of Arabidopsis thaliana are required for male gametogenesis and pollen competitiveness. Plant Cell Physiol 55:801–810

Mayer MP, Bukau B (2005) Hsp70 chaperones: cellular functions and molecular mechanism. Cell Mol Life Sci 62:670–684

Mittler R, Finka A, Goloubinoff P (2012) How do plants feel the heat? Trends Biochem Sci 37:118–125

Montero-Barrientos M, Hermosa R, Cardoza RE, Gutierrez S, Nicolas C, Monte E (2010) Transgenic expression of the Trichoderma harzianum hsp70 gene increases Arabidopsis resistance to heat and other abiotic stresses. J Plant Physiol 167:659–665

Mutava RN, Prince SJ, Syed NH, Song L, Valliyodan B, Chen W, Nguyen HT (2015) Understanding abiotic stress tolerance mechanisms in soybean: a comparative evaluation of soybean response to drought and flooding stress. Plant Physiol Biochem 86:109–120

Neuer A, Spandorfer SD, Giraldo P, Jeremias J, Dieterle S, Korneeva I, Liu HC, Rosenwaks Z, Witkin SS (1999) Heat shock protein expression during gametogenesis and embryogenesis. Infect Dis Obstet Gynecol 7:10–16

Ngara R, Ndimba BK (2014) Understanding the complex nature of salinity and drought-stress response in cereals using proteomics technologies. Proteomics 14:611–621

Ohta M, Wakasa Y, Takahashi H, Hayashi S, Kudo K, Takaiwa F (2013) Analysis of rice ER- resident J-proteins reveals diversity and functional differentiation of the ER-resident Hsp70 system in plants. J Exp Bot 64:5429–5441

Ono K, Hibino T, Kohinata T, Suzuki S, Tanaka Y, Nakamura T, Takabe T (2001) Overexpression of DnaK from a halotolerant cyanobacterium Aphanothece halophytica enhances the high-temperature tolerance of tobacco during germination and early growth. Plant Sci 160:455–461

Potocky M, Jones MA, Bezvoda R, Smirnoff N, Zarsky V (2007) Reactive oxygen species produced by NADPH oxidase are involved in pollen tube growth. New Phytol 174:742–751

Qin F, Shinozaki K, Yamaguchi-Shinozaki K (2011) Achievements and challenges in understanding plant abiotic stress responses and tolerance. Plant Cell Physiol 52:1569–1582

Renaut J, Lutts S, Hoffmann L, Hausman JF (2004) Responses of poplar to chilling temperatures: proteomic and physiological aspects. Plant Biol (Stuttg) 6:81–90

Ritossa F (1962) A new puffing pattern induced by temperature shock and DNP in drosophila. Experientia 18:571–573

Rizhsky L, Liang H, Mittler R (2002) The combined effect of drought stress and heat shock on gene expression in tobacco. Plant Physiol 130:1143–1151

Rodriguez-Celma J, Rellan-Alvarez R, Abadia A, Abadia J, Lopez-Millan AF (2010) Changes induced by two levels of cadmium toxicity in the 2-DE protein profile of tomato roots. J Proteomics 73:1694–1706

Rodziewicz P, Swarcewicz B, Chmielewska K, Wojakowska A, Stobiecki M (2014) Influence of abiotic stresses on plant proteome and metabolome changes. Acta Physiol Plant 36:1–19

Saidi Y, Finka A, Goloubinoff P (2011) Heat perception and signalling in plants: a tortuous path to thermotolerance. New Phytol 190:556–565

Sarkar NK, Kundnani P, Grover A (2013) Functional analysis of Hsp70 superfamily proteins of rice (Oryza sativa). Cell Stress Chaperones 18:427–437

Sarry JE, Kuhn L, Ducruix C, Lafaye A, Junot C, Hugouvieux V, Jourdain A, Bastien O, Fievet JB, Vailhen D, Amekraz B, Moulin C, Ezan E, Garin J, Bourguignon J (2006) The early responses of Arabidopsis thaliana cells to cadmium exposure explored by protein and metabolite profiling analyses. Proteomics 6:2180–2198

Scafaro AP, Haynes PA, Atwell BJ (2010) Physiological and molecular changes in Oryza meridionalis Ng., a heat-tolerant species of wild rice. J Exp Bot 61:191–202

Scharf KD, Berberich T, Ebersberger I, Nover L (2012) The plant heat stress transcription factor (Hsf) family: structure, function and evolution. Biochim Biophys Acta 1819:104–119

Schroda M, Vallon O, Wollman FA, Beck CF (1999) A chloroplast-targeted heat shock protein 70 (HSP70) contributes to the photoprotection and repair of photosystem II during and after photoinhibition. Plant Cell 11:1165–1178

Schroda M, Kropat J, Oster U, Rudiger W, Vallon O, Wollman FA, Beck CF (2001) Possible role for molecular chaperones in assembly and repair of photosystem II. Biochem Soc Trans 29:413–418

Sharma SS, Dietz KJ (2009) The relationship between metal toxicity and cellular redox imbalance. Trends Plant Sci 14:43–50

Shu L, Lou Q, Ma C, Ding W, Zhou J, Wu J, Feng F, Lu X, Luo L, Xu G, Mei H (2011) Genetic, proteomic and metabolic analysis of the regulation of energy storage in rice seedlings in response to drought. Proteomics 11:4122–4138

Sobhanian H, Aghaei K, Komatsu S (2011) Changes in the plant proteome resulting from salt stress: toward the creation of salt-tolerant crops? J Proteomics 74:1323–1337

Song A, Zhu X, Chen F, Gao H, Jiang J, Chen S (2014) A chrysanthemum heat shock protein confers tolerance to abiotic stress. Int J Mol Sci 15:5063–5078

Su PH, Li HM (2008) Arabidopsis stromal 70-kD heat shock proteins are essential for plant development and important for thermotolerance of germinating seeds. Plant Physiol 146:1231–1241

Su PH, Li HM (2010) Stromal Hsp70 is important for protein translocation into Pea and Arabidopsis chloroplasts. Plant Cell 22:1516–1531

Sung DY, Guy CL (2003) Physiological and molecular assessment of altered expression of Hsc70–1 in Arabidopsis. Evidence for pleiotropic consequences. Plant Physiol 132:979–987

Suzuki N, Rivero RM, Shulaev V, Blumwald E, Mittler R (2014) Abiotic and biotic stress combinations. New Phytol 203:32–43

Taylor NL, Heazlewood JL, Day DA, Millar AH (2005) Differential impact of environmental stresses on the pea mitochondrial proteome. Mol Cell Proteomics 4:1122–1133

Timperio AM, Egidi MG, Zolla L (2008) Proteomics applied on plant abiotic stresses: role of heat shock proteins (HSP). J Proteomics 71:391–411

Uchida A, Hibino T, Shimada T, Saigusa M, Takabe T, Araki E, Kajita H, Takabe T (2008) Overexpression of DnaK chaperone from a halotolerant cyanobacterium Aphanothece halophytica increases seed yield in rice and tobacco. Plant Biotechnol 25:141–150

Umezawa T, Nakashima K, Miyakawa T, Kuromori T, Tanokura M, Shinozaki K, Yamaguchi-Shinozaki K (2010) Molecular basis of the core regulatory network in ABA responses: sensing, signaling and transport. Plant Cell Physiol 51:1821–1839

Virdi AS, Singh S, Singh P (2015) Abiotic stress responses in plants: roles of calmodulin-regulated proteins. Front Plant Sci 6:809

Wakasa Y, Yasuda H, Oono Y, Kawakatsu T, Hirose S, Takahashi H, Hayashi S, Yang L, Takaiwa F (2011) Expression of ER quality control-related genes in response to changes in BiP1 levels in developing rice endosperm. Plant J 65:675–689

Wang W, Vinocur B, Altman A (2003) Plant responses to drought, salinity and extreme temperatures: towards genetic engineering for stress tolerance. Planta 218:1–14

Wang WX, Vinocur B, Shoseyov O, Altman A (2004) Role of plant heat-shock proteins and molecular chaperones in the abiotic stress response. Trends in Plant sci 9:244–252

Wang MC, Peng ZY, Li CL, Li F, Liu C, Xia GM (2008) Proteomic analysis on a high salt tolerance introgression strain of *Triticum aestivum/Thinopyrum ponticum*. Proteomics 8:1470–1489

Wang Y, Lin S, Song Q, Li K, Tao H, Huang J, Chen X, Que S, He H (2014) Genome-wide identification of heat shock proteins (Hsps) and Hsp interactors in rice: Hsp70s as a case study. BMC Genomics 15:344

Wu CH, Caspar T, Browse J, Lindquist S, Somerville C (1988) Characterization of an HSP70 cognate gene family in Arabidopsis. Plant Physiol 88:731–740

Xiao X, Yang F, Zhang S, Korpelainen H, Li C (2009) Physiological and proteomic responses of two contrasting Populus cathayana populations to drought stress. Physiol Plant 136:150–168

Xin H, Zhang H, Chen L, Li X, Lian Q, Yuan X, Hu X, Cao L, He X, Yi M (2010) Cloning and characterization of HsfA2 from Lily (Lilium longiflorum). Plant Cell Rep 29:875–885

Zhou W, Zhou T, Li MX, Zhao CL, Jia N, Wang XX, Sun YZ, Li GL, Xu M, Zhou RG, Li B (2012) The Arabidopsis J-protein AtDjB1 facilitates thermotolerance by protecting cells against heat-induced oxidative damage. New Phytol 194:364–378

Zhu B, Ye C, Lu H, Chen X, Chai G, Chen J, Wang C (2006) Identification and characterization of a novel heat shock transcription factor gene, GmHsfA1, in soybeans (Glycine max). J Plant Res 119:247–256

Chapter 8
The Role of Plant 90-kDa Heat Shock Proteins in Plant Immunity

Mariana G. Corigliano and Marina Clemente

Abstract Heat shock proteins of 90 kDa (Hsp90) are present in most of the living organisms. Their main function is to preserve cell viability by participating in the assembly, stabilization, maturation and activation of key signaling proteins. Most of the research on Hsp90 functions was made using mammal and yeast models, which has enable considerable progress in the current knowledge about their contribution to maintain cell homeostasis. In this trend, it has been shown that mammal and parasitic Hsp90s may function as immune regulators. Interestingly, recombinant plant Hsp90 is also able to stimulate murine cells and modulate their immune response, suggesting that common regulatory pathways might be evolutionarily conserved. Although plants and mammals present obvious differences in the development of the immune responses against pathogens, several coincidences have been found. Thus, several works demonstrated that plant Hsp90s has an active role in the activation of the defense mechanisms against pathogens, since the lack or deficit of these proteins impair defense against pathogenic bacteria and fungus. This chapter contains an overview of the molecular roles played by plant Hsp90s in the defense against pathogens, and ends with an example of how recombinant plant Hsp90s could be used as adjuvant in vaccine formulations.

Keywords Chaperones • Defense response • ETI • Hsp90 • Hypersensitive response • PAMP • Plants • PTI

Abbreviations

ETI	effector trigger immunity
GDA	Geldanamycin A
HR	hypersensitive response
Hsp90	heat shock proteins of 90 kDa
MAMP	microbial associated molecular patterns

M.G. Corigliano • M. Clemente (✉)
Laboratorio de Biotecnología Vegetal, IIB-INTECH, CONICET-UNSAM,
Avenida Intendente Marino Km 8.2, Chascomús 7130, Provincia de Buenos Aires, Argentina
e-mail: marianacorigliano@intech.gov.ar; mclemente@intech.gov.ar

A.A.A. Asea et al. (eds.), *Heat Shock Proteins and Plants*,
Heat Shock Proteins 10, DOI 10.1007/978-3-319-46340-7_8

PAMP	pathogen associated molecular patterns
PR genes	pathogenesis-related genes
PRR	pattern recognition receptors
PTI	PAMP trigger immunity
PVX	potato virus x
R protein	plant Resistance proteins
ROS	reactive oxygen species
TMV	tobacco mosaic virus
VIGS	virus-induced gene silencing

8.1 Introduction

Heat shock proteins of 90 kDa (Hsp90s) are highly conserved in most of the living organisms including bacteria, yeast, mammals and plants (Nover and Miernyk 2001; Pearl and Prodromou 2006). The story of its name is related to the fact that these proteins were discovered in cells after heat shock (Ritossa 1962). However, Hsp90 expression can be enhanced not only by temperature, but also by non-physiologic pH, nutrient deprivation, salinity, wounding, UV radiation and bacterial or viral infection, among others stimuli (Chaudhury et al. 2006; Swindell et al. 2007; Wallin et al. 2002). Nowadays, there are very useful websites for the search, identification, comparison and alignment of different genes, such as the National Center for Biotechnology Information (NCBI) at http://www.ncbi.nlm.nih.gov and Basic Local Alignment Sequence Tool (BLAST) at http://www.ncbi.nlm.nih.gov/BLAST. Gene information from various organisms is collected in databases available for free and also some of these sites have other resources like vectors or seeds for research proposal (Table 8.1).

In addition, Hsp90s are found in all cellular compartments examined to date. Plant Hsp90s have been detected in cytosol, nucleus, endoplasmic reticulum, mitochondria and chloroplasts (Conner et al. 1990; Felsheim and Das 1992; Koning et al. 1992; Krishna and Gloor 2001; Song et al. 2009; Takahashi et al. 1992; Yabe et al. 1994). Several Hsp90 from higher plants and also algae have been identified, cloned and sequenced (Table 8.2). Even though substantial progress has been achieved in the last years (Breiman 2014), there is still a gap in our understanding on many topics related to their molecular functions.

8.2 Expression of HSP90 in Plants

Homologous genes for Hsp90 have been identified among many eukaryotic organisms (Johnson 2012). Most of these species have at least two genes codifying for cytoplasmic Hsp90 (Gupta 1995) and it is believed that there is a redundancy of

Table 8.1 Summary of database and website information

Organism	Database	Website link	Reference
Arabidopsis thaliana	TAIR (The Arabidopsis Information Resource)	https://www.arabidopsis.org/	
	RIKEN	http://epd.brc.riken.jp/en/	
Nicotiana spp.	Sol Genomics Network	http://solgenomics.net/	
	RIKEN	http://epd.brc.riken.jp/en/	
Potato	Sol Genomics Network	http://solgenomics.net/	
Chlamydomonas reinhardtii		http://www.kazusa.or.jp/en/plant/chlamy/EST/blast.html	Schroda (2004)
		http://genome.jgi-psf.org/chlre2/chlre2.home.html	Willmund and Schroda (2005)
Zea mays	The TIGR Maize Database	http://maize.jcvi.org/	De La Fuente Van Bentem et al. (2005)
	Maize Cellgenomics Database	http://maize.jcvi.org/cellgenomics/index.php	
Poplar	Phytozome	http://www.phytozome.net/poplar.php	Zhang et al. (2013)
	RIKEN	http://epd.brc.riken.jp/en/	
Physcomitrella patens	Phytozome	http://www.phytozome.net	Zhang et al. (2013)
Brachypodium distachyon	Phytozome	http://www.phytozome.net	Zhang et al. (2013)
Medicago truncatula	Phytozome	http://www.phytozome.net	Zhang et al. (2013)
Vitis vinifera	Phytozome	http://www.phytozome.net	(Zhang et al. (2013)
Sorghum bicolor	Phytozome	http://www.phytozome.net	Zhang et al. (2013)
Oryza sativa	Rice Genome Annotation Project	http://rice.plantbiology.msu.edu/	Zhang et al. (2013)
	RIKEN	http://rice.plantbiology.msu.edu/analyses_search_locus.shtml	Zai et al. (2015)
		http://epd.brc.riken.jp/en/	
Solanum lycopersicum	Sol Genomics Network	http://solgenomics.net/	Zai et al. (2015)
	Tomato National BioResource Project	http://tomato.nbrp.jp/indexEn.html	
	Tomato Genetics Resource Center	http://tgrc.ucdavis.edu/	
		http://solcap.msu.edu/	
	Solanaceae Coordinated Agricultural Project	http://ted.bti.cornell.edu/	
		http://epd.brc.riken.jp/en/	
	Tomato Functional Genomics Database		
	RIKEN		
Matricaria recutita		http://www.ebi.ac.uk/interpro/	Ling et al. (2014)

Table 8.2 Summary of identified, cloned and/or sequenced Hsp90 from plant

Organism	Species	Reference
Arabidopsis	*Arabidopsis thaliana*	Takahashi et al. (1992)
		Milioni and Hatzopoulos (1997)
		Krishna and Gloor (2001)
Tomato	*Solanum lycopersicum*	Koning et al. (1992)
		Zai et al. (2015)
Japanese morning glory	*Pharbitis nil*	Felsheim and Das (1992)
Barley	*Hordeum vulgarum*	Walther-Larsen et al. (1993)
		Zhang et al. (2008)
Madagascar periwinkle	*Catharanthus roseus*	Schroder et al. (1993)
Maize	*Zea mayz*	Marrs et al. (1993)
Oilseed rape	*Brassica napus*	Krishna et al. (1995)
Rye	*Secale cereale*	Schmitz et al. (1996)
		Krishna et al. (1997)
Tobacco	*Nicotiana benthamiana*	Kanzaki et al. (2003)
Rice	*Oryza sativa*	Liu et al. (2006)
		Liu et al. (2009)
		Ye et al. (2012)
Soybean	*Glycine max*	Fu et al. (2009)
		Xu et al. (2013)
Brown seaweed	*Fucus serratus*	Pearson et al. (2010)
Brown seaweed	*Fucus vesiculosus*	Pearson et al. (2010)
Green macroalga	*Ulva fasciata*	Sung et al. (2011)
Sorghum	*Sorghum bicolor*	Virdi et al. (2011)
		Zhang et al. (2013)
Pearl millet	*Pennisetum glaucum*	Reddy et al. (2011)
Wheat	*Triticum aestivum*	Wang et al. (2011)
Grape	*Vitis vinifera*	Banilas et al. (2012)
		Zhang et al. (2013)
Microalga	*Dunaliella salina*	Wang et al. (2012)
		Chen et al. (2015)
Sweet orange	*Citrus sinensis*	Mendonça and Ramos (2012)
Medicago	*Medicago truncatula*	Zhang et al. (2013)
Moss	*Physcomitrella patens*	Zhang et al. (2013)
Poplar	*Populus trichocarpa*	Zhang et al. (2013)
Purple false brome	*Brachypodium distachyon*	Zhang et al. (2013)
Sugarcane	*Saccharum* spp.	da Silva et al. (2013)
German chamomile	*Matricaria recutita*	Ling et al. (2014)
Strawberry	*Fragaria x ananassa cv. camarosa*	Christou et al. (2014)

Hsp90 functions to guarantee cell viability, since the absence of a given isoform hardly ever causes deleterious effects on the cell (Borkovich et al. 1989; Hubert et al. 2003; Takahashi et al. 2003). In addition, the expression patterns of Hsp90s are diverse in different tissues, so it is easy to speculate that each isoform may play a specific role in cell growth and development as well in the response to many types of stresses. In plants, transcriptional patterns of Hsp90 expression have been widely analyzed in different organs. The cytosolic Hsp90 isoforms of *Arabidopsis thaliana*, Hsp90 (AtHsp90.1), AtHsp90.2, AtHsp90.3 and AtHsp90.4 have been cloned and characterized. *AtHsp90.1* mRNA was detected only in roots, while *AtHsp90.2* and *AtHsp90.3* were present in all tissues analyzed (Yabe et al. 1994). Besides, heat shock treatment revealed that these proteins present different patterns of induction. Thus, the evidence showed low levels of expression of *AtHsp90.1* in the absence of heat shock, but it is strongly induced by elevated temperatures (Takahashi et al. 1992; Yabe et al. 1994; Yamada et al. 2007). In turn, *AtHsp90.2*, *AtHsp90.3* and *AtHsp90.4* were shown to be constitutively expressed under normal growth temperatures with a moderated increment after heat treatment (Takahashi et al. 1992; Yabe et al. 1994; Yamada et al. 2007). In accordance with the role of these proteins on cell development, abnormal phenotypes were observed in Hsp90-RNAi *Arabidopsis* lines, characterized by loss of apical dominance and emergence of multiple shoots (Sangster et al. 2007).

In the same line of evidences, four isoforms of *Nicotiana benthamiana* Hsp90 (NbHsp90) were described. *NbHsp90.2* and *NbHsp90.3* have very low expression levels under normal growth conditions but they show to be heat inducible. On the other hand, *NbHsp90.1* and *NbHsp90.4* are constitutively expressed showing a slight heat-induction response (Kanzaki et al. 2003), suggesting they might play a role in homeostasis maintenance. Gene silencing of NbHsp90.1 by virus-induced gene silencing (VIGS) caused stunted phenotypes (Kanzaki et al. 2003), supporting the idea that this isoform makes a notable contribution to growth and development. This line does not show altered expression levels of cytosolic NbHsp90.4, suggesting that there is not a complement of function between these isoforms. So far, the effect of silencing NbHsp90.1 on the expression levels of NbHsp90.2 and NbHsp90.3 has not been studied (Kanzaki et al. 2003).

The number of isoforms present in plant genome may vary considerably in other plant species. For instance, 12 members of Hsp90 were identified in soybean (*Glycine max*) (Xu et al. 2013). Isoforms *Hsp90A1* (*GmHsp90A1*), *GmHsp90A2*, *GmHsp90B2* and *GmHsp90C2.2* were shown to be strongly induced by heat, whereas no responses were observed at the levels of *GmHsp90B1* and *GmHsp90C2.1*. So far, the expression patterns of the other isoforms has not been studied (Xu et al. 2013). In addition to the differences observed in response to stress, these isoforms were shown to be expressed in a tissue-specific manner. In this regard, the isoform *GmHsp90C1* is highly expressed in flowers, whereas *GmHsp90A3*, *GmHsp90A6* and *GmHsp90B* genes are highly transcribed in roots (Xu et al. 2013). Soybean plants with GmHsp90.1- and GmHsp90.2-silenced genes were severely stunted and showed profuse axillary growth (Fu et al. 2009), strengthening the idea that Hsp90 are key for plant growth and normal development.

Recently work on wheat reinforces the tissue-specific expression of Hsp90 previously observed (Wang et al. 2011). Thus, *Triticum aestivum Hsp90.1* (*TaHsp90.1*) is highly transcribed in reproductive organs compared to vegetative organs. In turn, *TaHsp90.2* and *TaHsp90.3* are ubiquitously transcribed, although transcription levels of *TaHsp90.2* are lower in roots and those of *TaHsp90.3* are higher in leaves (Wang et al. 2011). VIGS-silencing of TaHsp90.1 does not alter the expression levels of TaHsp90.2 or TaHsp90.3. Similar results were observed when *TaHsp90.2* or *TaHsp90.3* genes were silenced separately (Wang et al. 2011). Notably, there was a strong impact on wheat growth when *TaHsp90.1* or *TaHsp901/2/3* were silenced at the same time. The phenotype of this plant is dwarf, with the consequent reduction on its average weight, it also has strong stunting in their leaves and it die before than controls. None of these effects are observed in silenced TaHsp90.2, TaHsp90.3 or TaHsp90.2/3 plants (Wang et al. 2011) suggesting that the lack of *TaHsp90.1* gene would be the responsible of the phenotype described previously.

All of these examples taken together indicate that Hsp90 isoforms are specifically expressed and localized under normal growth conditions, and each isoform expression after stress exposure or treatment is independently induced. Even though Hsp90s are time and tissue differentially expressed, their protein sequences are highly similar. For instance, percentages of identity among Hsp90s from different plant species vary between 88 and 93 % (Krishna and Gloor 2001). The high homology among these proteins from different species demonstrates a remarkable degree of conservation, which is also related to conserved structural and functional aspects (Breiman 2014; Mayor et al. 2007; Nürnberger et al. 2004). Although plant Hsp90 shares most of the features observed in their mammal, bacterial and yeast counterparts, plant Hsp90s performs unique tasks related to plant physiology, such as the transport of pre-proteins into chloroplast (Breiman 2014).

8.3 HSP90 and the Plant Defense Response

Plants are sessile organisms that are continuously exposed to a considerable amount of potential pathogenic microorganisms. Nevertheless, plants hardly ever get sick, which is mostly due to the existence of pre-formed barriers constitutively present on the plant surface and cytosol, such as the cell walls, wax layers, antimicrobial proteins and secondary metabolites (Nürnberger et al. 2004). As most of the eukaryotic organisms, plants are able to recognize potential pathogens effectors and consequently induce several defense responses in order to prevent disease. The activation of these plant defense responses is a consequence of the perception of effector molecules derived from the pathogen or other molecules associated to tissue damage (Jones and Dangl 2006).

In this trend, the plant innate immune system possesses two different mechanisms to detect harmful microorganisms (Fig. 8.1). The first one recognizes and responds to molecules common to many classes of microbes, including non-pathogenic.

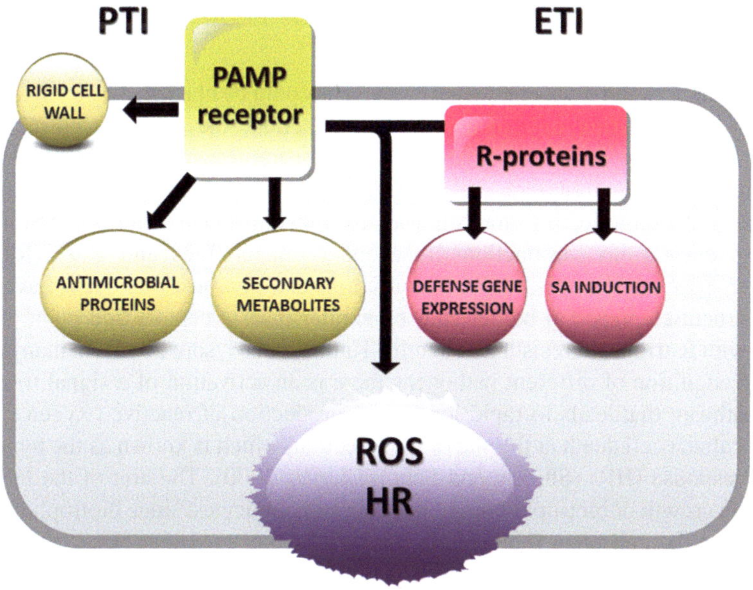

Fig. 8.1 Schematic representation of the mechanisms involved in plant defense response. PTI is triggered after PAMP recognition by PAMP receptors while ETI is triggered after effectors recognition by R-protein. PTI and ETI induce defense gene expression and culminate in HR

These molecules, known as M/PAMPs (Microbial/Pathogen Associated Molecular Patterns), are conserved microbial molecules absent in the host, widely distributed among different microorganisms and indispensable for microbes life (Chisholm et al. 2006; Jones and Dangl 2006; Nürnberger et al. 2004; Schwessinger and Zipfel 2008). The perception of M/PAMPs is performed by transmembrane pattern recognition receptors (PRR) present in plants and results in the activation of PAMP Trigger Immunity (PTI) (Jones and Dangl 2006) (Fig. 8.1). The second mechanism is triggered after recognition of specific pathogen-derived proteins that are usually injected into the plant cell cytoplasm to abolish the PTI (Jones and Dangl 2006). In this case, plants are able to respond to these effectors by the so-called Effector Trigger Immunity (ETI). The host resistance is elicited when a plant Resistance proteins (R proteins) specifically recognize pathogen effectors or effector-derived molecules (Jones and Dangl 2006; Shirasu and Schulze-Lefert 2003). The specificity of this defense mechanism depends on the R proteins, which act as immune sensors (Jones and Dangl 2006) (Fig. 8.1).

In this sense, Hsp90s plays a dual role in the immune response during the perception of microorganisms and when maintaining signaling-competent state of complexes necessary for the defense response against harmful organisms.

8.4 HSP90 and Non-host Resistance

Interaction between plants and microorganisms could be classified as compatible or incompatible. During any incompatible interaction, the microorganism cannot cause the disease since it is repealed by plant preformed or inducible defenses (Senthil-Kumar and Mysore 2013). Some of the defenses to restrict pathogen growth and multiplication are cell wall fortification due to callose, lignine and suberin depositions, production of secondary metabolites and ROS (Reactive Oxygen Species) (De Wit 2007). Because host and non-host pathogens have common structures, they can be detected by similar plant receptors and plant overlap basal with R-triggered resistance (Senthil-Kumar and Mysore 2013). In many cases, after recognition of different pathogens there is an activation of a signal transduction pathway that leads to rapid ion fluxes, production of reactive oxygen species and localized cell death at the site of the infection, which is known as the hypersensitive response (HR) (Shirasu and Schulze-Lefert 2000). The aim of the HR is to limit the growth of biotrophic fungi, bacteria and oomicetes, since biotrophic pathogens need the cell alive (Holt et al. 2005). It is supposed that HR is an induced resistance mechanism that is activated only when preformed mechanisms described above failed to arrest the pathogen completely (Lu et al. 2003). Several proteins are required to activate the HR during non-host resistance and several line of evidences demonstrated that the expression of Hsp90s is one of them (Chen et al. 2010; Garavaglia et al. 2008; Kanzaki et al. 2003; Swindell et al. 2007; Thao et al. 2007). For instance, Hsp90 expression is induced in *A. thaliana* leaves inoculated with *Phytophtora infestans* (Swindell *et al.* 2007). Similar results were observed in pepper (*Capsicum annuum* cv. *grossum*); in this case, a progressive induction of CaHsp90 was observed between 24 and 48 h after inoculation with the non-host pathogen *Xanthomonas axonopodis*. The expression of CaHsp90 was increased about 70 % compared to control (Garavaglia et al. 2008). Finally, the role of NbHsp90 during non-host response was also studied. Silencing of the NbHsp90.1 isoform blocked the activation of HR after inoculation with the non-host pathogen *Pseudomonas cichorii*. Besides, the titer of bacteria was higher than controls and the expression of pathogenesis-related genes (*PR* genes) were reduced compared to control plants, suggesting that Hsp90 could be directly or indirectly involved in the trigger of HR (Kanzaki et al. 2003). In this context, Shimamoto et al. characterized the role of Hsp90 in the rice innate immunity response. These authors studied OsCERK1, a PAMP receptor which recognizes two fungal PAMPs: chitin (N-acetylchitooligosacharide) and sphingolipids and OsRac1, a small GTPase that interacts with NADPH oxidase to regulate ROS production at the beginning of cell death in the defense signaling. They observed that cytoplasmic OsHsp90 interact with the rice chitin receptor OsCERK1 and that this chaperone is required for the transport of the receptor from the ER to the plasma membrane (Chen et al. 2010). Likewise, the inhibition of OsHsp90 by Geldanamycin A (GDA), a specific inhibitor of all Hsp90, impairs the defense response. These results demonstrate that OsHsp90 is essential for Rac1-mediated enhancement in PAMP-triggered immune responses in rice (Thao et al. 2007). Besides, *PR* genes were not induced, proving the requirement of OsHsp90 for Rac1-mediated enhancement of PAMP signaling (Thao et al. 2007). Furthermore, the authors proposed that OsCERK1 and OsRac1 are part of a larger complex called "defensome" at the plasma membrane (PM) and Hsp90 play a critical role in the defensome function (Chen et al. 2010).

Even though few examples involving Hsp90s in the PTI have been described so far, there seems to be an overlapping between non-host and PAMP-mediated innate immune response due to Hsp90 silencing impairs *PR* genes expression and HR defense mechanism. So far, a small number of examples related to the role of Hsp90 in innate response were found in the bibliography (Chen et al. 2010; Garavaglia et al. 2008; Kanzaki et al. 2003; Swindell et al. 2007; Thao et al. 2007), but there is no doubt about the importance of Hsp90s in plant defense response in non-host response described above.

8.5 HSP90 and Host Resistance

During compatible interactions, the plant cannot detect the pathogen by PAMP receptors or the defense mechanisms developed are ineffective. Therefore, pathogens succeed in installing the disease and it is said that the pathogen is virulent (Hammond-Kosack and Jones 1996). Several works have demonstrated that some virulent pathogens possess the effectors to overcome PTI. If one effector is recognized by R proteins codified by the plant, ETI-associated mechanisms are successful in avoiding pathogen invasion and this form of resistance is referred to as host resistance (Jones and Dangl 2006). The majority of the R proteins shares common features among different subfamilies (Elmore et al. 2012; Shirasu 2008). Most of R protein-triggered resistance is associated with HR development, which allows a rapid host cell death (Shirasu and Schulze-Lefert 2000). According to this, it is believed that common signaling pathway is used in response to different pathogens (Shirasu and Schulze-Lefert 2003; Shirasu 2008).

Given to the fact that resistance mediated by R protein is conditioned to Hsp90 function (Chapman et al. 2014; Hein et al. 2005; Lu et al. 2003; Scofield et al. 2005), the loss of HR activation was established as a primary criterion for identifying the extent of Hsp90 involvement in disease resistance in compatible interactions. In this sense, different approaches have been carried out, like VIGS screens, inhibitor treatments or mutations of cytosolic Hsp90 isoforms (Chapman et al. 2014; Lu et al. 2003; Scofield et al. 2005; Wang et al. 2011). In addition, several works have demonstrated by yeast two-hybrid analysis, co-immunoprecipitation and *pull-down* assay that cytosolic Hsp90 acts physically close to R proteins, supporting the idea that Hsp90 have a specific role in R protein-triggered immunity (Bieri et al. 2004; Hubert et al. 2003; De La Fuente Van Bentem et al. 2005; Liu et al. 2004; Lu et al. 2003). Besides, knockout or silencing of Hsp90 or treatment with GDA impairs the resistance against various pathogens and reduces the levels of R proteins (Holt et al. 2005; Hubert et al. 2003; Takahashi et al. 2003). The current model is that Hsp90 associated with other two proteins (RAR1 or/and SGT1) stabilizes R proteins to keep them in a controlled and competent conformation state in order to detect pathogen signals (Holt et al. 2005; Hubert et al. 2003; Schulze-Lefert 2004).

8.6 Specific Requirement of HSP90 Isoforms by R Proteins

As mentioned before, it is known that there are different isoforms of Hsp90 in several species and the evidence have shown that there is a degree of specificity associated to expression patterns of each isoform (Fu et al. 2009; Takahashi et al. 1992; Wang et al. 2011; Xu et al. 2013; Yabe et al. 1994). Interestingly, there is an apparent specific requirement of different Hsp90 isoforms for R protein-mediated disease resistance (Table 8.3). In this regard, several research labs have evaluated the role of different Hsp90 isoforms in the immune response against pathogens.

As a first approach, Takahashi *et al.* aimed to identify whether the expression of any of the AtHsp90 isoforms was specifically induced upon challenge with various *Pseudomonas syringae* pv. *tomato* DC3000 transgenic strains. They used two avirulent pathogens, *Pst* DC3000 (*avrRpt2*) and *Pst* DC3000 (*avrRmp1*) which codify effectors that are recognized in plants by the R proteins RPS2 and RPM1, respectively. They observed that only *AtHsp90.1* expression was induced upon *Pst* DC3000 (*avrRpt2*) challenge, but RPS2-dependent HR response was impaired when leaves were infiltrated with GDA. To test the extent of Hsp90-dependent resistance, the authors used AtHsp90.1 mutant lines. The results showed that when these mutant lines were infected with *Pst* DC3000 (*avrRpt2*), the bacteria grew between 5- and 20-fold more compared to control, indicating the AtHsp90.1 requirement for RPS2-dependent resistance (Takahashi et al. 2003). By contrast, AtHsp90.3 mutant lines infected with *Pst* DC3000 (*avrRpt2*) exhibited no significant difference compared to wild-type plants, demonstrating that this isoform has no effect on RPS2-mediated resistance (Bao et al. 2014). The characterization of the plant defense response mediated by RPM1 is the best described so far (Bao et al. 2014; Hubert et al. 2003; Takahashi et al. 2003). The challenge of *Arabidopsis* plants with *Pst* DC3000 (*avrRmp1*) resulted in an induced *AtHsp90.1* expression. However, the level of infection with *Pst* DC3000 (*avrRmp1)* were not affected compared to control, concluding that AtHsp90.1 is not required for RPM1-dependent resistance (Takahashi et al. 2003). To add more light to the current knowledge on pathogen resistance, Hubert *et al.* studied the participation of AtHsp90.2 by using *Arabidopsis* mutant plants. AtHsp90.2 mutant lines were challenged with *Pst* DC3000 (*avrRmp1*) or *Pst* DC3000 (*avrB*), both effectors were recognized by RPM1. RPM1-mediated HR was altered, but not completely abolished, suggesting that there exist a preferential use of AtHsp90.2 in RPM1 accumulation (Hubert et al. 2003). As there is not a full loss of RPM1 function, the authors believe that any of the other three cytosolic AtHsp90s compensate AtHsp90.2 absence, as previously observed in *S. cerevisiae* (Borkovich et al. 1989).

Table 8.3 Requirement of Hsp90 in plant defense response

Plant species	Plant R gene	Hsp90 isoform requirement	Pathogen	Matching pathogen gene	Reference
Arabidopsis thaliana	RPS2	Hsp90.1	*Pseudomonas syringae*	*AvrRpt2*	Takahashi et al. (2003)
	RPM1	Hsp90.2	*Pseudomonas syringae*	*AvrB*	Hubert et al. (2003)
	RPM1	Hsp90.2	*Pseudomonas syringae*	*AvrRpm1*	Hubert et al. (2003)
	RPM1	Hsp90.3	*Pseudomonas syringae*	*AvrRpm1*	Bao et al. (2014)
	RPS4	Hsp90.3	*Pseudomonas syringae*	*AvrRps4*	Bao et al. (2014)
Tobacco	Rx	Not specified	*Tobacco Mosaic Virus*	Not specified	Lu et al. (2003)
	N	Not specified	*Potato Virus X*	Not specified	Lu et al. (2003)
	N	Hsp90.1	*Potato Virus X*	Not specified	Liu et al. (2004)
	N	Hsp90.2	*Potato Virus X*	Not specified	Liu et al. (2004)
Potato	R3a	Not specified	*Phytophtora infestans*	Avr3a	Chapman et al. (2014)
Wheat	Lr21	Not specified	*Puccinia triticina*	Not specified	Scofield et al. (2005)
	Not specified	Hsp90.2	stripe rust fungus	Not specified	Wang et al. (2011)
	Not specified	Hsp90.3	stripe rust fungus	Not specified	Wang et al. (2011)
Barley	Mla13	Not specified	*Blumeria graminis*	Not specified	Hein et al. (2005)
Tomato	I-2	Hsp90.2	*Fusarium oxysporum*	Not specified	De La Fuente Van Bentem et al. (2005)

In a recent work, Bao *et al.* brought to light the isoform responsible of the redundant effect observed previously. Experiments carried out with *Arabidopsis* with point mutations in the AtHsp90.3 isoform, showed that the R protein-mediated resistance was impaired. The authors determined that, similarly to that observed for the AtHsp90.2 isoform, AtHsp90.3 also contribute to RPM1-mediated resistance when infected with *Pst* DC3000 (*avrRmp1*) (Bao et al. 2014). In addition, the authors demonstrated that RPS4 function depends also on AtHsp90.3 function since the infection using *Pst* DC3000 (*avrRps4*) in AtHsp90.3 mutant lines exhibits compromised HR defense response (Bao et al. 2014).

Other interesting model to study the involvement of the Hsp90 isoforms in plant immune response are the members of the genus *Nicotiana*, as it has been described that different species are resistant to the oomycete plant pathogen *Phytophtora infestans* (Kamoun et al. 1998). Although the HR is associated with the resistance response of *Nicotiana* to *P. infestans*, the timing, severity, and extent of the HR varied considerably, depending on the examined genotype (Kamoun 1998; Kamoun et al. 1998). To evaluate the role of NbHsp90 in this plant-pathogen interaction, a known elicitor from *P. infestans* (INF1) was assessed (Kanzaki et al. 2003). Results showed that in NbHsp90.1-silenced plants HR was totally absent after challenge with INF1, while control plants were able to mount HR (Kanzaki et al. 2003), concluding that the NbHsp90.1 isoform is required for HR defense plant in this model like it was observed in *A. thaliana*. Since tobacco and potato belong to the solanaceus family, *N. benthamiana* is an appropriate model plant to assess the R protein function from potato. In this sense, Chapman *et al.* expressed the potato R protein R3a in *N. benthamiana* and evaluated the defense response against *P. infestans* after silencing endogenous NbHsp90. The authors silenced NbHsp90 by VIGS with a conserved Hsp90 sequence, leading to a multigene silencing of the Hsp90 family. They observed that NbHsp90-silenced plants showed no HR triggered response, concluding that the HR response depends on Hsp90 function (Chapman et al. 2014). Similar results were obtained by Lu *et al.*, they also used VIGS with a conserved Hsp90 sequence. Unlike Lu *et al.* performed the experiments with *N. benthamiana* transgenic lines carrying R proteins, either *N*, which confer resistance to potato virus x (PVX), or *Rx* that confer resistance against tobacco mosaic virus (TMV). After silencing Hsp90 in N and Rx plants, it was observed the complete or partial loss of disease response against PVX or TMV respectively (Lu et al. 2003). In an attempt to identify the Hsp90 isoform involved in those interactions, Liu *et al.* performed the silencing of NbHsp90.1 in plants, which resulted in severe plant development defects and plant death, making this model not suitable. The authors hypothesized that due to the high similarity among Hsp90 isoforms, other isoforms besides NbHsp90.1 and NbHsp90.2 might have been silenced. Therefore, they used the tomato ortholog *Solanum lycopersicum* Hsp90.1 (SlHsp90.1) and it was able to silence NbHsp90.1 and NbHsp90.2 isoforms, but not others. Finally, the authors observed that the silencing of NbHsp90.1 and NbHsp90.2 in *N. benthamiana* plants compromises *N*-mediated disease response against TMV (Liu et al. 2004).

There is far less information about the requirements of Hsp90 on immune responses in monocotyledonous plants. In these cases, genetic analyses are greatly complicated since most of the cultivated wheat is hexaploid. In this sense, VIGS protocols have proven to be very useful for gene expression analyzes (Scofield et al. 2005). This strategy was used to evaluate the requirement of Hsp90s for the R protein Lr21 from wheat against *Puccinia triticina* (Scofield et al. 2005). The authors confirmed that wheat TaHsp90-silenced plants become susceptible when infected with the pathogen (Scofield *et al.* 2005). Further, an attempt to study the role of TaHsp90-silenced isoforms was successfully achieved (Wang et al. 2011). Wang *et al.* demonstrated that TaHsp90.2 and TaHsp90.3, but not TaHsp90.1 isoforms were required for the HR resistance to stripe rust fungus (Wang et al. 2011). Finally, Hein *et al.* used the strategy to silence several member of the barley HvHsp90 family.

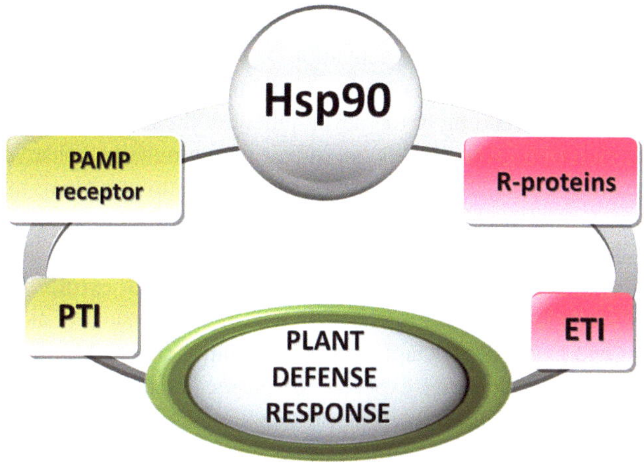

Fig. 8.2 Schematic representation of the requirement of Hsp90 in plant defense response. Hsp90 is involved in transport, maturation and stabilizing PAMP receptors and R-proteins. The non-active but ready-to use state of these receptors allow plant to trigger PTI and ETI, respectively, and mount the consequent plant defense response

They observed that silenced Hsp90 are required for the R protein Mla13-mediated resistance toward *Blumeria graminis* f. sp. *hordei* (Hein et al. 2005).

All the examples mentioned above have demonstrated that Hsp90 chaperone is an essential component of defense mechanism and that its absence causes an impairment of most of the defense responses. This suggests that cytosolic Hsp90 are indispensable components of plant defense signaling pathways (Kanzaki et al. 2003). During the last few years, it has become apparent that plant Hsp90 is involved in disease resistance since Hsp90 is able to link plant receptors to other components essential for mounting an effective defense responses against pathogens (Schulze-Lefert 2004). In accordance to this, mutation and silencing of plant cytosolic *Hsp90s* genes impairs the immune response triggered by host receptor. Besides, GDA abolish the resistance mediated by R proteins (Bao et al. 2014; Holt et al. 2005; Hubert et al. 2003; Takahashi et al. 2003; Thao et al. 2007). Similar to that observed in host resistance, Hsp90s participate in transport of M/PAMP receptor from the ER to the plasma membrane, thus Hsp90s are not only involved in ETI but also in PTI (Fig. 8.2).

8.7 Conclusion

Results showed in this chapter described that plant Hsp90s are an important component of the immune response making plants able to mount an appropriate defense against pathogens. The involvement of Hsp90s in either PTI or ETI has been plotted in the Fig. 8.2. Hsp90s participates in the transport of M/PAMP receptors to the plasma membrane and favors the activation of PTI (Chen et al. 2010; Thao et al. 2007). In addition, Hsp90 is able to stabilize and maintain R proteins in a non-active but ready-to use state, leading to the development of the ETI (Table 8.3). However, the lack or deficit of a given Hsp90 can impair the immune response and in consequence, plants are not capable of defend themselves. Although plants and mammals present obvious differences in the development of the immune response against pathogens, several coincidences have been found. So far, we talked about the function of intracellular Hsp90s, while it has been reported that extracellular mammal Hsp90s play a role in the activation of mammal immune system (Calderwood et al. 2007a; Kono and Rock 2009; Oura et al. 2011; Wallin et al. 2002). In addition, the evidence showed that artificial exposure of purified mammal Hsp90 to antigens leads to a phenomenon known as cross-presentation, essential to develop a specific immune response against intracellular pathogens. As a consequence, it has been demonstrated that mammal Hsp90s are able to chaperone and present antigens and modulate the immune response against them (Calderwood et al. 2007a, b; Murshid et al. 2010, 2012; Thériault et al. 2006). Thus, mammal and parasitic Hsp90s are emerging as immune regulators since they are able to modulate the immune response and they have been proposed as vaccine in cancer treatments (Hahn 2009; Murshid et al. 2012). In this sense, Corigliano et al. (2011, 2013) evaluated plant Hsp90 immunostimulatory properties in the murine model. As a first approach, they demonstrated that recombinant Hsp90 from *A. thaliana* and *N. benthamiana* behave as mitogens of B-lymphocyte and this stimulus is time- and doses-dependent (Corigliano et al. 2011). An interesting finding is that the proliferation observed in mammal cells after Hsp90 stimuli was impaired in a TLR4$^{-/-}$ mice, suggesting a possible role of M/PAMP receptor in the response observed (Corigliano et al. 2011). More recently, they assessed the ability of plant Hsp90 to be used as adjuvant in vaccine development. In this sense, plant Hsp90 immunization conferred immunity against the antigen assayed, making plant Hsp90 attractive to be use as adjuvant in vaccine formulations (Corigliano et al. 2013). On the other hand, a really interesting observation was made by Petriccione et al. (2013) since they demonstrated for the first time the presence of extracellular Hsp90 in plants. They identified Hsp90 in the apoplast of kiwiplant leaves infected with *P. syringae* pv. *Actinidiae.* Even though there is no evidence of Hsp90 function in the apoplast, the authors suggest that might be involved in defense mechanisms (Petriccione et al. 2013). Altogether, the results show Hsp90s play a critical role not only in the assembly, stabilization, maturation and activation of key signaling proteins (Young et al. 2001) but also in the plant immune response mechanisms. An open question is whether extracellular Hsp90 are playing an unknown role related to the modulation or the specificity of the defense response triggered or whether plant defense mechanisms manipulate the protein content in the apoplast in order to impede further disease development. Further research need to be done in order to answer these and other questions about the involvement of plant Hsp90s and plant immune response.

Acknowledgements We would like to acknowledge Dr. Andrés Gárriz (member of CONICET) for his suggestions, which greatly improved the English language of the manuscript. This work was supported by PICT 2014–3473 of the National Agency for Promotion of Science and Technology (ANPCyT, Argentina), Puente 2014-Marina Clemente and Puente 2014-Mariana G. Corigliano of National University of General San Martin (UNSAM). MC and MGC are members of CONICET and UNSAM.

References

Banilas G, Korkas E, Englezos V, Nisiotou AA, Hatzopoulos P (2012) Genome-wide analysis of the heat shock protein 90 gene family in grapevine (Vitis vinifera L.). Aust Soc Vitic Oenol Inc 18:29–38

Bao F, Huang X, Zhu C, Zhang X, Li X, Yang S (2014) Arabidopsis HSP90 protein modulates RPP4-mediated temperature-dependent cell death and defense responses. New Phytol 202:1320–1334

Bieri S, Mauch S, Shen QH, Peart J, Devoto A, Casais C, Ceron F, Schulze S, Steinbiss HH, Shirasu K, Schulze-Lefert P (2004) RAR1 positively controls steady state levels of barley MLA resistance proteins and enables sufficient MLA6 accumulation for effective resistance. Plant Cell 16:3480–3495

Borkovich KA, Farrelly FW, Finkelstein DB, Taulien J, Lindquist S (1989) Hsp82 is an essential protein that is required in higher concentrations for growth of cells at higher temperatures. Mol Cell Miol 9:3919–3930

Breiman A (2014) Plant Hsp90 and its co-chaperones. Curr Protein Pept Sci Pept Sci 15:1–13

Calderwood SK, Mambula SS, Gray PJ, Theriault JR (2007a) Extracellular heat shock proteins in cell signaling. FEBS Lett 581:3689–3694

Calderwood SK, Theriault J, Gray PJ, Gong J (2007b) Cell surface receptors for molecular chaperones. Methods 43:199–206

Chapman S, Stevens LJ, Boevink PC, Engelhardt S, Alexander CJ, Harrower B, Champouret N, McGeachy K, Van Weymers PSM, Chen X, Birch PRJ, Hein I (2014) Detection of the virulent Form of AVR3a from Phytophthora infestans following artificial evolution of potato resistance gene R3a. PLoS One 9:e110158

Chaudhury S, Welch TR, Blagg BSJ (2006) Hsp90 as a target for drug development. Chem Med Chem 1:1331–1340

Chen L, Hamada S, Fujiwara M, Zhu T, Thao NP, Wong HL, Krishna P, Ueda T, Kaku H, Shibuya N, Kawasaki T, Shimamoto K (2010) The Hop/Sti1-Hsp90 chaperone complex facilitates the maturation and transport of a PAMP receptor in rice innate immunity. Cell Host Microbe 7:185–196

Chen XJ, Wu MJ, Jiang Y, Yang Y, Yan YB (2015) Dunaliella salina Hsp90 is halotolerant. Int J Biol Macromol 75:418–425

Chisholm ST, Coaker G, Day B, Staskawicz BJ (2006) Host-microbe interactions: shaping the evolution of the plant immune response. Cell 124:803–814

Christou A, Filippou P, Manganaris GA, Fotopoulos V (2014) Sodium hydrosulfide induces systemic thermotolerance to strawberry plants through transcriptional regulation of heat shock proteins and aquaporin. BMC Plant Biol 14:42

Conner TW, LaFayette PR, Nagao RT, Key JL (1990) Sequence and expression of a HSP83 from Arabidopsis thaliana. Plant Physiol 94:1689–1695

Corigliano MG, Maglioco A, Laguía Becher M, Goldman A, Martín V, Angel SO, Clemente M (2011) Plant Hsp90 proteins interact with B-cells and stimulate their proliferation. PLoS One 6:e21231

Corigliano MG, Fenoy I, Sander V, Maglioco A, Goldman A, Clemente M (2013) Plant heat shock protein 90 as carrier-adjuvant for immunization against a reporter antigen. Vaccine 31:5872–5878

da Silva VCH, Cagliari TC, Lima TB, Gozzo FC, Ramos CHI (2013) Conformational and functional studies of a cytosolic 90 kDa heat shock protein Hsp90 from sugarcane. Plant Physiol Biochem 68:16–22

De La Fuente Van Bentem S, Vossen JH, De Vries KJ, Van Wees S, Tameling WIL, Dekker HL, de Koster CG, Haring MA, Takken FLW, Cornelissen BJC (2005) Heat shock protein 90 and its co-chaperone protein phosphatase 5 interact with distinct regions of the tomato I-2 disease resistance protein. Plant J 43:284–298

De Wit PJGM (2007) How plants recognize pathogens and defend themselves. Cell Mol Life Sci 64:2726–2732

Elmore JM, Lin ZD, Coaker G (2012) Plant NB-LRR signaling: upstreams and downstreams. Curr Opin Plant Biol 14:365–371

Felsheim RF, Das A (1992) Structure and expression of a heat-shock protein 83 gene of Pharbitis nil. Plant Physiol 100:1764–1771

Fu D, Ghabrial S, Kachroo A (2009) GmRAR1 and GmSGT1 are required for basal, R gene-mediated and systemic acquired resistance in soybean. Mol Plant-Microbe Interact 22:86–95

Garavaglia BS, Garofalo CG, Orellano EG, Ottado J (2008) Hsp70 and Hsp90 expression in citrus and pepper plants in response to Xanthomonas axonopodis pv. citri. Eur J Plant Pathol 123:91–97

Gupta RS (1995) Phylogenetic analysis of the 90 kD heat shock family of protein sequences and an examination of the relationship among animals, plants, and fungi species. *Mo*. Biol Evol 12:1063–1073

Hahn JS (2009) The Hsp90 chaperone machinery: from structure to drug development. BMB Rep 42:623–630

Hammond-Kosack KE, Jones JD (1996) Resistance gene-dependent plant defense responses. Plant Cell 8:1773–1791

Hein I, Barciszewska-Pacak M, Hrubikova K, Williamson S, Dinesen M, Soenderby IE, Sundar S, Jarmolowski A, Shirasu K, Lacomme C (2005) Virus-induced gene silencing-based functional characterization of genes associated with powdery mildew resistance in barley. Plant Physiol 138:2155–2164

Holt BF, Belkhadir Y, Dangl JL (2005) Antagonistic control of disease resistance protein stability in the plant immune system. Science 309:929–932

Hubert DA, Tornero P, Belkhadir Y, Krishna P, Takahashi A, Shirasu K, Dangl JL (2003) Cytosolic HSP90 associates with and modulates the Arabidopsis RPM1 disease resistance protein. EMBO J 22:5679–5689

Johnson JL (2012) Evolution and function of diverse Hsp90 homologs and cochaperone proteins. Biochim Biophys Acta 1823:607–613

Jones JDG, Dangl JL (2006) The plant immune system. Nat Rev 444:323–329

Kamoun S (1998) Dissection of nonhost resistance of Nicotiana to phytophthora infestans. Phytopathology 88:S45

Kamoun S, van West P, Vleeshouwers VGAA, de Groot KE, Govers F (1998) Resistance of Nicotiana benthamiana to phytophthora infestans is mediated by the recognition of the elicitor protein INF1. Plant Cell 10:1413–1425

Kanzaki H, Saitoh H, Ito A, Fujisawa S, Kamoun S, Katou S, Yoshioka H, Terauchi R (2003) Cytosolic HSP90 and HSP70 are essential components of INF1-mediated hypersensitive response and non-host resistance to Pseudomonas cichorii in Nicotiana benthamiana. Mol Plant Pathol 4:383–391

Koning AJ, Rose R, Comai L (1992) Developmental expression of tomato heat-shock cognate protein 80. Plant Physiol 100:801–811

Kono H, Rock KL (2009) How dying cells alert the immune system to danger. Nat Rev Immunol 8:279–289

Krishna P, Gloor G (2001) The Hsp90 family of proteins in Arabidopsis thaliana. Cell Stress Chaperones 6:238–246

Krishna P, Sacco M, Cherutti JF, Hill S (1995) Cold-induced accumulation of hsp90 transcripts in Brassica napus. Plant Physiol 107:915–923

Krishna P, Reddy RK, Sacco M, Frappier JR, Felsheim RF (1997) Analysis of the native forms of the 90 kDa heat shock protein (hsp90) in plant cytosolic extracts. Plant Mol Biol 33:457–466

Ling SP, Su SS, Zhang HM, Liu XY, Pan GF, Yuan Y (2014) Molecular cloning and characterization of the full-length Hsp90 gene from Matricaria recutita. Genet Mol Res 13:10994–11003

Liu Y, Burch-Smith T, Schiff M, Feng S, Dinesh-Kumar SP (2004) Molecular chaperone Hsp90 associates with resistance protein N and its signaling proteins SGT1 and Rar1 to modulate an innate immune response in plants. J Biol Chem 279:2101–2108

Liu D, Zhang X, Cheng Y, Takano T, Liu S (2006) rHsp90 gene expression in response to several environmental stresses in rice (Oryza sativa L.). Plant Physiol Biochem 44:380–386

Liu D, Lu Z, Mao Z, Liu S (2009) Enhanced thermotolerance of E. coli by expressed OsHsp90 from rice (Oryza sativa L.). Curr Microbiol 58:129–133

Lu R, Malcuit I, Moffett P, Ruiz MT, Peart J, Wu A-J, Rathjen JP, Bendahmane A, Day L, Baulcombe DC (2003) High throughput virus-induced gene silencing implicates heat shock protein 90 in plant disease resistance. EMBO J 22:5690–5699

Marrs KA, Casey ES, Capitant SA, Bouchard RA, Dietrich PS, Mettler IJ, Sinibaldi RM (1993) Characterization of two maize HSP90 heat shock protein genes: expression during heat shock, embryogenesis, and pollen development. Dev Genet 14:27–41

Mayor A, Martinon F, Smedt TD, Pétrilli V, Tschopp J (2007) A crucial function of SGT1 and HSP90 in inflammasome activity links mammalian and plant innate immune responses. Nat Immunol 8:497–503

Mendonça YA, Ramos CHI (2012) Cloning, purification and characterization of a 90 kDa heat shock protein from citrus sinensis (sweet orange). Plant Physiol Biochem 50:87–94

Milioni D, Hatzopoulos P (1997) Genomic organization of hsp90 gene family in Arabidopsis. Plant Mol Biol 35:955–961

Murshid A, Gong J, Calderwood SK (2010) Heat shock protein 90 mediates efficient antigen cross presentation through the scavenger receptor expressed by endothelial cells-I. J Immunol 185:2903–2917

Murshid A, Gong J, Calderwood SK (2012) The role of heat shock proteins in antigen cross presentation. Front Immunol 3:1–10

Nover L, Miernyk JA (2001) A genomics approach to the chaperone network of Arabidopsis thaliana. Cell Stress Chaperones 6:175–176

Nürnberger T, Brunner F, Kemmerling B, Piater L (2004) Innate immunity in plants and animals: stinking similarities and obvious differences. Immunol Rev 198:249–266

Oura J, Tamura Y, Kamiguchi K, Kutomi G, Sahara H, Torigoe T, Himi T, Sato N (2011) Extracellular heat shock protein 90 plays a role in translocating chaperoned antigen from endosome to proteasome for generating antigenic peptide to be cross-presented by dendritic cells. Int Immunol 23:223–237

Pearl LH, Prodromou C (2006) Structure and mechanism of the Hsp90 molecular chaperone machinery. Annu Rev Biochem 75:271–294

Pearson GA, Hoarau G, Lago-Leston A, Coyer JA, Kube M, Reinhardt R, Henckel K, Serrao ETA, Corre E, Olsen JL (2010) An expressed sequence tag analysis of the intertidal brown seaweeds Fucus serratus (L.) and F. vesiculosus (L.) (Heterokontophyta, Phaeophyceae) in response to abiotic stressors. Mar Biotechnol 12:195–213

Petriccione M, Di Cecco I, Arena S, Scaloni A, Scortichini M (2013) Proteomic changes in Actinidia chinensis shoot during systemic infection with a pandemic Pseudomonas syringae pv. actinidiae strain. J Proteomics 78:461–476

Reddy PS, Thirulogachandar V, Vaishnavi CS, Aakrati A, Sopory SK, Reddy MK (2011) Molecular characterization and expression of a gene encoding cytosolic Hsp90 from Pennisetum glaucum and its role in abiotic stress adaptation. Gene 474:29–38

Ritossa F (1962) A new puffing pattern induced by temperature shock and DNP in Drosophila. Experientia 18:571–573
Sangster TA, Bahrami A, Wilczek A, Watanabe E, Schellenberg K, McLellan C, Kelley A, Kong SW, Queitsch C, Lindquist S (2007) Phenotypic diversity and altered environmental plasticity in Arabidopsis thaliana with reduced Hsp90 levels. PLoS One 2:e648
Schmitz G, Schmidt M, Feierabend J (1996) Characterization of a plastid-specific HSP90 homologue: identification of a cDNA sequence, phylogenetic descendence and analysis of its mRNA and protein expression. Plant Mol Biol 30:479–492
Schroda M (2004) The Chlamydomonas genome reveals its secrets: chaperone genes and the potential roles of their gene products in the chloroplast. Photosynth Res 82:221–240
Schroder G, Beck M, Eichel J, Vetter H-P, Schroder J (1993) Hsp90 homolog from Madagascar periwinkle (Catharanthus roseus) – cDNA sequence, regulation of protein expression and location in the endoplasmic-reticulum. Plant Mol Biol 23:583–594
Schulze-Lefert P (2004) Plant immunity: the origami of receptor activation. Curr Biol 14:22–24
Schwessinger B, Zipfel C (2008) News from the frontline: recent insights into PAMP-triggered immunity in plants. Curr Opin Plant Biol 11:389–395
Scofield SR, Huang L, Brandt AS, Gill BS (2005) Development of a virus-induced gene-silencing system for hexaploid wheat and its use in functional analysis of the Lr21 -mediated leaf rust resistance pathway 1. Plant Physiol 138:2165–2173
Senthil-Kumar M, Mysore KS (2013) Nonhost resistance against bacterial pathogens: retrospectives and prospects. Annu Rev Phytopathol 51:407–427
Shirasu K (2008) The HSP90-SGT1 chaperone complex for NLR immune sensors. Annu Rev Plant Biol 60:139–164
Shirasu K, Schulze-Lefert P (2000) Regulators of cell death in disease resistance. Plant Mol Biol 44:371–385
Shirasu K, Schulze-Lefert P (2003) Complex formation, promiscuity and multi-functionality: protein interactions in disease-resistance pathways. Trends Plant Sci 8:252–258
Song H, Zhao R, Fan P, Wang X, Chen X, Li Y (2009) Overexpression of AtHsp90.2, AtHsp90.5 and AtHsp90.7 in Arabidopsis thaliana enhances plant sensitivity to salt and drought stresses. Planta 229:955–964
Sung MS, Hsu YT, Ho KL, Lee TM (2011) Implications of the up-regulation of genes encoding protein degradation enzymes and heat shock protein 90 for intertidal green macroalga Ulva fasciata against hypersalinity-induced protein oxidation. Mar Biotechnol 13:684–694
Swindell WR, Huebner M, Weber AP (2007) Transcriptional profiling of Arabidopsis heat shock proteins and transcription factors reveals extensive overlap between heat and non-heat stress response pathways. BMC Genomics 8:125
Takahashi T, Naito S, Komeda Y (1992) Isolation and analysis of the expression of two genes for the 81-kDa heat-shock proteins from Arabidopsis. Plant Physiol 99:383–390
Takahashi A, Casais C, Ichimura K, Shirasu K (2003) HSP90 interacts with RAR1 and SGT1 and is essential for RPS2-mediated disease resistance in Arabidopsis. Proc Natl Acad Sci U S A 100:11777–11782
Thao NP, Chen L, Nakashima A, Hara S, Umemura K, Takahashi A, Shirasu K, Kawasaki T, Shimamoto K (2007) RAR1 and HSP90 form a complex with Rac/Rop GTPase and function in innate-immune responses in rice. Plant Cell 19:4035–4045
Thériault JR, Adachi H, Calderwood SK (2006) Role of scavenger receptors in the binding and internalization of heat shock protein 70. J Exp Med 177:8604–8611
Virdi AS, Pareek A, Singh P (2011) Evidence for the possible involvement of calmodulin in regulation of steady state levels of Hsp90 family members (Hsp87 and Hsp85) in response to heat shock in sorghum. Plant Signal Behav 6:393–399
Wallin RP, Lundqvist A, Moré SH, Bonin A, Kiessling R, Ljunggren G (2002) Heat-shock proteins as activators of the innate immune system. Trends Immunol 23:130–135

Walther-Larsen H, Brandt J, Collinge DB, Thordal-Christensen H (1993) A pathogen-induced gene of barley encodes a HSP90 homologue showing striking similarity to vertebrate forms resident in the endoplasmic reticulum. Plant Mol Biol 21:1097–1108
Wang GF, Wei X, Fan R, Zhou H, Wang X, Yu C, Dong L, Dong Z, Wang X, Kang Z, Ling H, Shen Q-H, Wang D, Zhang X (2011) Molecular analysis of common wheat genes encoding three types of cytosolic heat shock protein 90 (hsp90): Functional involvement of cytosolic hsp90s in the control of wheat seedling growth and disease resistance. New Phytol 191:418–431
Wang S-J, Wu M-J, Chen X-J, Jiang Y, Yan Y-B (2012) DsHsp90 is involved in the early response of dunaliella salina to environmental stress. Int J Mol Sci 13:7963–7979
Willmund F, Schroda M (2005) Heat shock protein 90C is a bona fide Hsp90 that interacts with plastidic Hsp70B in Chlamydomonas reinhardtii. Plant Physiol 138:2310–2322
Xu J, Xue C, Xue D, Zhao J, Gai J, Guo N, Xing H (2013) Overexpression of GmHsp90s, a heat shock protein 90 (Hsp90) gene family cloning from soybean, decrease damage of abiotic stresses in Arabidopsis thaliana. PLoS One 8:1–14
Yabe N, Takahashi T, Komeda Y (1994) Analysis of tissue-specific expression of Arabidopsis thaliana HSP90-family gene HSP81. Plant Cell Physiol 35:1207–1219
Yamada K, Fukao Y, Hayashi M, Fukazawa M, Suzuki I, Nishimura M (2007) Cytosolic HSP90 regulates the heat shock response that is responsible for heat acclimation in Arabidopsis thaliana. J Biol Chem 282:37794–37804
Ye SF, Yu SW, Shu LB, Wu JH, Wu AZ, Luo LJ (2012) Expression profile analysis of 9 heat shock protein genes throughout the life cycle and under abiotic stress in rice. Chin Sci Bull 57:336–343
Young JC, Moarefi I, Hartl UF (2001) Hsp90: a specialized but essential protein-folding tool. J Cell Biol 154:267–273
Zai WS, Miao LX, Xiong ZL, Zhang HL, Ma YR, Li YL, Chen YB, Ye SG (2015) Comprehensive identification and expression analysis of Hsp90s gene family in Solanum lycopersicum. Genet Mol Res 14:7811–7820
Zhang M, Botër M, Li K, Kadota Y, Panaretou B, Prodromou C, Shirasu K, Pearl LH (2008) Structural and functional coupling of Hsp90- and Sgt1-centred multi-protein complexes. EMBO J 27:2789–2798
Zhang J, Li J, Liu B, Zhang L, Chen J, Lu M (2013) Genome-wide analysis of the Populus Hsp90 gene family reveals differential expression patterns, localization, and heat stress responses. BMC Genomics 14:532

Chapter 9
Chloroplasts Heat Shock Protein 70B as Marker of Oxidative Stress

Stephanie Chankova and Nadezhda Yurina

Abstract Organisms respond to environmental impact by developing a series of physiological, biochemical and molecular strategies. Heat shock proteins are one of them, possess a special place among defense systems and could contribute to cellular homeostasis. Here we address and try to discuss the following questions: 1. Whether species isolated from habitats with extreme environmental conditions are good model for studying cell resistance to oxidative stress? 2. Whether HSP70B could be used as a reliable marker for cell resistance to oxidative stress? 3. Whether HSP70B could be applied to evaluate the magnitude of environmentally induced stress? In this chapter we tried to throw more light on these three "whether". Briefly, results and discussion presented here contribute to the hypothesis of consistent functional properties of HSP70B as a mechanism of thermo-tolerance in plant species. Some assumptions are done concerning the role of constitutive and well-expressed overproduction of HSP70B as a part of *Chlorella* survival strategy against environmental stress stimuli. Here we show that overproduction of HSP70B could be used as an early warning marker for induced oxidative stress in the studied genotypes. Our experimental finding that HSP70B induction correlates with the magnitude of PQ-induced oxidative stress contributes to the still unresolved challenge for identification of reliable markers for screening of genotype resistance/susceptibility to oxidative stress. For the first time we identified homologue of chloroplast HSP70B in *Chlorella* chloroplasts. The development of plant-based biomarker test systems corresponds to the strategies for protection biodiversity preservation and genome stability of plant populations.

Keywords Abiotic stress • Antioxidant enzymes • Cell resistance • *Chlamydomonas* • *Chlorella* • Chloroplast chaperones • Glutathione-S-transferase • Heat shock proteins • Paraquat • Stress proteins

---

S. Chankova
Institute of Biodiversity and Ecosystem Research, BAS, Sofia, Bulgaria

N. Yurina (✉)
Laboratory of Bioenergetics, A.N. Bach Institute of Biochemistry Research Center of Biotechnology, RAS, 33, bld. 2 Leninsky Ave., Moscow 119071, Russia
e-mail: NadezhdaYurina@hotmail.com

A.A.A. Asea et al. (eds.), *Heat Shock Proteins and Plants*,
Heat Shock Proteins 10, DOI 10.1007/978-3-319-46340-7_9

Abbreviations

CAT	catalase
GST	glutathione-S-transferase
HSPs	heat shock proteins
PQ	paraquat
ROS	reactive oxygen species
SOD	superoxide dismutase

9.1 Introduction

Markers of oxidative stress can be used to monitoring toxic effects of various different damaging factors on plant organisms. Identifying markers of oxidative stress has been in the focus of many researchers because their potential role to act as an "integrator" of a multitude of processes that abiotic stress could drive. Plant biomarkers of oxidative stress can be classified as molecules that are modified by interactions with reactive oxygen species (ROS) in the microenvironment; and molecules of the antioxidant system that could be changed as a response to increased oxidative stress. In many experiments ROS-responsive genes were identified as biomarkers.

ROS are natural products formed during the oxygen metabolism and generation of H_2O. The main pro-oxidant agents are ROS formed by unstable forms of oxygen: superoxide, hydrogen peroxide, and hydroxyl radicals. Cells can tolerate moderate oxidative loads by increasing gene expression to upregulate their reductive defense systems and restore the oxidant/antioxidant balance (Mittler 2002). But when this increased synthesis cannot be achieved due to damage to enzymes, or substrate limitations, or when the increased or prolonged oxidative load is overwhelming, an imbalance persists and the result is oxidative stress. In this chapter we give an overview of our current understanding of plant tolerance to oxidative stress with a special focus on chloroplast HSP/chaperones as a biochemical marker of oxidative stress.

9.2 Chloroplast HSP

Abiotic stress is produced by the negative impact of environmental factors such as extreme temperatures, strong light, chemicals, drought, salinity, hypoxia etc. on plant. Organisms respond to such negative factors by developing a series of physiological and biochemical strategies, including synthesis of different stress proteins. Heat shock proteins being one of them and having a special place among defense systems could contribute to cellular homeostasis. It has shown that the expression of HSPs and other heat-responsive genes are regulated by heat shock factors (HSFs)

(Kotak et al. 2007) through their association to the heat shock element (HSE) in the promoter region of the heat-responsive genes (Pelham 1982; Scharf et al. 2012; Driedonks et al. 2015). Obviously, both the activation and production of HSFs/HSPs and the increase in ROS/scavenging activity belong to the major responses of plants to stress and play important roles in plant acclimation. A number of recent genetic and biochemical studies, however, indicate that there are complex interactions between these responses (Driedonks et al. 2015).

HSPs are obligatory components of nonspecific response to heat stress or other environmental stressful factors (Timperio et al. 2008; Barua and Heckathorn 2006; Niu et al. 2006; Huang and Xu 2008). HSPs are particularly important for prokaryotic and eukaryotic cells contributing to the cellular homeostasis under optimal or stress environmental conditions (Wang et al. 2004; Yu et al. 2015) as well as for the formation of heat tolerance (Shen and Leen 1997; Xu and Huang 2010). It is described that HSPs expression could occur in natural environment. HSPs genes are found in all species but they vary in patterns of expression. The threshold of HSPs expression in different species correlated with the strength of environmental stress and the expression of HSPs could be correlated with resistance to stress (Feder and Hofmann 1999; Al-Whaibi 2011). In last decades HSPs are considered as a very robust candidate for "early – warning indicator" of environmentally induced stress by different pollutants, because they could be activated very early as a result of toxic exposure or at concentrations below the lethal dose (Bierkens et al. 1998; Monari et al. 2011). The present state of knowledge that HSPs/chaperones play an important role in plant adaptation to environmental stress assisting in the folding, intracellular distribution, assembly and degradation of proteins, mainly by stabilizing partially unfolded states in many normal cellular processes and can assist in the protein refolding under stress conditions (Baniwal et al. 2004; Schroda 2004; Wang et al. 2004).

Among the various HSP families a special role in protecting cells from the action of different kind of stress belongs to HSP70 proteins. The induction of HSP70 has been confirmed in a wide range of organisms – from microorganisms to humans (Lindquist and Graig 1988; Al-Whaibi 2011). This chaperone system is relatively well studied in bacteria and in most compartments of eukaryotic cells (the cytosol, endoplasmic reticulum, mitochondria) (Nordhues et al. 2010; Al-Whaibi 2011). Misfolded proteins formed during stress, bind to HSP70. Hsp70 is associated with the ribosomal subunit (Zargar et al. 2006). It has been proposed that HSP70 could be considered as good candidates for bio-monitoring assay because Hsp70 gene is a sensitive biomarker for different classes of environmental pollutants in algae (Bierkens et al. 1998; Ireland et al. 2004; Tukaj and Tukaj 2010).

The significance of HSP70 for cellular response to thermal stress, thermotolerance and physiological adaptation has been shown in different biological objects (Zargar et al. 2006; Clark and Peck 2009; Shatilina et al. 2011; Yu et al. 2015). Plastid polyfunctional protein HSP70B together with co-chaperones proteins forms chloroplast "foldosome" and has an important role in protecting plants from oxidative and heat stress (Schroda and Mühlhaus 2009; Nordhues et al. 2010). There are indirect evidences of the involvement of chloroplast HSP70B/chaperone in refolding

Table 9.1 Plastid HSP70 and HSP90 of *C. reinhardtii*

HSP family	Localization	Proposed function	References
HSP70B	Stroma; membrane-associated	Protein folding; protein import into chloroplasts (dis)assembly of VIPP1 oligomers; protection/repair of PS2 from photoinhibition	Trösch et al. (2015)
HSP70D, HSP70F	Unknown (C-terminally truncated putative stromal HSP70s)	–	Veyel et al. (2014), Liu et al. (2007), and Schroda et al. (1999)
HSP90C	Stroma; membrane-associated	Maturation of proteins involved in signal transduction; VIPP1 oligomer disassembly; protein import	Trösch et al. (2015)

of denatured proteins (Drzymalla et al. 1996; Nordhues et al. 2010). It has been also shown that HSP70B is involved in the biogenesis of thylakoid membranes, the synthesis or assembly of components of the new reaction centers, which contributes to the protection and repair of photosystem II during and after photoinhibition (Schroda et al. 1999; Schroda 2004; Trösch et al. 2015). Heat shock protein 70B (HSP70B) is the major HSP70 in the *Clamydomonas* chloroplast. Plastidic HSP70B and HSP90C exist as multichaperone complex with escort proteins CDJ1 (chloroplast DnaJ-like protein) and CGE1 (chloroplast GrpE homologue) (Willmund et al. 2008; Trösch et al. 2015). The expression of these two genes from heat and light is a clear indication of this (Trösch et al. 2015) (Table 9.1).

The fact that these proteins are involved in the formation of the multi-enzyme complex could be understood as an indication of their possible similarity in gene expression regulation. Actually, a great similarity was found in the dynamics and accumulation of both proteins – HSP70B and HSP90C (Swindell et al. 2007; Chankova et al. 2009). Also, our work demonstrated that heating with temperatures in the range of 37–42 °C operated as a stress factor in *C. reinhardtii* cells and resulted in over accumulation of HSP70B and HSP90C. Therefore one of these proteins HSP70B was used as a marker of oxidative stress in our further experiments (Chankova et al. 2013, 2014). Additionally, it was observed that heat–heat treatment led to further accumulation of chloroplast chaperones which could indicate that an acquired thermotolerance was developed (Banzet et al. 1998; Mendez-Alvarez et al. 1999; Chankova et al. 2009). Working methods with HSP are described on the website Bioprotocols (http://www.bio-protocol.org/wenzhang.aspx?id=849; http://www.bio-protocol.org/wenzhang.aspx?id=850).

In many experiments, the increased expression of HSPs genes has been used as a marker for response to heat shock, although this over-expression could be seen rather as a marker for disruption of protein homeostasis, as a result of thermal shock than as a marker for complex heat shock response. (Trösch et al. 2015). Based on increased HSP expression as marker, a shock response was shown to be elicited in *Chlamydomonas* when cells were shifted from 20 °C to 39–41 °C (Tanaka et al.

2000). Kobayashi et al. (2014)) reported a shock response to be induced when *Chlamydomonas* cells were shifted from 24 °C to at least 36 °C, and found cell survival to be compromised when the temperature shift went beyond 42.4 °C. Schroda et al. (2015)) found HSP genes to be induced when *Chlamydomonas* cells were shifted from 25 °C to at least 37 °C and cells to tolerate a maximum temperature of 43.5 °C. The variations observed are best explained by differences in growth conditions and/or variations between the strains employed (Trösch et al. 2015).

9.3 Heat Shock Protein 70B (HSP70B) – Biomarker for Cell Resistance or Cell Susceptibility to Environmentally Induced Stress?

Present state of knowledge concerning heat shock protein's (HSPs) physiology, ecology and evolution has allowed drawing number of findings about HSPs importance for living organisms (Feder and Hofmann 1999; Al-Whaibi 2011). HSPs expression is classified as non-specific defense mechanism occurring in natural environment and mediated by various signaling path ways; the *hsp* genes are found in all species but interspecies variability has been obtained; good relationship is determined between HSPs expression, cell resistance to stress and the magnitude of stress prevailing in the environment (Wang et al. 2004; Xu et al. 2011; Chankova et al. 2013; Lipiec et al. 2013; Piterková et al. 2013; Sarkar et al. 2013; Wang et al. 2014). Regulatory role of ROS in Hsp70B production and accumulation as a result of environmental stimuli as well as the relationship between HSP70B induction and the level of oxidative stress was obtained (Ahn and Thiele 2003; Piterková et al. 2013; Chankova et al. 2013).

For a long time mechanisms involved in the formation of genotype and induced resistance are in the focus of our attention (Chankova et al. 1990, 2001, 2005; Chankova et al. 2009; Chankova and Yurina 2012; Chankova et al. 2013, 2014). During the last years we ask a few questions: whether species isolated from habitats with extreme environmental conditions are good model for studying cell resistance to oxidative stress; whether HSP70B could be used as a reliable marker for cell resistance to oxidative stress or HSP70B could be applied to evaluate the level of stress? (Chankova et al. 2009; Chankova and Yurina 2012; Chankova et al. 2013, 2014). To get answers to these questions we commonly use mutant strains of unicellular green algae *Chlorella, Chlamydomonas* with different radio-and chemo-resistance and *Chlorella* species isolated from habitats with different environmental conditions. Strains are available in our algae collection at the Institute of Biodiversity and Ecosystem Research, BAS. Why such approach was chosen? The advantages of these model organisms are well known and described previously (Chankova and Yurina 2012).

Here, an attempt is made to summarize and discuss in the light of the present state of knowledge, some of our previously published and not yet published results.

Table 9.2 Survival fraction (SF) after temperature treatment of three *Chlorella* species: *Chlorella vulgaris*- Antarctic, *Chlorella vulgaris* 8/1 and *Chlorella kesslery*

Species[a]	SF at 39 °C	SF at 42 °C	SF at 45 °C
Chlorella vulgaris	0.93 ± 0.02	0.88 ± 0.05	0.83 ± 0.01
Chlorella vulgaris 8/1	0.87 ± 0.02	0.80 ± 0,01	0.78 ± 0.01
Chlorella kesslery	0.91 ± 0.01	0.82 ± 0.02	0.70 ± 0.01

[a]Differences between species are statistically significant (P < 0001). Error represents standard errors of mean values

A speculation is done that species with different thermal tolerance and temperature preferences will respond in a different way to heat shock induced stress. This reasoning was inspired also by the concept that endemic species are more susceptible to environmental changes that cosmopolitan species and can be characterized by various HSPs properties than cosmopolitans (Wiencke 1991; Brennecke et al. 1998; Tomanek 2002; Shatilina et al. 2011).

To check this presumption we compare temperature-induced response in three Chlorella species that differ in terms of their geographical origin, temperature preferences and tolerance. Three species of *Chlorella* were studded. Two of them originate from contrasting habitats – Antarctica (*Chlorella vulgaris* – extremophilic) and hot spring in Bulgaria (*Chlorella vulgaris* strain 8/1 – thermophilic). The third one *Chlorella kesslery* – mesophilic was from Trebon collection, Czech Republic (Chankova and Yurina 2012). Cell survival and induction of chloroplast HSP70B were measured. Cultures were maintained at standard temperature and light. Stress response to three temperature regimes (t = 39 °C for 30 min and t = 42 °C for 5 min and 45 °C for 5 min) was analyzed on the basis of two endpoints – cell survival and chloroplast HSP70B accumulation.

Data presented in a Table 9.2 indicate that short -term heating with temperatures in the range of 39–45 °C moderately affects cell survival depending on the genotype and temperature conditions. Species could be arranged as follows: *Chlorella vulgaris* Antarctic > *Chlorella vulgaris* 8/1 > *Chlorella kesslery*. Mesophilic *Chlorella kesslery* exhibits a higher sensitivity to this temperature range.

Some studies (White et al. 1994; Heckathorne et al. 1996) support the presumption that heat-shock protein production can vary among closely related species as a function of habitat. In our work interspecies variability was also obtained (Chankova et al. 2013) concerning constitutive and induced levels of HSP70B. About 30 % higher content of constitutive HSP70B level was measured in *Chlorella vulgaris* isolated from Antarctic soil. No statistically significant difference between constitutive levels (control, non-stress) of chloroplast HSP70B in *Chlorella kesslery* and *Chlorella vulgaris* 8/1 was calculated. Our results are in a agreement that the higher constitutive expression of HSP proteins in the more tolerant genotype could represent a pre-formed tolerance mechanism (Witzel et al. 2009).

To clarify whether higher constitutive content of HSP70B *Chlorella vulgaris* (Antarctic*)* was genotypically determined or due to the temperatures conditions of cultivation (23 °C) the following experiment was performed – *Chlorella* species

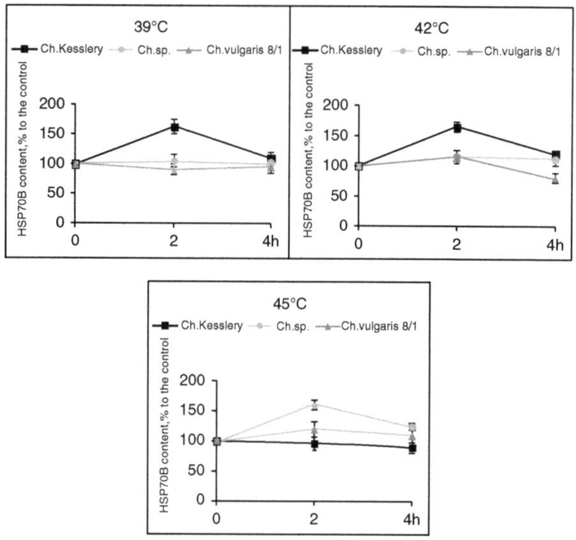

Fig. 9.1 Comparison of HSP70B level in *Chlorella* species after heat stress: 39 °C/30 min, 42 °C/5 min and 45 °C/5 min. Densitometry of HSP70B contents ($P \leq 0.05$); C – control sample (non stress)

were cultivated at 8 °C for 12 h. Under these experimental conditions of cultivation relative contents of HSP70B for all species were very similar to those measured when cells were cultivated at 23 °C. Again *Chlorella vulgaris* (Antarctic) was found to have the most pronounced production of constitutive HSP70B (127 % at 23 °C and 135 % at 8 °C). This result argues in a favor of the assumption that higher HSP70B contents in both Chlorella species isolated from habitats with more stressful environments compared with that in mesophilic Chlorella is evolutionary fixed sign promoting to the survival of cells under extreme environments. This is consistent with the idea of the key role of HSP70 in plant resistance to thermal stress. It is interesting that *Chlorella vulgaris* (Antarctic) has been also characterized by increased activity of superoxide dismutase and catalase at high temperatures, and increased efficiency of antioxidant system than other strains (Malanga et al. 2008; Nedeva and Pouneva 2009).

Further Western blot analysis was performed to investigate the relationship between increased heating and chloroplast HSP70B production (Fig. 9.1). Our data show that heating with t = 39 °C–42 °C for 5 min not alter in a statistically significant way the level of chloroplast HSP70B in both species isolated from habitats with more stressful environments. It could be assumed that temperatures in the range of 39–42 °C are not stressful factor for these species, although some authors suggested that t = 42 °C is a critical thermal threshold for plants (Knight 2010; Schroda et al. 2015). At the same temperature range, about 60 % increasing of the level of HSP70B for *Chlorella kesslery,* living in moderate environmental conditions was measured. Quite different stress response was defined as a result of heating with 45 °C for 5 min. The levels of HSP70B were enhanced with 55 % for Antarctic *Chlorella vulgaris* and 20 % *Chlorella vulgaris* 8/1. No any statistically significant changes in HSP70B content was measured for *Chlorella kesslery*. Our findings imply that the adaptations to life in habitats with more stressful environments resulted to different HSP70 expression that is consistent with the results for *Hydra oligactis* and *H. magnipapillata* (Brennecke et al. 1998) where differences in *hsp70* mRNA stability appear to be responsible for the habitat-correlated differences in the stress response in *Hydra* species. Our results could be also attributed to the inhibition of protein synthesis or photosynthetic performance than with the cell death, which is found to be about 30 % (SF = 0.70). Thus, interspecies variability found by us corresponds well with present understanding that genes coding HSPs are evolutionary conserved and can exhibit great interspecies variability leading to variation of tolerance ability and sensitivity to some diseases or environmental stimuli.

9.4 Conclusion

Let's ask ourselves: whether we have managed to answer to the questions?

Question 1
Whether species isolated from habitats with extreme environmental conditions are good model for studying cell resistance to oxidative stress?

Unfortunately we are still in the beginning. To be able definitely to reply this question further studies should be conducted, including more species and inductors of stress.

The Beginning Is Set Results presented here contribute to the hypothesis of consistent functional properties of HSP70B as a mechanism of thermo-tolerance in plant species. Some suggestion could be made that HSP70B overproduction is involved as a part of *Chlorella* species survival strategy against environmental stress stimuli. Higher constitutive content and well expressed overproduction of HSP70B are probably among of the factors that allow *Chlorella vulgaris* to survive at the extreme Antarctic environment.

Question 2
Whether HSP70B could be used as a reliable marker for cell resistance to oxidative stress?

Over-expression of chloroplast HSP70B could be considered as an early warning and sensitive short-lived biomarker of heat stress.

Commonly members of other HSP families have been recommended as biomarkers (Song et al. 2014; Wang et al. 2014). For the first time we identified homologue of chloroplast HSP70B in *Chlorella* chloroplasts. Earlier, HSP70B was detected and characterized in *Chlamydomonas reinhardtii* (Trösch et al. 2015), in *Arabidopsis thaliana* (Su and Li 2008), *Oryza sativa* (Kim and An 2013), *Triticum aestivum* (Duan et al. 2011) and in moss *Physcomitrella patens* (Shi and Theg 2010).

Question 3
Whether HSP70B could be applied to evaluate the level of stress?

Our data illustrate variations in magnitude of HSP70B response between investigated species that could be used for such purpose.

9.5 Genotype Resistance to Oxidative Stress and HSP70B Chaperone in *Chlamydomonas Reinhardtii*

Biota is subjected to increasingly strong pressure from the changing environment occurring as a result of natural global climate alterations or with the active intervention of man. At present abiotic stress is seen as a very important environmental factor acting via generation of reactive oxygen species (ROS) in cellular compartments (Jaspers and Kangasjärvi 2010) and inducing damages at different levels – DNA, proteins and membranes (Rochat et al. 2005; Driedonks et al. 2015). ROS act as molecules with dual function – signaling molecule and as molecule controlling gene expression and anti-stress systems activity (Vranová et al. 2002; Apel and Hirt 2004; Torres and Dangl 2005; Timperio et al. 2008; Galvez-Valdivieso and Mullineaux 2010; Mittler et al. 2011; Petrov and Van Breusegem 2012; Driedonks et al. 2015). A natural consequence of such pressure is the reduction of the yield of cultivated plants, genome degradation of cultivars and wild species and in some cases – human diseases.

9.6 How Organisms Cope with Oxidative Stress?

Evolutionary, organisms have evolved various defense mechanisms to protect themselves from environmentally induced stress, including expression of genes and synthesis of ROS scavenging proteins (Mittal et al. 2012; Suzuki et al. 2013; Driedonks et al. 2015), changes in plant transcriptome, proteome and metabolome and their corresponding metabolic pathways, activating DNA repair, antioxidant and chaperone systems etc. (Ahuja et al. 2010; Rodziewicz et al. 2014; Ramalingam et al. 2015).

9.7 How Men Can Cope with Oxidative Stress in Plants?

Recently, the idea was launched that the improvement of plant stress tolerance and introduction of genotypes with enhanced resistance to abiotic stress could be promising strategy to overcome this problem (Bianco and Defez 2012). A number of studies show that plant stress tolerance could be improved by increasing antioxidant enzymes activities (Rui et al. 1990; Badiani et al. 1993; Gupta et al. 1993; Sairam et al. 2000; Almeselmani et al. 2006; Wu et al. 2012; Chen et al. 2013), induction of ROS scavenging genes (Rainwater et al. 1996; Driedonks et al. 2015), activation of DNA repair, antioxidant and chaperone systems (Chankova et al. 2005, 2007; Chankova and Yurina 2012; Wegener and Jansen 2013; Chankova et al. 2014; Driedonks et al. 2015).

9.8 What Approach Should Be Used to Obtain Fast and Reliable Information?

Experimental data confirm that the strategy should include two main approaches: (i) model species, mutant/transgenic lines/strains with different resistance to oxidative stress; (ii) finding of robust plant markers for screening stress-tolerant/resistant genotypes (Badahur et al. 2011; Lipiec et al. 2013; Chankova and Yurina 2012; Chankova et al. 2013). As was shown before our teams in its long-term research programs have set as a goal to throw more light on the possible contribution of DSBs repair system, some antioxidant and detoxifying enzymes as well as chloroplast chaperone content in the formation of genotype resistance to oxidative stress. Commonly *Chlamydomonas reinhardtii* genotypes with different resistance to oxidative stress have been used. Detailed information concerning strains characterization could be found in (Chankova et al. 2001; Chankova et al. 2005; Dimova et al. 2008, 2009). Previously it was shown that most of the strains differ in their resistance to gamma-rays, radiomimetics, standard mutagens, pesticides, heating etc (Chankova et al. 2000, 2005, 2007, 2013, 2014; Chankova and Yurina 2012; Dimova et al. 2008).

We use strains with different resistance to various stress stimuli because our understanding that genotypes that differ in terms of their resistance to oxidative stress perhaps also differ in their antioxidant and HSP70B responses. In the example presented here Methyl Viologen (1,1′-Dimethyl –4,4′- bipyridinium dichloride (PQ – paraquat) was used as an oxidative-stress-inducing agent (Vermeulen et al. 2005; Paraquat Information Center, www.paraquat.com). Stress response of 4 strains with different genotype resistance – 137C, wild type (WT); CW15(+), cell-wall-less strain sensitive to PQ; hybrid strain H-3 with moderate level of PQ-resistance; and highly PQ-resistant mutant strain AK-9-9 was analyzed. Strains resistance to PQ was previously evaluated on the basis of colony forming ability (Dimova et al. 2008) and DSBs repair capacity (Chankova and Yurina 2012).

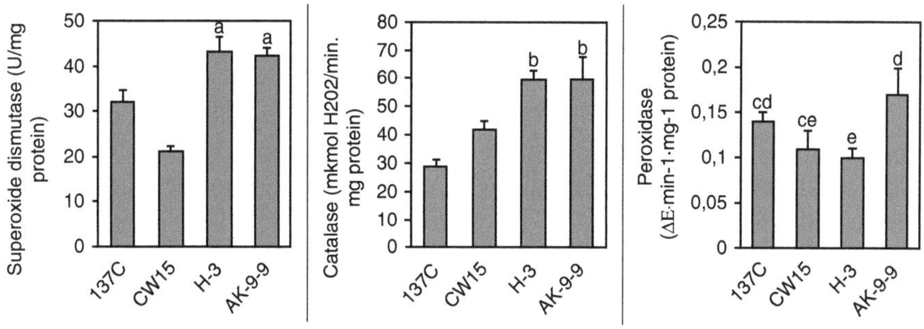

Fig. 9.2 Specific enzyme activities of superoxide dismutase, catalase and peroxidase of *C. reinhardtii* strains cultivated in physiological conditions ^{a,b,c,d,e} – values denoted with one and the same letter are not significantly different ($P > 0.05$) according to t test for equality of means: ^{a} ($t = 0.304$; $t_k = 2.0345$); ^{b} ($t = 0.041$; $t_k = 2.0244$); ^{c} ($t = 1.369$; $t_k = 2.0423$); ^{d} ($t = 1.011$; $t_k = 2.0395$); ^{e} ($t = 0.724$; $t_k = 2.0452$)

The herbicide PQ was applied as a free-radical-generating herbicide for three main reasons described in (Chankova et al. 2014). At the beginning endogenous SOD and CAT activities were analyzed in physiological conditions. Data, presented on Fig. 9.2, demonstrate higher endogenous SOD and CAT activities in the PQ-resistant and the moderately PQ-resistant genotypes. Our finding is in a good agreement with results of (Asker et al. 2007; Badahur et al. 2011; Zaka et al. 2002) where genotype resistance to oxidative stress correlates with antioxidant and non-antioxidant capacities.

Previously we have described several characteristics typical for radio – resistant strains – higher content of endogenous thiols (Chankova et al. 1990), higher levels of SOD activity (Chankova et al. 2001), increased DSBs rejoining capacity (Chankova et al. 2005; Dimova et al. 2009), higher constitutive HSP70B content (Chankova and Yurina 2012; Chankova et al. 2013). Some of these characteristics are found for other species, e.g. *Drosophila* (Vermeulen et al. 2005). At the same time opposite data – no statistically significant difference is obtained between the SOD activity of PQ-susceptible and PQ-resistant biotypes of *Conyza canadensis* (Szigeti and Lehoczki 2003) or between the activity of antioxidant enzymes in WT and PQ-resistant strains of *C. reinhardtii* (Vartak and Bhargava 1999). Here it should be noted that most probably a number of other factors should not be excluded – chromatin structure, mutations, cell cycle, altered DNA damage checkpoint response, defective apoptosis, cell wall maintenance, disturbance of the integrity of morphological, metabolic and genetic components and pathways, differences in the ultrastructure organization of cells and their cell etc. (Badahur et al. 2011; Bao et al. 2006; Chankova et al. 2005; Chalmers 2007; Costantini et al. 2013; Strasser et al. 2007).

Data concerning the antioxidant response to PQ in different plants and algae vary from the lack of enhancement of antioxidant enzyme activities (Szigeti and Lehoczki 2003) to stimulation of SOD, CAT, POX, GPX, APX (Ekmekci and Terzioglu 2005; Sood et al. 2011; Tanaka et al. 2011). In our experiments with PQ the effect of concentrations was very clearly manifested. A pronounced enhancement of SOD, CAT, GST activities was determined for PQ-sensitive genotype. Despite this increase, the elevated SOD and CAT activities in the PQ-sensitive genotype did not exceed the ones characteristic of the more resistant genotypes in physiological conditions (Chankova et al. 2014). The lack of statistically significant response measured as SOD, CAT, GST activities in WT and resistant genotypes could be considered as an indication of absence of strong oxidative stress and could be related to higher levels of endogenous SOD and CAT activities (Chankova et al. 2014). Although the WT genotype responded with minor increasing in APX activity, in general, the levels of peroxidases in the four *C. reinhardtii* genotypes were low. Therefore, it is likely that the major role in the response of *C. reinhardtii* cells to PQ treatment is played by defense mechanisms other than POX and APX (Chankova et al. 2014).

Current state of knowledge considered heat shock proteins (HSPs) overproduction as non-specific defense mechanism mediated by various signaling pathways (Lipiec et al. 2013; Piterková et al. 2013; Sarkar et al. 2013; Wang et al. 2004; Xu et al. 2011). Today, the relationship between increased ROS levels, expression of heat- responsive genes and accumulation of HSPs is proven (Ahn and Thiele 2003; Volkov et al. 2006; Wahid et al. 2007; Banti et al. 2008). The involvement of ROS in the regulation of Hsp70 production and accumulation under abiotic and biotic stresses was also confirmed (Piterková et al. 2013). Hypotheses promoted in recent years suggested that stress can indirectly activate HSFs via the action of ROS (Driedonks et al. 2015).

In order to clarify HSP70B differences among strains after PQ treatment chloroplast HSP70B content in *C. reinhardtii* strains was measured 1 h and 4 h following PQ treatment (Fig. 9.3). Our Western – blot data provided new experimental evidence that apparently HSP70B induction correlated with the magnitude of PQ-induced oxidative stress. HSP70B levels were higher in PQ-susceptible genotypes than in PQ-resistant genotypes and HSP70B accumulation is proportional to the PQ concentration. It is well known that in PQ-treated cells, hydrogen peroxide can be generated during the dismutation of superoxide radical anions by SOD. It is proposed, that H_2O_2 is signal molecule for the induction of Hsp genes expression. Possibly the diminished HSP70B level in PQ-resistant genotypes comparative to the PQ-susceptible genotypes correlated with lower level of H_2O_2 in these cells after PQ treatment.

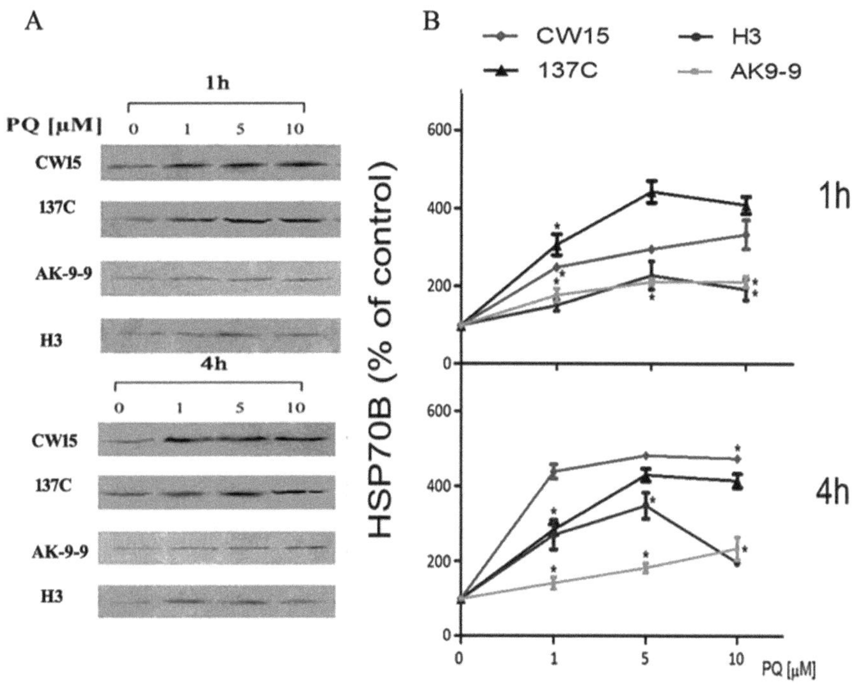

Fig. 9.3 Chloroplast HSP70B content in *C. reinhardtii* strains CW15, 137C, AK-9-9 and H-3 1 h and 4 h following PQ treatment. Western blot analysis using specific antisera against HSP70B (**a**). Results of densitometric estimations of the contents HSP70B after 1 h (**b**) and 4 h (**c**) PQ treatment; values (as percent of non-treated control) are means of triplicate ± S.E.; where no error bars are visible, errors were equal to or less than the symbols. ANOVA significant (treatment vs. control) at $P < 0.05$. * – $P < 0.01$

9.9 Conclusion

Here we show that overproduction of HSP70B could be used as an early marker for induced oxidative stress in the studied genotypes. Our experimental finding that HSP70B induction correlates with the magnitude of PQ-induced oxidative stress contributes to the still unresolved challenge for identification of reliable markers for screening of genotype resistance/susceptibility to oxidative stress. In the light of current understanding, oxidative stress triggered by different natural and anthropogenic environmental stimuli could induce excessive production of reactive oxygen species (ROS). Affecting expression of various genes ROS can influence and control many processes like stress response, growth, cell cycle, systemic signaling, development etc. and could cause disturbances in cell homeostasis, mutations, genome degradation and even death. These consequences have been well described previously by Dat et al. (2000) and Gill and Tuteja (2010).

During the evolution, plants have developed a number of mechanisms for protection against damaging factors present in the environment (Driedonks et al. 2015)

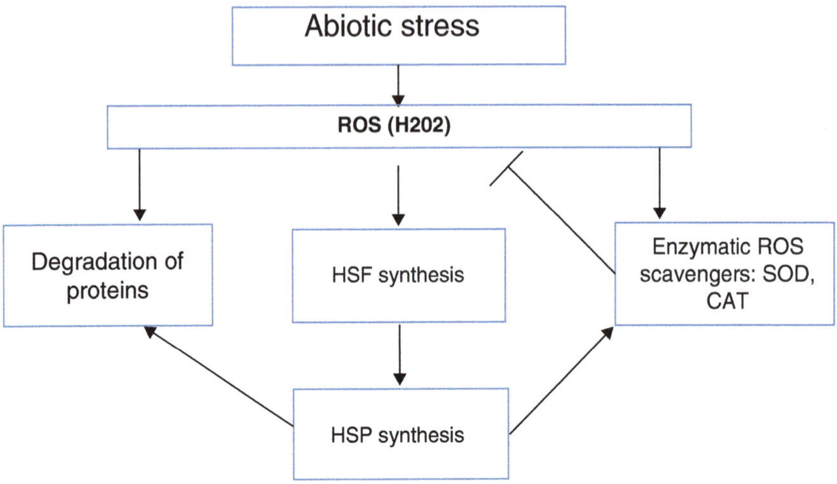

Fig. 9.4 The sequence of responses in plant biosynthesis induced by abiotic stress

and several typical futures for genotypes with well-expressed radio-and chemo-resistant were shown earlier by us (Fig. 9.4) (Chankova et al. 2000; Chankova and Yurina 2012). The mechanisms involved in the formation of genotypic resistance are of interest to scientists over the last 20 years (Badiani et al. 1993; Gupta et al. 1993; Rainwater et al. 1996; Sairam et al. 2000; Chankova et al. 2005, 2007; Chankova and Yurina 2012; Wu et al. 2012; Chen et al. 2013; Chankova et al. 2014; Driedonks et al. 2015). Here in this review we have made an attempt to answer at least two of them using some our own and other authors data. Of course, it is impossible to mentioned and interpreted the huge database and probably we have overlooked certain major studies for which we apologies in advance.

Let's ask ourselves: whether we have managed to answer to the following questions?

Question 1
Whether species isolated from habitats with extreme environmental conditions are good model for studying cell resistance to oxidative stress?

Unfortunately we are still in the beginning. To be able definitely to reply this question further studies should be conducted, including more species and inductors of stress.

The beginning is set. Results presented here contribute to the hypothesis of consistent functional properties of HSP70B as a mechanism of thermo-tolerance in plant species. Some suggestion could be made that HSP70B overproduction is involved as a part of Chlorella species survival strategy against environmental stress stimuli. Higher constitutive content and well expressed overproduction of HSP70B are probably among of the factors that allow *Chlorella vulgaris* to survive at the extreme Antarctic environment.

Question 2
Whether HSP70B could be used as a reliable marker for cell resistance to oxidative stress?

Here we show that overproduction of HSP70B could be used as an early marker for induced oxidative stress in the studied genotypes.

Our experimental finding that HSP70B induction correlates with the magnitude of PQ-induced oxidative stress contributes to the still unresolved challenge for identification of reliable markers for screening of genotype resistance/susceptibility to oxidative stress.

Over-expression of chloroplast HSP70B could be considered as an early warning and sensitive short-lived biomarker of heat stress.

Commonly members of other HSP families have been recommended as biomarkers (Wang et al. 2014). For the first time we identified homologue of chloroplast HSP70B in Chlorella chloroplasts.

Earlier, HSP70B was detected and characterized in *Chlamydomonas reinhardtii* (Trösch et al. 2015), in *Arabidopsis thaliana* (Su and Li 2008), *Oryza sativa* (Kim and An 2013), *Triticum aestivum* (Duan et al. 2011) and in moss *Physcomitrella patens* (Shi and Theg 2010).

Question 3
Whether HSP70B could be applied to evaluate the level of stress?

Our data illustrate variations in magnitude of HSP70B response between investigated species that could be used for such purpose.

Acknowledgements We thank Prof. Michael Schroda (Molecular Biotechnology & Systems Biology, TU Kaiserslautern, Kaiserslautern, Germany) for the antiserum against HSP70B and HSP90C. This work was supported by joint research projects between the BAS and RAS ('Biochemical markers of plant resistance to oxidative stress' and 'Molecular mechanisms of induced resistance to oxidative stress in plants'), the Bulgarian Ministry of Education and Science (projects K-1204, B-1520, BioCORE), Russian Foundation for Basic Research (project № 16-04-01626a), Program MCB RAS.

References

Ahn S-G, Thiele DJ (2003) Direct redox regulation of mammalian heat shock factor 1 is essential for Hsp gene activation and cellular protection from stress. Genes Dev 17:516–528

Ahuja I, de Vos RC, Bones AM, Hall RD (2010) Plant molecular stress responses face climate change. Trends Plant Sci 15:664–674

Almeselmani M, Deshmukh P, Sairam R, Kushwaha S, Singh T (2006) Protective role of antioxidant enzymes under high temperature stress. Plant Sci 171:382–388

Al-Whaibi MH (2011) Plant heat shock proteins: a mini review. J King Saud Univ Sci 23:139–150

Apel K, Hirt H (2004) Reactive oxygen species: metabolism, oxidativestress, and signal transduction. Annu Rev Plant Biol 55:373–399

Asker D, Beppu T, Ueda K (2007) Unique diversity of carotenoid-producing bacteria isolated from Misasa, a radioactive site in Japan. Appl Microbiol Biotechnol 77:383–392
Badahur A, Antra C, Rajesh K, Major S, Naik P (2011) Physiological and biochemical basis of drought tolerance in vegetables. Veg Sci 38:1–16
Badiani M, Schenone G, Paolacci AR, Fumagalli I (1993) Daily Fluctuations of antioxidants in bean (*Phaseolus vulgaris* L.) leaves as affected by the presence of ambient air pollutants. Plant Cell Physiol 34:271–279
Baniwal SK, Bharti K, Chan KY, Fauth M, Ganguli A, Kotak S, Mishra SK, Nover L, Port M, Scharf KD, Tripp J, Weber C, Zielinski D, von Koskull-Döring P (2004) Heat stress response in plants: a complex game with chaperones and more than twenty heat stress transcription factors. J Biosci 29:471–487
Banti V, Loreti E, Novi G, Santaniello A, Alpi A, Perata P (2008) Heat acclimation and cross-tolerance against anoxia in Arabidopsis. Plant Cell Environ 31:1029–1037
Banzet N, Richaud C, Deveaux Y, Kazmaier M, Gagnon J, Triantaphylides C (1998) Accumulation of small heat shock proteins, including mitochondrial HSP22, induced by oxidative stress and adaptive response in tomato cells. Plant J 13:519–527
Bao S, Wu Q, McLendon RE, Hao Y, Shi Q, Hjelmeland AB, Dewhirst MW, Bigner DD, Rich JN (2006) Glioma stem cells promote radioresistance by preferential activation of the DNA damage response. Nature 444(7120):756–760
Barua D, Heckathorn SA (2006) The interactive effects of light and temperature on heat shock protein accumulation in *Solidago altissima* (Asteracea) in the field and laboratory. Am J Bot 93:102–109
Bianco C, Defez R (2012) Soil bacteria support and protect plants against abiotic stresses. In: Venkateswarlu B, Shanker A (eds) Abiotic stress in plants – mechanisms and adaptations. Agricultural and biological sciences. Springer, Dordrecht, pp 143–171
Bierkens J, Van dePerre WV, Maes J (1998) Effect of different environmental variables on the synthesis of Hsp70 in *Raphidocelis subcapitata*. Comp Biochem Physiol A Mol Integr Physiol 120:29–34
Brennecke T, Gellner K, Bosch TC (1998) The lack of stress response in *Hydra oligactis* is due to reduce hsp70 mRNA stability. Eur J Biochem 255:703–709
Chalmers AJ (2007) Radioresistant glioma stem cells – therapeutic obstacle or promising target? DNA Repair 6:1391–1394
Chankova S, Yurina N (2012) Micro-algae as a model system for studying of genotype resistance to oxidative stress and adaptive response. In: Mothersill CE, Korogodina V, Seymour CB (eds) Radiobiology and environmental security. Springer, Dordrecht, pp 19–30
Chankova SG, Vinarova KM, Nikolov SH, Mehandjiev AD, Sergeeva SA, Ptitzina SN, Semov AB, Shevchenko VA (1990) Changes in *Chlorella vulgaris* B. populations after chronic influence with chemical and physical mutagenic factors. Biol Plant 32:35–41
Chankova SG, Kapchina VM, Stoyanova DP (2000) Some aspects of the plant radioresistance. Radiats Biol Radioecol 40:535–543
Chankova SG, Kapchina-Toteva VM, Frolova O, Stoilov LM, Blagoeva ED (2001) Characterization of new radioresistant strains of *Chlamydomonas reinhardtii*. C R Acad Bulg Sci 54:51–54
Chankova SG, Matos JA, Simões F, Bryant PE (2005) Adaptive response of a new radioresistant strain of *Chlamydomonas reinhardtii* and correlation with increased DNA double-strand break rejoining. Int J Radiat Biol 81:509–514
Chankova SG, Dimova E, Dimitrova M, Bryant PE (2007) Induction of DNA double-strand breaks by zeocin in *Chlamydomonas reinhardtii* and the role of increased DNA double-strand breaks rejoining in the formation of an adaptive response. Radiat Environ Biophys 46:409–416
Chankova SG, Yurina NP, Dimova EG, Ermohina OV, Oleskina YP, Dimitrova MT, Bryant PE (2009) Pretreatment with heat does not affect double strand DNA rejoining in *Chlamydomonas reinhardtii*. J Therm Biol 34:332–336
Chankova SG, Mitrovska Z, Miteva D, Oleskina YP, Yurina NP (2013) Heat shock protein HSP70B as a marker for genotype resistance to environmental stress in *Chlorella* species from contrasting habitats. GENE 516:184–189

Chankova SG, Dimova EG, Mitrovska Z, Miteva D, Mokerova DV, Yonova PA, Yurina NP (2014) Antioxidant and HSP70B responses in *Chlamydomonas reinhardtii* genotypes with different resistance to oxidative stress. Ecotoxicol Environ Saf 101:131–137

Chen S, Liu A, Zhang S, Li C, Chang R, Liu D, Ahammed GJ, Lin X (2013) Overexpression of mitochondrial uncoupling protein conferred resistance to heat stress and *Botrytis cinerea* infection in tomato. Plant Physiol Biochem 73:245–253

Clark M, Peck LS (2009) Triggers of the HSP70 stress response: environmental responses and laboratory manipulation in an Antarctic marine invertebrate (*Nacella concinna*). Cell Stress Chaperones 14:649–660

Costantini D, Monaghan P, Metcalfe NB (2013) Loss of integration is associated with reduced resistance to oxidative stress. J Exp Biol 216:2213–2220

Dat J, Vandenabeele S, Vranová E, Van Montagu M, Inzé D, Van Breusegem F (2000) Dual action of the active oxygen species during plant stress responses. Cell Mol Life Sci 57:779–795

Dimova E, Dimitrova M, Miteva D, Mitrovska Z, Chankova S (2008) Paraquat induced adaptive response in *Chlamydomonas reinhardtii* strains depends on the genotype. Compt Rend Acad Bulg Sci 61:911–918

Dimova E, Dimitrova M, Miteva D, Mitrovska Z, Yurina NP, Bryant PE, Chankova S (2009) Does single-dose cell resistance to the radio-mimetic zeocin correlate with a zeocin-induced adaptive response in *Chlamydomonas reinhardtii* strains? Radiat Environ Biophys 48:77–84

Driedonks N, Xu J, Peters JL, Park S, Rieu I (2015) Multi-level interactions between heat shock factors, heat shock proteins, and the redox system regulate acclimation to heat. Front Plant Sci 6:999

Drzymalla C, Schroda M, Beck CF (1996) Light-inducible gene HSP70B encodes a chloroplast-localized heat shock protein in *Chlamydomonas reinhardtii*. Plant Mol Biol 31:1185–1194

Duan Y-H, Guo J, Ding K, Wang SJ, Zhang H, Dai XW, Chen YY, Govers F, Huang LL, Kang ZS (2011) Characterization of a wheat HSP70 gene and its expression in response to stripe rust infection and abiotic stresses. Mol Biol Rep 38:301–307

Ekmekci Y, Terzioglu S (2005) Effects of oxidative stress induced by paraquat on wild and cultivated wheats. Pestic Biochem Physiol 83:69–81

Feder ME, Hofmann GE (1999) Heat shock proteins, molecular chaperones, and stress response: evolutionary and ecological physiology. Annu Rev Physiol 61:243–282

Galvez-Valdivieso G, Mullineaux P (2010) The role of reactive oxygen species in signaling from chloroplasts to the nucleus. Physiol Plant 138:430–439

Gill SS, Tuteja N (2010) Reactive oxygen species and antioxidant machinery in abiotic stress tolerance in crop plants. Plant Physiol Biochem 48:909–930

Gupta AS, Webb RP, Holaday AS, Allen RD (1993) Overexpression of superoxide dismutase protects plants from oxidative stress (induction of ascorbate peroxidase in superoxide dismutase-overexpressing plants). Plant Physiol 103:1067–1073

Heckathorne SA, Poeller GJ, Colemann JS, Hallberg RL (1996) Nitrogen availability alters patterns of accumulation of heat stress-induced proteins in plants. Oecologia (Berl) 105:413–418

Huang B, Xu C (2008) Identification and characterization of proteins associated with plant tolerance to heat stress. J Integr Plant Biol 50:1230–1237

Ireland HE, Harding SJ, Bonwick GA, Jones M, Smith CJ, Williams JH (2004) Evaluation of heat shock protein 70 as a biomarker of environmental stress in *Fucus serratus* and *Lemna minor*. Biomarkers 9:139–155

Jaspers P, Kangasjärvi J (2010) Reactive oxygen species in abiotic stress signaling. Physiol Plant 138:405–413

Kim S-R, An G (2013) Rice chloroplast-localized heat shock protein 70, OsHsp70CP1, is essential for chloroplast development under high-temperature conditions. J Plant Physiol 170:854–863

Knight CA (2010) Small heat shock protein responses differ between chaparral shrubs from contrasting microclimates. J Bot ID 171435, 7 p

Kobayashi Y, Harada N, Nishimura Y, Saito T, Nakamura M, Fujiwara T, Kuroiwa T, Misumi O (2014) Algae sense exact temperatures: small heat shock proteins are expressed at the survival

threshold temperature in *Cyanidioschyzon merolae* and *Chlamydomonas reinhardtii*. Genome Biol Evol 6:2731–2740
Kotak S, Larkindale J, Lee U, von Koskull-Döring P, Vierling E, Scharf KD (2007) Complexity of the heat stress response in plants. Curr Opin Plant Biol 10:310–316
Lindquist S, Graig EA (1988) The heat-shock proteins. Annu Rev Genet 22:651–677
Lipiec J, Doussan C, Nosalevicz A, Kondracka K (2013) Effect of drought and heat stresses on plant growth and yield: a review. Int Agrophys 27:463–477
Liu C, Willmund F, Golecki JR, Cacace S, Hess B, Markert C, Schroda M (2007) The chloroplast HSP70B-CDJ2-CGE1 chaperones catalyse assembly and disassembly of VIPP1 oligomers in *Chlamydomonas*. Plant J 50:265–277
Malanga G, Gonzalez PM, Estevez MS, Abele D, Puntarulo S (2008) Oxidative stress in Antarctic algae and mollusks. Ber Polarforsch Meeresforsch 571:208–215
Mendez-Alvarez S, Leisinger U, Eggen RI (1999) Adaptive responses in *Chlamydomonas reinhardtii*. Int Microbiol 2:15–22
Mittal D, Madhyastha D, Grover A (2012) Genome-wide transcriptional profiles during temperature and oxidative stress reveal coordinated expression patterns and overlapping regulons in rice. PLoS One 7:e40899
Mittler R (2002) Oxidative stress, antioxidants and stress tolerance. Trends Plant Sci 7:405–410
Mittler R, Vanderauwera S, Suzuki N, Miller G, Tognetti VB, Vandepoele K, Gollery M, Shulaev V, Van Breusegem F (2011) ROS signaling: the new wave? Trends Plant Sci 16:300–309
Monari M, Fosch J, Rosmini R, Marin MG, Serrazanetti GP (2011) Heat shock protein 70 response to physical and chemical stress in *Chamelea galina*. J Exp Mar Biol Ecol 397:71–78
Nedeva D, Pouneva I (2009) Changes in electrophoretic profiles of proteins and some antioxidant enzymes in Antarctic alga *Choricystis minor* and *Chlorella sp*. as affected by temperature and oxidative stress. Biotechnol Biotechnol Equip 23:233–236
Niu P, Liu L, Gong Z, Tan H, Wang F, Yuan J, Fen Y, Wei Q, Tanguag RM, Wu T (2006) Overexpressed heat shock protein 70 protects cells against DNA damage caused by ultraviolet C in a dose-dependent manner. Cell Stress Chaperones 11:162–169
Nordhues A, Miller SM, Muehlhaus T, Schroda M (2010) New insights into the roles of molecular chaperones in *Chlamydomonas* and *Volvox*. Int Rev Cell Mol Biol 285:75–113
Pelham HR (1982) A regulatory upstream promoter element in the Drosophila hsp70 heat-shock gene. Cell 30:517–528
Petrov V D, Van Breusegem F (2012) Hydrogen peroxide – a central hub for information flow in plant cells. AoB Plants pls014.
Piterková J, Luhová L, Mieslerová B, Lebeda A, Petřivalský M (2013) Nitric oxide and reactive oxygen species regulate the accumulation of heat shock proteins in tomato leaves in response to heat shock and pathogen infection. Plant Sci 207:57–65
Rainwater DT, Gossett DR, Millhollon EP, Hanna HY, Banks SW, Lucas MC (1996) The relationship between yield and the antioxidant defense system in tomatoes grown under heat stress. Free Radic Res 25:421–35
Ramalingam A, Kudapa H, Pazhamala LT, Weckwerth W, Varshney RK (2015) Proteomics and metabolomics: two emerging areas for legume improvement. Front Plant Sci 24:1116
Rochat T, Miyoshi A, Gratadoux JJ, Duwat P, Sourice S, Azevedo V, Langella P (2005) High-level resistance to oxidative stress in *Lactococcus lactis* conferred by *Bacillus subtilis* catalase KatE. Microbiology 151:3011–3018
Rodziewicz P, Swarcewicz B, Chmielewska K, Wojakowska A, Stobiecki M (2014) Influence of abiotic stress on plant proteome and metabolome changes. Acta Physiol Plant 36:1–19
Rui R, Nie Y, Tong H (1990) SOD activity as a parameter for screening stress-tolerant germplasm resources in sweet potato (*Ipomoea batatas* L.). Jiangsu J Agric Sci 6:52–56
Sairam R, Srivastava G, Saxena D (2000) Increased antioxidant activity under elevated temperatures: a mechanism of heat stress tolerance in wheat genotypes. Biol Plant 43:245–251
Sarkar NK, Kundnani P, Grover A (2013) Functional analysis of Hsp70 superfamily proteins of rice (*Oryza sativa*). Cell Stress Chaperones 18:427–437

Scharf K-D, Berberich T, Ebersberger I, Nover L (2012) The plant heat stress transcription factor (Hsf) family: structure, function and evolution. Biochim Biophys Acta 1819:104–119
Schroda M (2004) The *Chlamydomonas* genome reveals its secrets: chaperone genes and the potential roles of their gene products in the chloroplast. Photosynth Res 82:221–240
Schroda M, Mühlhaus T (2009) A 'foldosome' in the chloroplast? Plant Signal Behav 4:301–303
Schroda M, Vallon O, Wollman FA, Beck CF (1999) A chloroplast-targeted heat shock protein 70 (HSP70) contributes to the photoprotection and repair of photosystem II during and after photoinhibition. Plant Cell 11:1165–1178
Schroda M, Hemme D, Mühlhaus T (2015) The *Chlamydomonas* heat stress response. Plant J 82:466–480
Shatilina ZM, Riss HW, Protopopova MV, Trippe M, Meyer EI, Pavlichenko VV, Bedulina DS, Axenov-Gribanov DV, Timofeyev MA (2011) The role of the heat shock proteins (HSP70 and sHSP) in the thermotolerance of freshwater amphipods from contrasting habitats. J Therm Biol 36:142–149
Shen H, Leen YK (1997) Thermotolerance induced by heat shock in *Chlorella*. J Appl Phycol 9:471–475
Shi LX, Theg SM (2010) A stromal heat shock protein 70 system functions in protein import into chloroplasts in the moss Physcomitrella patens. Plant Cell 22:205–220
Song A, Zhu X, Chen F, Gao H, Jiang J, Chen S (2014) A chrysanthemum heat shock protein confers tolerance to abiotic stress. Int J Mol Sci 15:5063–5078
Sood A, Pabbi S, Uniyal P (2011) Effects of paraquat on lipid peroxidation and antioxidant enzymes in aquatic fern *Azolla microphylla*. Russ J Plant Physiol 58:667–673
Strasser H, Grabenbauer GG, Sprung CN, Sauer R, Distel LVR (2007) DNA double-strand break induction and repair in irradiated lymphoblastoid, fibroblast cell lines and white blood cells from ATM, NBS and radiosensitive patients. Strahlenther Onkol 183:447–453
Su P-H, Li H-M (2008) Arabidopsis stromal 70-kD heat shock proteins are essential for plant development and important for thermotolerance of germinating seeds. Plant Physiol 146:1231–1241
Suzuki N, Miller G, Sejima H, Harper J, Mittler R (2013) Enhanced seed production under prolonged heat stress conditions in *Arabidopsis thaliana* plants deficient in cytosolic ascorbate peroxidase 2. J Exp Bot 64:253–263
Swindell WR, Huebner M, Weber AP (2007) Transcriptional profiling of Arabidopsis heat shock proteins and transcription factors reveals extensive overlap between heat and non-heat stress response pathways. BMC Genomics 22(8):125
Szigeti Z, Lehoczki E (2003) A review of physiological and biochemical aspects of resistance to atrazine and paraquat in Hungarian weeds. Pest Manag Sci 59:451–458
Tanaka Y, Nishiyama Y, Murata N (2000) Acclimation of the photosynthetic machinery to high temperature in *Chlamydomonas reinhardtii* requires synthesis *de novo* of proteins encoded by the nuclear and chloroplast genomes. Plant Physiol 124:441–449
Tanaka S, Ikeda K, Miyasaka H, Shioi Y, Suzuki Y, Tamoi M, Takeda T, Shigeoka S, Harada K, Hirata K (2011) Comparison of three *Chlamydomonas* strains, which show distinctive oxidative stress tolerance. J Biosci Bioeng 112:462–468
Timperio AM, Egidi MG, Zolla L (2008) Proteomics applied on plant abiotic stresses: role of heat shock proteins (HSP). J Proteomics 7:391–411
Tomanek L (2002) The heat shock response: its variation, regulation and ecological importance in intertidal gastropods (genus Tegula). Integr Comp Biol 42:797–807
Torres MA, Dangl JL (2005) Functions of the respiratory burst oxidase in biotic interactions, abiotic stress and development. Curr Opin Plant Biol 8:397–403
Trösch R, Muehlhaus T, Schroda M, Willmund F (2015) ATP-dependent molecular chaperones in plastids – More complex than expected. Biochem et Biophys Acta 1847:872–888
Tukaj S, Tukaj Z (2010) Distinct chemical contaminants induce the synthesis of Hsp70 proteins in green microalgae *Desmodemus subspicatus*: heat treatment increases cadmium resistance. J Therm Biol 35:239–244

Vartak V, Bhargava S (1999) Photosynthetic performance and antioxidant metabolism in a paraquat-resistant mutant of *Chlamydomonas reinhardtii* L. Pestic Biochem Physiol 64:9–15

Vermeulen SJ, Van de Zande L, Bijlsma R (2005) Resistance to oxidative stress induced by paraquat correlates well with both decrease and increased lifespan in *Drosophila melanogaster*. Biogerontology 6:387–395

Veyel D, Erban A, Fehrle I, Kopka J, Schroda M (2014) Rationales and approaches for studying metabolism in eukaryotic microalgae. Metabolites 4:184–217

Volkov RA, Panchuk II, Mullineaux PM, Schöffl F (2006) Heat stress-induced H_2O_2 is required for effective expression of heat shock genes in *Arabidopsis*. Plant Mol Biol 61:733–746

Vranová E, Inzé D, Van Breusegem F (2002) Signal transduction during oxidative stress. J Exp Bot 53:1227–1236

Wahid A, Gelani S, Ashraf M, Foolad M (2007) Heat tolerance in plant: an overview. Environ Exp Bot 61:199–223

Wang W, Vinocur B, Shoseyov O, Altman A (2004) Role of plant heat-shock proteins and molecular chaperones in the abiotic stress response. Trends Plant Sci 9:244–252

Wang K, Zhang X, Goatley M, Ervin E (2014) Heat shock proteins in relation to heat stress tolerance of creeping bentgrass at different N levels. PLoS One 9:e102914

Wegener CB, Jansen G (2013) Antioxidants in different potato genotypes: effect of drought and wounding stress. Agriculture 3:131–146

White CN, Hightower LE, Schultz RJ (1994) Variations in heat shock proteins among species of desert fishes (Poecilidae, Poeciliopsis). Mol Biol Evol 11:106–119

Wiencke T (1991) Photosynthesis, dark respiration and light independent carbon fixation of endemic Antarctic microalgae. Polar Biol 11:329–337

Willmund F, Dorn KV, Schulz-Raffelt M, Schroda M (2008) The chloroplast DnaJ homolog CDJ1 of *Chlamydomonas reinhardtii* is part of a multichaperone complex containing HSP70B, CGE1, and HSP90C. Plant Physiol 148:2070–2082

Witzel K, Weidner A, Surabhi GK, Börner A, Mock HP (2009) Salt stress-induced alterations in the root proteome of barley genotypes with contrasting response towards salinity. J Exp Bot 60:3545–3557

Wu Q, Lin J, Liu J-Z, Wang X, Lim W, Oh M, Park J, Rajashekar CB, Whitham SA, Cheng NH, Hirschi KD, Park S (2012) Ectopic expression of *Arabidopsis* glutaredoxin AtGRXS17 enhances thermotolerance in tomato. Plant Biotechnol J 10:945–955

Xu C, Huang B (2010) Differential proteomic responses to water stress induced by PEG in two creeping bentgrass cultivars differing in stress tolerance. J Plant Physiol 67:1477–1485

Xu Y, Zhan C, Huang B (2011) Heat shock proteins on association with heat tolerance in grasses. Int J Proteomics 2011:11

Yu A, Li P, Tang T, Wang J, Chen Y, Liu L (2015) Roles of Hsp70s in stress responses of microorganisms, plants and animals. BioMed Res Int 2015, ID 510319

Zaka R, Vandecasteele CM, Misset MT (2002) Effects of low chronic doses of ionizing radiation on antioxidant enzymes and G6PDH activities in *Stipa capillata* (Poaceae). J Exp Bot 53:1979–1987

Zargar S, Krishnamurthi K, Saravana Devi S, Ghosh TK, Chakrabarti T (2006) Temperature-induced stress on growth and expression of HSP in freshwater alga *Scenedesmus quadricauda*. Biomed Environ Sci 19:414–421

Chapter 10
The Involvement of HSP70 and HSP90 in *Tomato Yellow Leaf Curl Virus* Infection in Tomato Plants and Insect Vectors

Rena Gorovits, Yule Liu, and Henryk Czosnek

Abstract Cellular chaperones HSP70 and HSP90 are involved in *Tomato yellow leaf curl virus* (TYLCV) infection. TYLCV is a begomovirus transmitted by the whitefly *Bemisia tabaci* to tomato and other crops. In infected tomato and *B. tabaci* vector, chaperones are redistributed, from soluble to aggregated state. Together with chaperones and viral proteins, ubiquitin, 26S proteasome subunits and autophagy protein ATG8, all were found in large protein aggregates. The appearance of these aggregates containing protein quality control elements and infectious virions can be considered as markers of a successful virus invasion. Capturing of HSP70/HSP90 in aggregates results in a decrease of the free chaperones pool, which triggers the transcription of *HSP* encoding genes under the control of heat stress transcription factors. Indeed, TYLCV infection downregulates the heat stress response of plants grown at high temperatures, and alleviates cell death caused by the other stresses. Stress response mitigation is used by TYLCV for successful multiplication. Even though HSP70 and HSP90 are similarly recruited in TYLCV aggregates, their roles in viral multiplication are different. HSP70, but not HSP90, is important for the viral coat protein shuttling from cytoplasm into nuclei. HSP70 impairment leads to decreased viral amounts, while HSP90 inhibition causes an inactivation of cellular protein degradation and consequently promotes the accumulation of viral proteins.

Keywords Plant chaperones • Protein aggregation • Stress response

R. Gorovits (✉) • H. Czosnek
Institute of Plant Sciences and Genetics in Agriculture, Robert H. Smith Faculty of Agriculture, Food 6 and Environment, Hebrew University of Jerusalem, Rehovot, Israel
e-mail: rena.gorovits@mail.huji.ac.il

Y. Liu
School of Life Sciences, Tsinghua 7 University, Beijing, China

A.A.A. Asea et al. (eds.), *Heat Shock Proteins and Plants*,
Heat Shock Proteins 10, DOI 10.1007/978-3-319-46340-7_10

Abbreviations

Co-IP	co-immunoprecipitation
HSF	heat stress transcription factor
HSP	heat stress protein
PQC	protein quality control
TYLCV	*Tomato yellow leaf curl virus*
VF	viral factory

10.1 Introduction

Viruses are intracellular obligate parasites that recruit the host biosynthetic machinery for their replication and multiplication. Given the functional complexity of many viral proteins, it is not surprising that they are dependent on cellular chaperones for their function. Viruses co-opt chaperones from the host cell to assist the synthesis, correct folding and shuttling of abundant viral proteins, to regulate viral multiplication via activation of replication proteins. The most frequently subverted host chaperones are heat shock protein 70 (HSP70) and HSP90. These chaperones together with co-chaperones affect many aspects of virus multiplication, through regulating translation, replication, virion assembly and counter-defense against cell apoptosis or native immune responses (Nagy et al. 2011 and reference therein). The recruited HSPs play major roles in assembly of the virus-induced protein aggregates, named virus factories (VFs) and in their activity. Detailed characteristics of VFs in the infected mammalian cells have been described elsewhere (Wileman 2006, 2007; Netherton and Wileman 2011; Netherton and Wileman 2013; Burch and Weller 2005; Livingston et al. 2009; Weller 2010).

In plant and insect cells, the knowledge concerning the involvement of cellular protein quality control (PQC) elements, including chaperones, in the virus life cycle is restricted. Plant viruses can recruit HSP70 to assist the synthesis, folding and localization of viral proteins, to regulate virus replication and to interfere with the antiviral host response (Nagy et al. 2011). HSP90 interacts with *Bamboo mosaic virus* replicases in promoting viral accumulation (Huang et al. 2012). Coordinate function of HSP70 together with HSP90 in multicomponent complex assembly was revealed for the positive-strand RNA plant virus *Red clover necrotic mosaic virus* (RCNMV), which replicase complex can be detected readily as a 480-kDa functional protein complex. HSP70 and HSP90 are required for RCNMV RNA replication, and they interact with p27, a viral-encoded component of the 480-kDa replicase complex, on the endoplasmic reticulum membrane (Mine et al. 2012). HSP90 and HSP70 function together in the process of the incorporation of small RNAs into Argonaute proteins, which play central roles in post-transcriptional gene silencing in *Drosophila* (Miyoshi et al. 2010; Iwasaki et al. 2010). The association of HSP70/HSC70 and HSP90 is known to involve interaction with HSP90 co-chaperone, SGT1 (for Suppressor of G2 allele of skp1) (Noel et al. 2007). In plants and animals, SGT1 is essential to the function of many NLR proteins, which contribute to host immunity by activating plant defense responses (Liu et al. 2004). SGT1was shown to contribute to the systemic accumulation of *Plantago asiatic mosaic virus* (PlAMV) and *Potato virus X* (PVX) infection in susceptible hosts (Komatsu et al. 2010; Ye et al. 2012). While SGT1 enhanced PVX multiplication, *SGT1* silencing led to an increased accumulation of PlAMV in *Nicotiana benthamiana*. Such opposite effects of plant SGT1 on virus multiplication points to the intricate role of SGT1-chaperone complexes in virus biotic stress. Indeed, SGT1 has multiple functions in association with the Skp1-Cullin/F-box type E3 ubiquitin ligase complexes that target regulatory proteins for degradation by the 26S proteasome (Muskett and Parker 2003). Furthermore, the SGT1 complex also interacts with two COP9 signalosome components, indicating a strong involvement of SGT1 in PQC (Liu et al. 2002; Azevedo et al. 2002).

The aim of this review is to present the current knowledge of the roles of HSP70 and HSP90 in *Tomato yellow leaf curl virus* (TYLCV) infection in plant and insect cells. TYLCV is a representative of geminiviruses (genus *Begomovirus*, family *Geminiviridae*) transmitted exclusively by the whitefly *Bemisia tabaci* in a persistent-circulative manner (Czosnek 2007). In the insect host, TYLCV seems to replicate, and levels of virus steadily increase, especially upon stress (Pakkianathan et al. 2015; Ghanim and Czosnek 2016). TYLCV possesses a single-stranded circular DNA genome of 2787 nucleotides encapsidated in a geminate particle of approximately 20 × 30 nm (Díaz-Pendón et al. 2010). The viral sense genome encodes two genes: V1 or coat protein (CP) and V2. The complementary sense genomes encodes four genes C1 to C4: C1 is a replicase-associated protein (Rep), C2 a transcription activator (TrAP), C3 a replication enhancer (REn) and C4. The viral gene products interact with plant and insect host proteins to perform their function. In insect vector, once acquired, virions pass along the food canal in the stylet and reach the guts. Virions can cross to the haemolymph via the filter chamber and the midgut to reach the salivary gland, from which it is transmitted to plants with the saliva during feeding (Czosnek et al. 2002). It is likely that along the viral path different cellular chaperones are present the translocation of the virions in the insect. In the haemolymph, TYLCV virions interact with a 63 KDa GroEL protein produced by the primary endosymbiotic bacteria of *B. tabaci*, which protects the virions from proteolysis by the insect's immune system (Morin et al. 1999; Gottlieb et al. 2010).

10.2 TYLCV Infection Leads to Reorganization of PQC Elements Including HSP70 and HSP90 in Host Cells

10.2.1 TYLCV-Induced Aggregation in Plant and Insect Host Organisms

CP is a major structural TYLCV protein, essential for cell-to-cell movement, systemic infection and transmission by the whitefly vector. TYLCV CP patterns in plant and insect cells, obtained by separating proteins in sucrose gradient ultracentrifugation,

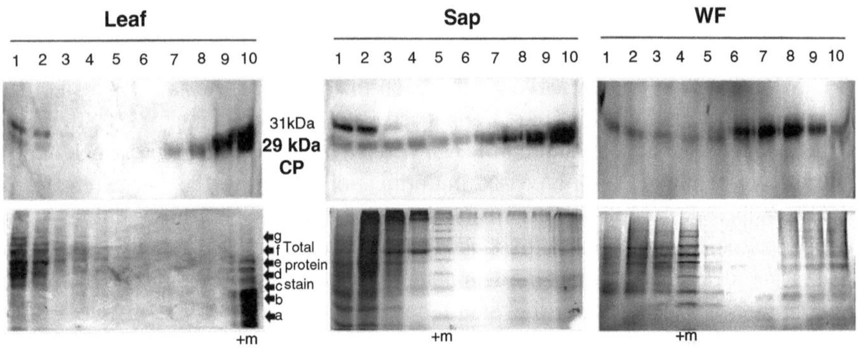

Fig. 10.1 Distribution of viral CP following sedimentation on 10–50 % sucrose gradients of native proteins extracted from tomato leaf/stem tissues and *B. tabaci* cells. Leaf and stem homogenates were prepared from tomato at 28 (dpi) days after the onset of TYLCV infection (B), whitefly extracts at the 7th day of TYLCV presence. Gradients were divided into 10 fractions, 1 (*top*) to 10 (*bottom*) and concentrated about 20 times by TCA precipitation. Aliquots were subjected to SDS-PAGE. The gels were stained with Coomassie blue (total protein) and Western blotted using anti-CP antibodies. Anti-CP recognized an additional minor 31 kDa polypeptide. Molecular weight markers (noted as + m) a–g (in kDa): (a) 17, (b) 26, (c) 34, (d) 43, (e) 55, (f) 72, and (g) 95

showed CP in soluble states (fractions 1–3), in complexes (fractions from 3 to 5; see sap gradient), but mostly as small, mid-sized and large aggregates (fractions from 6 to 10). In whitefly protein extracts, CP was found in all gradient fractions, in the soluble state (1–3), in large aggregates (10), but mostly in small/midsized aggregates (fractions 6–9) (Fig. 10.1). In addition to the structural CP, TYLCV V2 was the only non-structural protein detectable in the gradient fractions. V2 patterns resembled those of CP (Moshe et al. 2015a; Gorovits et al. 2016).

The immuno-histological detection of TYLCV proteins in tomatoes and *B. tabaci* cells confirmed their association with protein aggregates (Fig. 10.2). At the early stages of TYLCV infection in tomatoes, all the six viral proteins were observed in protein intermediate-sized aggregates (Fig. 10.2a); Gorovits et al. 2016). Along the progress of viral infection, CP and V2 appeared in aggregates of increasing size localized first in cytoplasm then in the nucleus, while the C1, C2, C3 and C4 aggregates disappeared. The appearance of CP and V2 aggregates in whitefly was confirmed by *in vivo* immunostaining of those proteins in dissected out *B. tabaci* midguts (Fig. 10.2b). Most of CP and V2- aggregates were observed in the cytoplasm, but nuclear CP and V2 aggregations were abundant as well.

So, TYLCV proteins were detected not only in soluble, but also in aggregated state in both viral hosts. The analysis of the tomato and whitefly capacity to degrade the six proteins encoded by the TYLCV genome demonstrated the highest proteolytic activity in the fractions containing soluble proteins, less – in large protein aggregates; a significant decrease of TYLCV proteolysis was detected in the intermediate-sized aggregates. We suggested that TYLCV could confront host degradation by sheltering in small/midsized aggregates, where viral proteins are less exposed to proteolysis, and, indeed, TYLCV protein aggregation has been revealed in both host organisms (Gorovits et al. 2016).

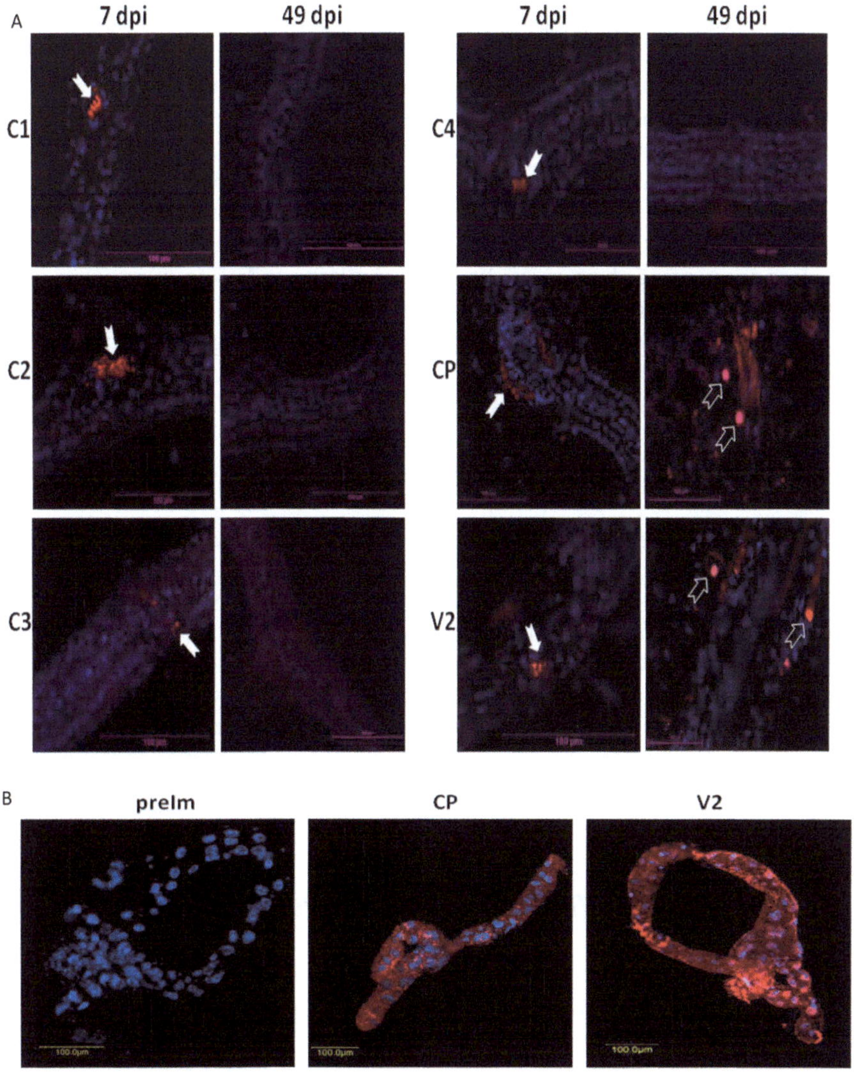

Fig. 10.2 *In vivo* immunodetection of TYLCV proteins in the infected host tissues. (**a**) Visualization under the fluorescence microscope of the six TYLCV proteins in cross sections of midribs of infected tomato leaves using primary antibodies and a Cy3-labeled secondary antibody (appear as *red*). Nuclei DAPI stained appear as *blue*, viral proteins localized in nuclei appear as *pink*. Full *white* arrows point to small and midsized aggregates; empty *white* arrows points to large aggregates. Bar: 100 μm. (**b**) Immunostaining of TYLCV CP and V2 in viruliferous *B. tabaci* midguts. Preimmune serum (preIm) was used in control staining. Viral proteins were labeled by Cy3 and appear as *red*; nuclei DAPI stained appear as *blue*. Viral proteins localized in nuclei appear as *pink*. Bar: 100 μm

The appearance of large, particularly, nuclear CP/V2-aggregates, was a feature of a successful TYLCV invasion in plants, while the prolonged maintenance of mid-sized aggregates was detected in TYLCV resistant tomatoes and was considered as part of the plant defense responses to virus stress. At the beginning of infection in resistant plants, viral proteins are captured in intermediate-sized aggregates, slowing down the formation of new virions until the plant has grown enough to sustain the deleterious effects of the virus. Moreover, TYLCV genomic dsDNA replicative form together with CP-DNA complexes and infectious particles were found in large nuclear aggregates, which pointed to their similarity to VFs in mammalian cells (Moshe and Gorovits 2012; Gorovits et al. 2013a; Moshe et al. 2015a).

Plant virus aggregation is much less known than aggregation of mammalian viruses, even though cylindrical inclusions induced by the potyvirus *Tobacco etch virus* (Edwardson 1966), and crystalline arrays caused by TMV (Willison 1976), were described decades ago. Another typical characteristic of plant virus invasion known for a long time is the alteration of the morphology of host organelles and membranes, caused for example by *Cowpea chlorotic mottle virus* (Kim 1977) and *Cymbidium ringspot virus* (Russo et al. 1983). In some instances, tubules containing virus-like particles were identified in or near the cell walls of infected cells (by *Cowpea mosaic virus*, for example (van der Scheer and Groenewegen 1971). Virus-induced membrane structures, mostly shown for plant RNA viruses, house the RNA replication complex and were compared with VFs in infected mammalian cells (Vogel et al. 2007; Netherton and Wileman 2011), even though plant cells lack MTOC, the main VFs' component. Studies in plant virology emphasize the absolute requirement for the formation of virus inclusions, VFs or viroplasms for the successful virus multiplication (reviewed in Laliberte and Sanfacon 2010).

10.2.2 Interaction Between Cellular HSPs and TYLCV Proteins

10.2.2.1 Co-localization of Host HSP70 and TYLCV CP

TYLCV does not induce the expression of HSPs in plant tissues, while transcriptome analysis using a whitefly cDNA microarray showed that *hsp70* was slightly upregulated during the ingestion of TYLCV (Götz et al. 2012). Oppositely, during prolonged TYLCV infection, the accumulation of viral DNA and CP was accompanied by a progressive decrease in the amounts of several plant chaperones such as HSP60, members of HSP70 family and HSP90 (Gorovits et al. 2007; Gorovits and Czosnek 2007; Moshe et al. 2012). The spatial distribution of HSP70 and viral CP and their possible co-localization in tomatoes and whiteflies was investigated by immunological detection of both proteins with a fluorescent microscopy. In plants, HSP70 and CP co-localized in aggregates of increasing size, first in the cytoplasm then in the nucleus of infected leaves and stems (Gorovits et al. 2013b) (Fig. 10.3a). At the late infection (49 dpi), CP and HSP70 co-localized in aggregates of

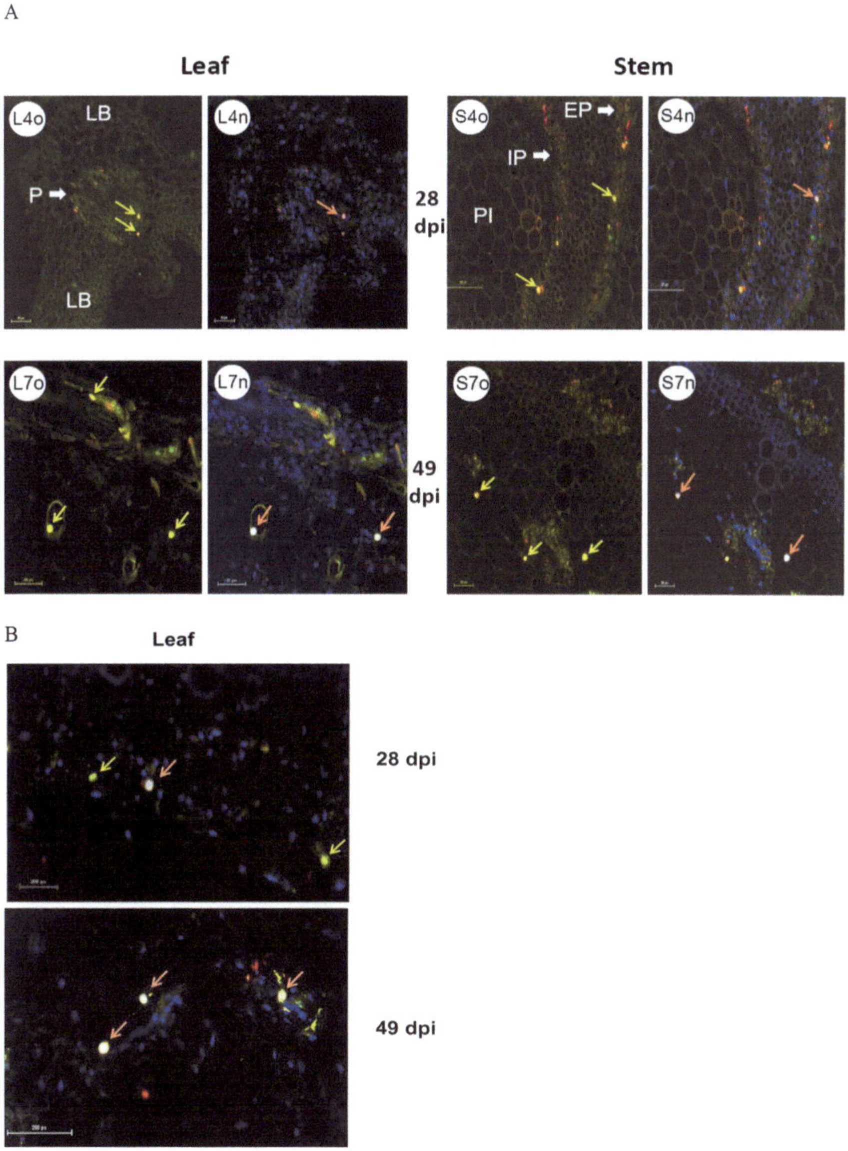

Fig. 10.3 Co-localization of CP and HSP70 (HSP90) in cytoplasm and nucleus of infected tomato cells at 28 and 49 dpi. (**a**): Leaf: cross-section through the midrib (P, phloem; LB, leaf blade). Stem: cross section between true leaves 2 and 3 (IP and EP, internal and external phloem; PI, pith). Fluorescent microscopy using primary anti-CP antisera and Cy3-labeled secondary antibody, primary anti-HSP70 antisera and Cy2-labeled secondary antibody; nuclei were DAPI stained. L4 and L4n: leaves 4 weeks after the onset of infection, without and with nuclei stain, respectively; L7 and L7n: same but 7 weeks after the onset of infection. S4 and S4n: stems 4 weeks after the onset of infection, without and with nuclei stain, respectively; S7 and S7n: same but 7 weeks after the onset of infection. Viral CP appears as *red*, cellular HSP70 as *green*, nuclei as *blue*; CP in nuclei appears as *violet*, HSP70 in nuclei as *light blue*; CP co-localizing with HSP70 appears as *yellow*, CP co-localizing with HSP70 in nuclei – as *pink*. Bar: 50 µm, except for S4 and S4n, 100 µm. (**b**): In leaf cross-section TYLCV CP appears as *red*, cellular HSP90 as *green*, nuclei as *blue*; CP in nuclei appears as *violet*, HSP70 in nuclei as *light blue*; CP co-localizing with HSP90 appears as *yellow*, CP co-localizing with HSP90 in nuclei – as *pink*

increasing sizes in cytoplasm and nucleus. Most of the large aggregates/inclusion bodies contained both proteins. HSP70 did not appear in large aggregates/inclusion bodies in non-infected tomatoes (Gorovits et al. 2013b). HSP70-CP co-localization was visualized in aggregated states in whitefly midgut (Ghanim and Czosnek 2016). The immuno-histological images demonstrated the double staining of CP and HSP70 in infected tomato leaves and in midguts of viruliferous whiteflies, moreover, co-localization happened in various aggregates.

The potential co-localization of viral CP with HSP90 was investigated in TYLCV infected tomato leaf tissue. Figure 10.3b demonstrates the fluorescent immunodetections of HSP90 (green color) and CP (red color), the two proteins co-localized in cytoplasmic (yellow color) and in nuclear (pink or light pink) aggregates, while nuclei were stained with DAPI. In non-infected tomatoes, HSP90 was not detected in large aggregates (Moshe et al. 2015b). Hence, HSP70 and HSP90 co-localized with viral CP in cytoplasmic aggregates and, later, in large nuclear aggregates, operating as VFs.

10.2.2.2 Complexes Between HSPs and TYLCV Proteins

The possible interaction between CP and HSP70 was tested by co-immunoprecipitation (Co-IP). CP-HSP70 complexes were pulled-down using anti-CP and anti-HSP70 antibodies and identified using anti-HSP70 and anti-CP, respectively (Fig. 10.4). That *in vitro* assay was used to detect potential interaction of viral CP with HSP70 in plant and insect protein extracts. The results showed that TYLCV CP formed complexes with cellular HSP70 in the tissues analyzed. Co-immunoprecipitation assay revealed interaction between plant HSP90 and CP in nuclear, but not cytoplasmic protein fractions (not shown), the complexes between *B. tabaci* HSP90 and viral proteins have not been investigated yet.

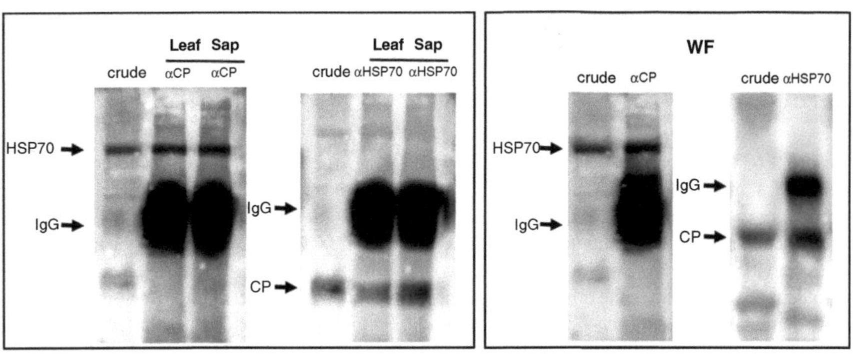

Fig. 10.4 Co-immunoprecipitations of cellular HSP70 with viral CP and *vice versa* in tomato tissues and *B. tabaci*. Co-immunoprecipitation of HSP70 with anti-TYLCV CP specific antibody and TYLCV CP with anti-HSP70 specific antibody in leaf, stem and whitefly (WF) protein extracts. The direct immunodetection was designed as "crude"

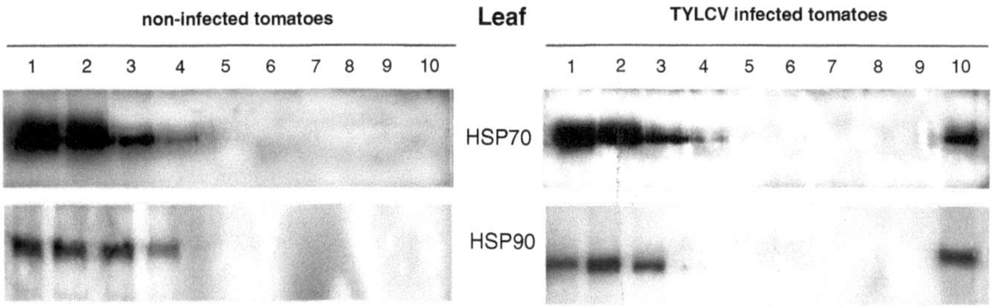

Fig. 10.5 Distribution of plant HSP70 and HSP90 following sedimentation on 10–50 % sucrose gradients of native proteins extracted from TYLCV infected tomato leaves

10.2.2.3 TYLCV Infection Causes the Re-localization of PQC Elements in Host Cells

HSP70 and HSP90 were not found in aggregates in uninfected tomatoes. Indeed, western blot analysis of proteins, separated by centrifugation in sucrose gradient, demonstrated the presence of HSP70 and HSP90 in fraction (N10) containing large aggregates/inclusion bodies only in TYLCV infected plants (Fig. 10.5). The "conversion" of HSP70 and HSP90 from soluble into insoluble states was found not only in leaf tissue, but also in sap of infected tomatoes and in viruliferous whiteflies (not shown).

It is important to note, that in the gradients containing large protein aggregates (9–10th) were detected such PQC markers as ubiquitin, regulatory subunit of 26S proteasome (Gorovits et al. 2014) and ATG8 (Gorovits et al. 2016), one of the key autophagy proteins. In fractions 9–10 of gradients of whitefly separated proteins, 26S proteasome was present, while ATG8 was not immunodetected. Proteolytic activity against all the six TYLCV proteins was present in large aggregates. Moreover, *in vivo* treatments of detached tomato leaves by MG132 (26S proteasome inhibitor) and by wortmannin (autophagy inhibitor)/rapamycin (autophagy inducer) caused changes in CP and V2 aggregation patterns, confirming the involvement of these degradation mechanisms in the TYLCV life cycle during plant infection. All the six TYLCV proteins restored their abundances at various extents in the presence of MG132 or wortmannin, confirming the involvement of 26S proteasome and autophagy in begomovirus proteolysis (Gorovits et al. 2016). The presence of PQC elements, including key chaperones, together with the proteolytic activity appearance in large aggregates proved the suggestion that these aggregates could be TYLCV factories, comparable with those of animal DNA viruses (Netherton et al. 2007; Netherton and Wileman 2011, 2013).

To confirm the association of TYLCV-induced aggregation with PQC, we addressed the question of whether V2 aggregates could be degraded by the 26S proteasome. TYLCV V2 fused to the fluorescent GFP protein was examined in tomato and in *N. benthamiana* epidermal cells as cytoplasmic aggregates of different sizes throughout the cell (Fig. 10.6). The V2:GFP fluorescent aggregates resembled the V2 aggregates seen in tomato inoculated with viruliferous whiteflies

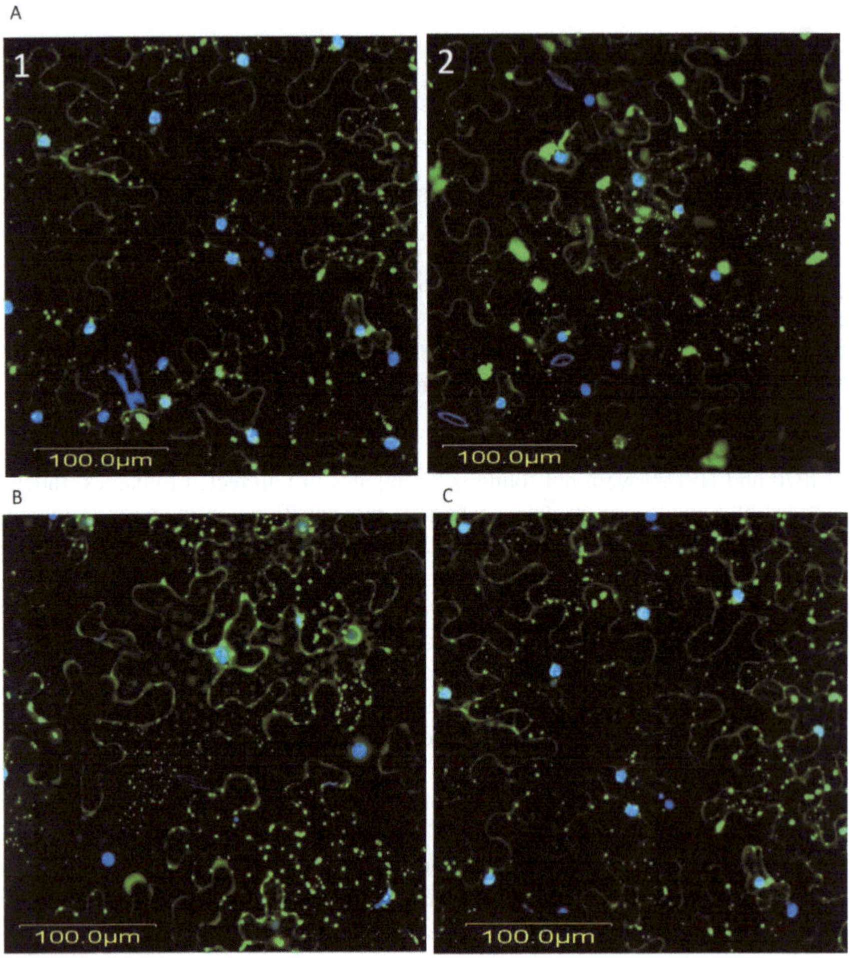

Fig. 10.6 Inhibition of the 26S proteasome results in the formation of large aggregates; disruption of microtubules or of actin filaments inhibits the delivery of V2 to the proteasome. (**a**): *N. benthamiana* epidermal cells expressing V2:GFP in untreated (1) and MG132 treated (2) leaf; DAPI stained nuclei appear as blue; V2:GFP appears as green. (**b**) and (**c**): *N. benthamiana* epidermal cells expressing V2:GFP in MG132 (proteasome inhibitor) and latB (latrunculin B is a toxin that inhibits the polymerization of actin microfilaments) (B), and in MG132 and oryzalin (oryzalin is a herbicide known to depolymerize microtubules) (C) treated leaves. DAPI stained nuclei appear as blue; V2:GFP appears as green. Bar: 100 µm

(Moshe et al. 2015a), suggesting that V2:GFP expression in *N. benthamiana* is suitable for the analysis of V2 aggregation. V2 aggregates were associated with the cytoskeleton; furthermore, intact actin filaments and microtubules are needed to deliver V2 to the 26S proteasome for degradation (Fig. 10.6). Hence, the cytoskeleton, which was shown to be important PQC element in aggresome and VFs development (Wileman 2006, 2007), was required for the formation of TYLCV aggregates and their subsequent degradation via 26S proteasome.

Based on the results obtained in both TYLCV-infected plants and viruliferous whiteflies, we suggest that in TYLCV aggregation and consequently, in viral mobilization, movement, accumulation and degradation the key chaperones (HSP70 and HSP90) play a major role together with the other PQC components, such as ubiquitin, 26S proteasome and autophagy proteins, and cytoskeleton elements. TYLCV-induced reorganization of host PQC is a complicated multilevel process, which has just been started to be investigated.

10.3 Downregulation of HSP70 and HSP90 Expression/Activity Differently Affect TYLCV Accumulation in Tomatoes

Relying on the similar virus-dependent recruitment of both chaperones in large aggregates/VFs, promoting viral successful invasion, we suggested a role of HSP70 and HSP90 in TYLCV multiplication. Tomato HSP70 expression was inactivated by quercetin, a bioflavonoid known to inhibit *HSP70* transcription in plant cells (Wang et al. 2009). Decreased HSP70 amounts were accompanied by reduced amounts of nuclear CP aggregates and re-localization of the GFP-CP fusion from the nucleus to the cytoplasm (Fig. 10.7). HSP70 inactivation resulted in a decrease of TYLCV DNA levels, demonstrating the role of HSP70 in TYLCV multiplication *in planta* (Gorovits et al. 2013b). Reduced nuclear CP transportation could be one of the consequences in HSP70 pattern shifts.

In contrast to HSP70, inhibition of HSP90 activity by geldanamycin (GDA) did not affect the subcellular localization of CP in tomato and tobacco cells; hence, HSP90 is not required for nuclear transportation of TYLCV CP. Silencing of *HSP90* and *SGT1*, encoding the integrative protein SGT1 of HSP90 complexes, positively affected TYLCV accumulation during prolonged infection. HSP90 inactivation also led to the impaired degradation of V2 through the 26S proteasome machinery (Moshe et al. 2015a). Altogether, inhibition of HSP70 and HSP90 had opposite effects on TYLCV. While HSP70 impairment reduced TYLCV propagation, GDA-treated and *HSP90* silenced tomatoes showed the inactivation of the ubiquitin–proteasome system, followed by increased viral CP/DNA levels.

Interesting, membrane feeding of whiteflies with anti-HSP70 antibodies and TYLCV virions induced an increase in TYLCV transmission, suggesting that HSP70 inhibits virus transmission (Götz et al. 2012), a role that might be related to protection against begomoviruses deleterious effects in the whitefly (Rubinstein and Czosnek 1997). However, binding of TYLCV CP with another chaperone, a GroEL homologue produced by *B. tabaci* endosymbionts, was shown to protect the virus from degradation in the haemolymph (Morin et al. 1999). In contrast, other GroEL proteins produced by secondary endosymbionts in both B and Q biotypes did not

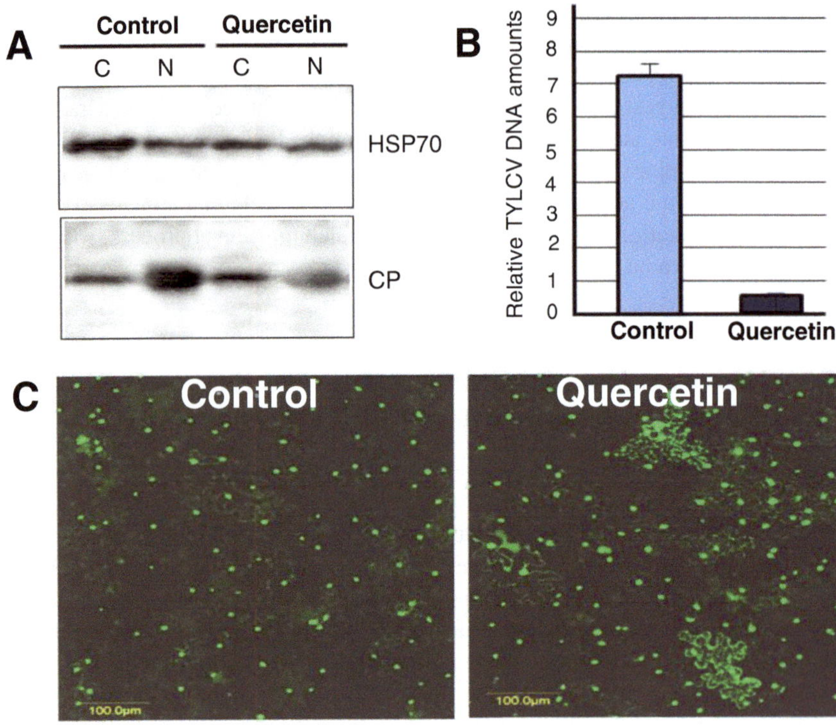

Fig. 10.7 Changes affecting TYLCV and HSP70 in plants treated with quercetin. (**a**) Western blot analysis of leaf cytoplasmic and nuclear proteins from 28 dpi tomatoes before and after quercetin treatment (400 μM for 4 days). (**b**) Tomato leaflets at 28 dpi were incubated for 4 days with 400 μM quercetin, TYLCV DNA amounts were estimated by qPCR analysis. (**c**) *N. benthamiana* epidermal cells transiently expressing GFP-CP following infiltration with quercetin (800 μM) or DMSO (control). Bar: 100 μm

interact with the TYLCV CP, indicating that the secondary endosymbionts play a minor role, if at all, in TYLCV transmission (Gottlieb et al. 2010). It will be interesting to find how HSP70 or HSP90 inactivation in the infected plants interferes with *B. tabaci* acquisition of virions.

10.4 TYLCV Infection Interferes with Plant Stress Response Through Modifications in the Availability of HSPs

The combination of extreme temperature and viral infection occurs quite often in the field in many tropical regions, impairing crop productivity. It is well known that increasing temperatures facilitate pathogen spread in major food crops (Scharf et al. 2012). Temperature was identified as the dominant abiotic factor directly affecting

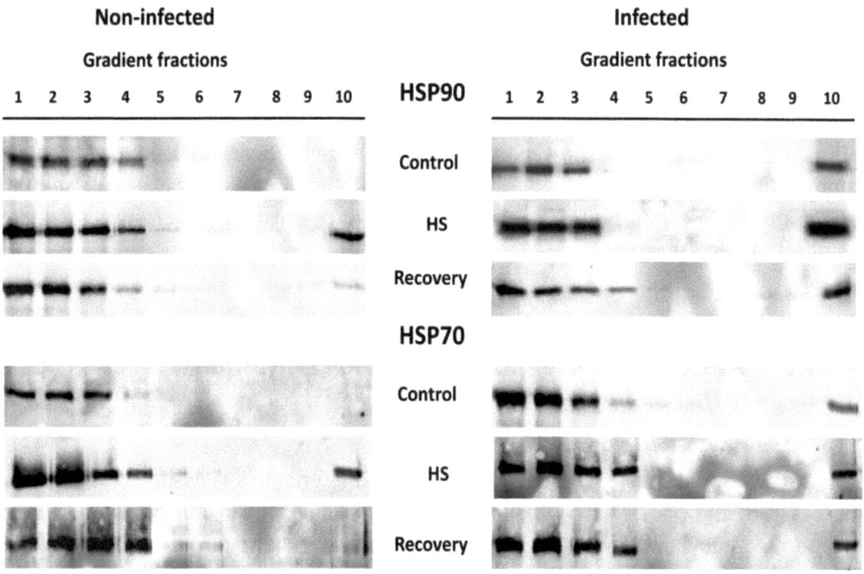

Fig. 10.8 Influence of TYLCV infection on the pattern of HSP90 and HSP70 upon heat shock and recovery. Infected and uninfected tomato leaves were incubated at room temperature (control), and at 42–43°C for 2 h (heat shock), followed by 2 h recovery at room temperature (recovery). Extracts of native proteins were subjected to ultracentrifugation on sucrose gradients, which were subsequently divided in ten fractions; aliquots were analyzed by western blots with anti-HSP90 and HSP70 antibodies

herbivorous insects by changing their development, survival, range and abundance (Luck et al. 2011). Many abiotic stresses were shown to weaken the defense mechanisms of plants and enhance their susceptibility to pathogen infection (Bale et al. 2002; Atkinson and Urwin 2012). How does TYLCV infection influence plant heat stress response?

10.4.1 Combined Heat and Viral Stresses Induce the Efficient Aggregation of the Cellular Chaperones HSP70 and HSP90

TYLCV-dependent conversion of HSP70 and HSP90 from soluble into aggregated state is a highlight of the current review. Does heat shock together with TYLCV infection influence the degree of aggregation of these chaperones? The results of the ultracentrifugation through sucrose gradients showed that in extracts of heat shock treated leaves were observed increased amounts of HSPs and their appearance in large aggregates (Fig. 10.8). Recovery from heat shock to normal temperature led to the restoration of initial chaperone patterns in uninfected tomatoes. By comparison, in TYLCV-infected leaves that recovered from heat shock, significant amounts of aggregated HSPs remained in large aggregates (sucrose gradient fraction 10), after recovery, there is more HSPs in fraction 10 of infected plants than before heat shock (Fig. 10.8). These results showed that TYLCV enhances heat shock-induced protein aggregation. These aggregates contained viral CP together with at least two cellular chaperones. Restoration of initial chaperone patterns during recovery was impaired in viral infected plants.

10.4.2 TYLCV Downregulates Heat Stress Response in Infected Plants

Capturing of HSP70/HSP90 in TYLCV-induced aggregates is assumed to result in decrease of the pool of free chaperones, which triggers the transcription of *HSP* encoding genes under the control of heat stress transcription factors. Indeed, TYLCV infection downregulates the heat stress response of plants grown at high temperatures (Fig. 10.9). The transcription levels of heat stress transcription factors HsfA2, HsfB1 and HSF regulated genes, such as *Hsp17*, *Apx1*, *Apx2*, *HSP90*, were at the lowest levels at 14 dpi, when TYLCV is actively accumulated, while some release of suppression was observed later during infection (49 dpi). Special interest aroused during the recovery period, when tomato leaf samples were returned to room temperature (23–25 °C) after heat shock. *HsfA2*, *HsfB1*, *Hsp17*, *Apx1*, *Apx2*, *HSP90* genes were still less expressed in leaves of infected than in uninfected tomatoes (Fig. 10.9). Decreased amounts of HSFA2 coincided with a significant reduction in the expression of several heat-inducible genes, as demonstrated for *Hsp17*, *Apx1/2*.

Begomoviruses are long-lived viruses, which need a healthy host environment for prolonged time. To ensure a successful infection cycle, geminiviruses must restrain their destructive effect on the host cells and prevent drastic plant responses, heat stress response, for example (Anfoka et al. 2016). TYLCV infection not only was incapable to cause hypersensitive response/cell death (HR/CD) in tomatoes, but on the contrary, suppressed CD, induced by the other factors (Gorovits and Czosnek 2007; Moshe et al. 2015b). CD suppression was associated with a downregulation of protein degradation.

10.5 Conclusion

TYLCV stress does not induce HSP70/HSP90 expression, but causes their shift from soluble proteins into aggregates. During the development of plant infection, HSP70 and HSP90 aggregates re-localize from the cytoplasm to the nucleus in the

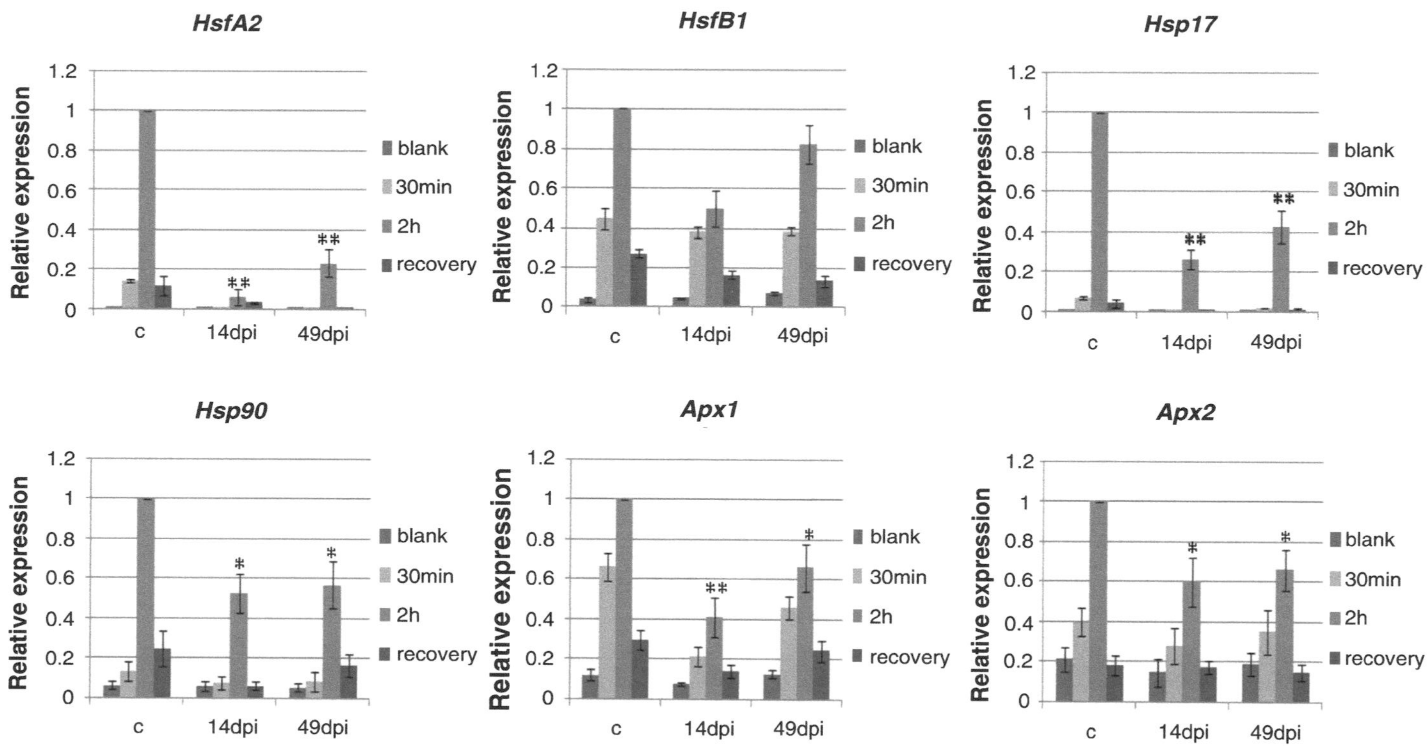

Fig. 10.9 Analyses of expression profiles of heat stress-depended genes upon TYLCV infection. Detached leaves from uninfected and TYLCV-infected (14, 21 and 49 dpi) tomato plants were incubated at ambient (23–25°C) and high (42–43°C) temperatures for 2 h followed by 2 h recovery at ambient temperature. qPCR transcription profile in uninfected (c) and infected (14 and 49 dpi) leaf samples. The expression level of each gene was calculated in relation to uninfected leaves after 2 h heat shock. The results were normalized using the *β-actin* gene as an internal marker. Results were analyzed using analysis of variance (ANOVA). One asterisk denotes $p < 0.05$. Two asterisks denotes $p < 0.01$. Bars represent the average and standard deviation of the relative expression from five independent biological repeats. Pooled leaves of three different plants were taken for each sample

cells associated with the vascular system. In the large nuclear VFs, the other PQC elements such as ubiqutin, 26S proteasome subunits, autophagy ATG8 are present together with TYLCV proteins, mainly CP, viral DNA, DNA-protein complexes and infectious virions. Host PQC could be used by virus for correct folding and maintenance of multiple virus proteins to promote virions assembly. Active HSP70 and HSP90 with the complex network of co-chaperones may define the balance of protein assembly and degradation by the ubiquitin-proteasome system in nuclear TYLCV factories. The co-localization of HSP70 with the major TYLCV protein CP was detected in aggregates in plants and in the TYLCV insect vector. CP is essential not only as the building brick of the virion, but also for intracellular and intercellular movement, systemic infection and transmission by the insect vector. CP specifically interacts with HSP70 in tomato and *B. tabaci*. Moreover, in infected plant cells, HSP70 has been shown to be involved in nuclear-cytoplasmic shuttling of viral CP. Inhibition of plant HSP70 expression by quercetin caused a decrease in the amount of nuclear CP and its re-localization from the nucleus to the cytoplasm, followed by a decrease of TYLCV DNA levels. On the contrary, inactivation of tomato HSP90 machinery by GDA treatment or by silencing of the *HSP90* and *SGT1* genes leads to accumulation TYLCV DNA. TYLCV enrichment was suggested to be caused by HSP90-dependent decline of 26S proteasome degradation in the infected tomatoes. The influence of HSP70 and HSP90 activity on TYLCV acquisition by *B. tabaci* is the subject of future investigation. The relations between *B. tabaci*, TYLCV and the environment are intricate (especially at increasingly high temperatures). Whiteflies thrive at high temperatures and their heat tolerance may have contributed to their success as a major insect pest to agriculture. On the other hand, it seems that TYLCV decreases its fitness at elevated temperatures. *B. tabaci HSP70* and *HSP90* expression are upregulated upon the passage from 25 to 40 °C (Mahadav et al. 2009; Díaz et al. 2015), and the fitness was increased upon exposure to heat. Understanding the thermal tolerance of the various *B. tabaci* biotypes in the various environments worldwide, and their capacity to transmit begomoviruses adds to the complexity of developing efficient management strategies of whiteflies in the field.

TYLCV could induce a mechanism to sequester virus-induced misfolded or modified cellular proteins in aggregates to prevent the triggering of innate antiviral responses. For example, sequestration of certain signals may represent a mechanism to stop the induction of cell death or the unfolded protein response pathway, consequently, to prevent an activation of HSFs and their substrates, which suppress virus successful multiplication. Indeed, low levels of cell death in TYLCV infected tomatoes have been shown, even more, TYLCV is able to alleviate cell death, induced by the other stress, and stress (heat, for example) response itself.

Acknowledgements This research was supported by a grant from the Israel Science Foundation Award 1037/13.

References

Anfoka G, Moshe A, Fridman L et al (2016) Tomato yellow leaf curl virus infection mitigates the heat stress response of plants grown at high temperatures. Sci Rep 6:19715. doi:10.1038/srep19715

Atkinson NJ, Urwin PE (2012) The interaction of plant biotic and abiotic stresses: from genes to the field. J Exp Bot 63:3523–3543

Azevedo C, Sadanandom A, Kitagawa K, Freialdenhoven A, Shirasu K, Schulze-Lefert P (2002) The RAR1 interactor SGT1, an essential component of R gene-triggered disease resistance. Science 295:2073–2076

Bale JS, Masters GJ, Hodkinson ID et al (2002) Herbivory in global climate change research: direct effects of rising temperature on insect herbivores. Glob Chang Boil 8:1–16

Burch AD, Weller SK (2005) Herpes simplex virus type 1 DNA polymerase requires the mammalian chaperone hsp90 for proper localization to the nucleus. J Virol 79:10740–10749

Czosnek H (2007) Tomato yellow leaf curl virus disease. In: Czosnek H (ed) Management, molecular biology, breeding for resistance. Springer, Dordrecht, p 440

Czosnek H, Ghanim M, Ghanim M (2002) Circulative pathway of begomoviruses in the whitefly vector *Bemisia tabaci* – insights from studies with *Tomato yellow leaf curl virus*. Ann Appl Biol 140:215–231

Díaz F, Orobio RF, Chavarriaga P, Toro-Perea N (2015) Differential expression patterns among heat-shock protein genes and thermal responses in the whitefly *Bemisia tabaci* (MEAM1). J Therm Biol 52:199–207

Díaz-Pendón JA, Cañizares MC, Moriones E, Bejarano ER, Czosnek H, Navas-Castillo J (2010) T*omato yellow leaf curl viruses*: ménage à trois between the virus complex, the plant and the whitefly vector. Mol Plant Pathol 11:441–450

Edwardson JR (1966) Cylindrical inclusions in the cytoplasm of leaf cells infected with tobacco etch virus. Science 153:883–884

Ghanim M, Czosnek H (2016) Interactions between the whitefly *Bemisia tabaci* and begomoviruses: biological and genomic perspectives. In: Czosnek H, Ghanim M (eds) Management of insect pests to agriculture: lessons learned from deciphering their genome, transcriptome and proteome. Springer, Cham

Gorovits R, Czosnek H (2007) Biotic and abiotic stress responses in breeding tomato lines resistant and susceptible to tomato yellow leaf curl virus. In: Czosnek H (ed) Management, molecular biology, breeding for resistance. Springer, Dordrecht, pp 223–237

Gorovits R, Akad F, Beery H, Vidavsky F, Mahadav A, Czosnek H (2007) Expression of stress-response proteins upon whitefly-mediated inoculation of tomato yellow leaf curl virus (TYLCV) in susceptible and resistant tomato plants. Mol Plant Microbe Interact 20:1376–1383

Gorovits R, Moshe A, Kolot M, Sobol I, Czosnek H (2013a) Progressive aggregation of *tomato yellow leaf curl virus* coat protein in systemically infected tomato plants, susceptible and resistant to the virus. Virus Res 171:33–43

Gorovits R, Moshe A, Ghanim M, Czosnek H (2013b) Recruitment of the host plant heat shock protein 70 by *tomato yellow leaf curl virus* coat protein is required for virus infection. PLoS One 8(7):e70280

Gorovits R, Moshe A, Ghanim M, Czosnek H (2014) Degradation mechanisms of the *tomato yellow leaf curl virus* coat protein following inoculation of tomato plants by the whitefly *Bemisia tabaci*. Pest Manag Sci 70:1632–1639

Gorovits R, Fridman L, Kolot M et al (2016) Tomato yellow leaf curl virus confronts host degradation by sheltering in small/midsized protein aggregates. Virus Res 213:304–313. doi:10.1016/j.virusres

Gottlieb Y, Zchori-Fein E, Mozes-Daube N et al (2010) The transmission efficiency of tomato yellow leaf curl virus by the whitefly Bemisia tabaci is correlated with the presence of a specific symbiotic bacterium species. J Virol 84:9310–9317

Götz M, Popovski S, Kollenberg M et al (2012) Implication of Bemisia tabaci heat shock protein 70 in begomovirus-whitefly interactions. J Virol 86:13241–13252

Huang YW, Hu CC, Liou MR et al (2012) Hsp90 interacts specifically with viral RNA and differentially regulates replication initiation of Bamboo mosaic virus and associated satellite RNA. PLoS Pathog 8(5):e1002726. doi:10.1371/journal.ppat.1002726

Iwasaki S, Kobayashi M, Yoda M et al (2010) Hsc70/Hsp90 chaperone machinery mediates ATPdependent RISC loading of small RNA duplexes. Mol Cell 39:292–299

Kim KS (1977) An ultrastructural study of inclusions and disease development in plant cells infected by cowpea chlorotic mottle virus. J Gen Virol 35:535–543

Komatsu K, Hashimoto M, Ozeki J, Yamaji Y, Maejima K, Senshu H (2010) Viral-induced systemic necrosis in plants involves both programmed cell death and the inhibition of viral multiplication, which are regulated by independent pathways. Mol Plant Microbe Interact 23:283–293

Laliberte J-F, Sanfacon H (2010) Cellular remodeling during plant virus infection. Annu Rev Phytopathol 48:69–91

Liu Y, Schiff M, Serino G, Deng XW, Dinesh-Kumar SP (2002) Role of SCF ubiquitin-ligase and the COP9 signalosome in the N gene-mediated resistance response to Tobacco mosaic virus. Plant Cell 14:1483–1496

Liu Y, Burch-Smith T, Schiff M, Feng S, Dinesh-Kumar SP (2004) Molecular chaperone Hsp90 associates with resistance protein N and its signaling proteins SGT1 and Rar1 to modulate an innate immune response in plants. J Biol Chem 279:2101–2108

Livingston CM, Ifrim MF, Cowan AE, Weller SK (2009) Virus-Induced Chaperone-Enriched (VICE) domains function as nuclear protein quality control centers during HSV-1 infection. PLoS Pathog 10:e1000619

Luck J, Spackman M, Freeman A et al (2011) Climate change and diseases of food crops. Plant Pathol 60:113–121

Mahadav A, Kontsedalov S, Czosnek H, Ghanim M (2009) Thermotolerance and gene expression following heat stress in the whitefly *Bemisia tabaci* B and Q biotypes. Insect Biochem Mol Biol 39:668–676

Mine A, Hyodo K, Tajima Y et al (2012) Differential roles of Hsp70 and Hsp90 in the assembly of the replicase complex of a positive-strand RNA plant virus. J Virol 86:12091–12104. doi:10.1128/JVI.01659-12

Miyoshi T, Takeuchi A, Siomi H, Siomi MC (2010) A direct role for Hsp90 in pre-RISC formation in Drosophila. Nat Struct Mol Biol 17:1024–1026

Morin S, Ghanim M, Zeidan M, Czosnek H, Verbeek M, van den Heuvel JFJM (1999) A GroEL homologue from endosymbiotic bacteria of the whitefly Bemisia tabaci is implicated in the circulative transmission of Tomato yellow leaf curl virus. Virol 256:75–84

Moshe A, Gorovits R (2012) Virus-induced aggregates in infected cells. Viruses 4:2218–2232. doi:10.3390/v4102218

Moshe A, Pfannstiel J, Brotman Y, et al (2012) Stress responses to Tomato yellow leaf curl virus (TYLCV) infection of resistant and susceptible tomato plants are different. Metabolom S1:006.

Moshe A, Belausov E, Niehl A, Heinlein M, Czosnek H, Gorovits R (2015a) The *tomato yellow leaf curl virus* V2 protein forms aggregates depending on the cytoskeleton integrity and binds viral genomic DNA. Sci Rep 5:9967

Moshe A, Gorovits R, Liu Y and Czosnek H (2015b) Tomato plant cell death induced by inhibition of HSP90 is alleviated by *Tomato yellow leaf curl virus* infection. Mol Plant Pathol 17(2): 247–260. doi: 10.1111/mpp

Muskett P, Parker J (2003) Role of SGT1 in the regulation of plant R gene signalling. Microbiol Infect 5(11):969–976

Nagy PD, Wang RY, Pogany J, Hafren A, Makinen K (2011) Emerging picture of host chaperone and cyclophilin roles in RNA virus replication. Virol 411:374–382
Netherton C, Wileman T (2011) Virus factories, double membrane vesicles and viroplasm generated in animal cells. Curr Opin Virol 1:381–387. doi:10.1016/j.coviro
Netherton CL, Wileman TE (2013) African swine fever virus organelle rearrangements. Virus Res 173(1):76–86. doi:10.1016/j.virusres
Netherton C, Moffat K, Brooks E, Wileman T (2007) A guide to viral inclusions, membrane rearrangements, factories, and viroplasm produced during virus replication. Adv Virus Res 70:101–182
Noel LD, Cagna G, Stuttmann J et al (2007) Interaction between SGT1 and cytosolic/nuclear HSC70 chaperones regulates Arabidopsis immune responses. Plant Cell 19:4061–4076
Pakkianathan BC, Kontsedalov S, Lebedev G et al (2015) Replication of *tomato yellow leaf curl* in its whitefly vector *Bemisia tabaci*. J Virol 89:9791–9803
Rubinstein G, Czosnek H (1997) Long-term association of tomato yellow leaf curl virus with its whitefly vector Bemisia tabaci: effect on the insect transmission capacity, longevity and fecundity. J Gen Virol 78:2683–2689
Russo M, Di Franco A, Martelli GP (1983) The fine structure of Cymbidium ringspot virus infections in host tissues. III. Role of peroxisomes in the genesis of multivesicular bodies. J Ultrastruct Res 82:52–63
Scharf KD, Berberich T, Ebersberger I, Nover L (2012) The plant heat stress transcription factor (Hsf) family: structure, function and evolution. Biochim Biophys Acta 1819:104–119
van der Scheer C, Groenewegen J (1971) Structure in cells of Vigna unguiculata infected with cowpea mosaic virus. Virol 46:493–497
Vogel F, Hofius D, Sonnewald U (2007) Intracellular trafficking of potato leafroll virus movement protein in transgenic Arabidopsis. Traffic 8:1205–1214
Wang RY, Stork J, Nagy PD (2009) A key role for heat shock protein 70 in the localization and insertion of tombusvirus replication proteins to intracellular membranes. J Virol 83:3276–3287
Weller SK (2010) Herpes simplex virus reorganizes the cellular DNA repair and protein quality control machinery. PLoS Pathog 6(11):e1001105
Wileman T (2006) Aggresomes and autophagy generate sites for virus replication. Science 312:875–878
Wileman T (2007) Aggresomes and pericentriolar sites of virus assembly: cellular defense or viral design? Annu Rev Microbiol 61:149–617
Willison JH (1976) The hexagonal lattice spacing of intracellular crystalline tobacco mosaic virus. J Ultrastruct Res 54:176–182
Ye CM, Kelly V, Payton M, Dickman MB, Verchot J (2012) SGT1 is induced by the Potatovirus X TGBp3and enhances virus accumulation in *Nicotiana benthamiana*. Mol Plant 5:1151–1153

Part III
Heat Shock Proteins and Plant Therapeutics

Chapter 11
Heat Shock Proteins and Phytochemicals: Role in Human Health and Disease

Xinyu Wang, Srujana Rayalam, and Vishakha Bhave

Abstract The use of plants for therapeutic purpose has a long history in many continents. However, the potential of phytochemicals, the bioactive components in the plant, in disease prevention and treatment is recognized and supported only recently by the scientific community. A lot of human diseases and health-related issues are related to the disruption of redox homeostasis by various stressors encountered constantly throughout human life. As a consequence, it leads to accumulation of abnormal proteins, inflammatory response, and oxidative stress, which are common denominators shared by many degenerative disorders. Heat shock proteins are stress proteins known to provide cytoprotection and play important roles in protein folding/unfolding. This chapter briefly describes recent findings in the effects of phytochemicals on oxidative stress-involved ischemia/reperfusion injury, obesity, and liver diseases through regulation of heat shock proteins. These phytochemicals include caffeic acid phenethyl ester from bee glue, synthetic oleanane triterpenoid CDDO-Im, curcumin from Indian spice turmeric, resveratrol from red grapes, naringin found in grapefruit, epigallocatechin-3-gallate from green tea, anthocyanins from pomegranate, and flavonoids.

Keywords Heat shock proteins • Liver disease • Obesity • Oxidative stress • Phytochemicals

Abbreviations

ARE	antioxidant response element
ASH	alcoholic steatohepatitis
CA	caffeic acid
CAPE	caffeic acid phenethyl ester
CCl_4	carbon tetrachloride
CDDO	2-cyano-3,12-dioxooleana-1,9-dien-28-oic acid

X. Wang (✉) • S. Rayalam • V. Bhave
Department of Pharmaceutical Sciences, School of Pharmacy, Philadelphia College of Osteopathic Medicine, Georgia Campus, Suwanee, GA, USA
e-mail: xinyuwa@pcom.edu

A.A.A. Asea et al. (eds.), *Heat Shock Proteins and Plants*,
Heat Shock Proteins 10, DOI 10.1007/978-3-319-46340-7_11

CDDO-Im	1-[2-cyano-3,12-dioxooleana-1,9(11)-dien-28-oyl]imidazole
DENA	diethylnitrosamine
EGCG	Epigallocatechin-3-Gallate
HCC	hepatocellular carcinoma
HO-1	heme oxygenase-1
HSF-1	heat shock factor 1
HSPs	heat shock proteins
HUVEC	human umbilical vein endothelial cells
I/R	ischemia/reperfusion
Keap1	Kelch-like ECH-associated protein 1
MB	mallory bodies
NASH	nonalcoholic steatohepatitis
NF-kB	nuclear factor kappa B
Nrf2	nuclear factor, erythroid 2-like 2
PPAR-γ	peroxisome proliferator-activated receptor gamma
ROS	reactive oxygen species

11.1 Introduction

Phytochemicals are bioactive components found in natural plants. In the Eastern world, plants with therapeutic activities have long been utilized for disease prevention and treatment. The record of using plants for therapeutic purpose can be traced back to 2800 BC when the first manuscript of medicinal herbs was prepared by the ancient Chinese emperor, *Shen Nung* (Wang et al. 2012). In the Western world, the beneficial effects of phytochemicals from vegetables and fruits are well recognized. The well-known saying of "an apple a day keeps the doctor away" is a good example. Recently, more and more scientific evidence from research publications demonstrated the therapeutic potential of phytochemicals for the treatment of various human diseases (Arumuggam et al. 2015; Sirerol et al. 2016; Zhang et al. 2015). Interestingly, some of these beneficial effects are found to be regulated through a family of molecular chaperones known as heat shock proteins (HSPs) (Calabrese et al. 2008; Maiti et al. 2014). HSPs belong to a highly conserved group of proteins responding to various factors inducing cellular stress by preventing or reducing protein misfolding and aggregation. Here, we reviewed the effects of phytochemicals through the regulation of HSPs on oxidative stress-involved ischemia/reperfusion injury, obesity, and liver diseases.

11.2 Oxidative Stress – Involved Ischemia/Reperfusion Injury

Ischemia/reperfusion (I/R) injury is a potentially serious pathologic state involved in a number of cardiovascular diseases including myocardial infarction and stroke. This condition also occurs following septic or hemorrhagic shock, surgery, organ transplantation, and traumatic injury (Carden and Granger 2000; Garcia et al. 1996; Tapuria et al. 2008; Turer and Hill 2010). All these disorders cause great economic loss and may lead to death not only in the civilian sector but also on the battlefield because hemorrhage is one of the major causes of morbidity and mortality during combat (Kauvar et al. 2006). Therefore, there is an urgent need to identify new therapeutic strategies that ameliorate I/R injury and related disorders. Understanding I/R injury and corresponding molecular mechanism will help achieve this goal.

Ischemia refers to the interruption of blood flow to a tissue or an organ, which leads to the depletion of oxygen and nutrients to the constituent cells. Consequently, it can cause cell damage and death of subsequent cells, tissue, or organ. Reperfusion is a paradox with double-edged sword. On one side, it is required to restore oxygen and nutrients to maintain normal cell function. On the other hand, the reintroduction of oxygen to an ischemia area could result in further cellular and vascular injury (Granger et al. 1986). There is a large body of information indicating oxidative stress accompanies I/R injury (Granger et al. 1986). Oxidative stress is a condition where the sustained intracellular production of pro-oxidants (reactive oxygen species) exceeds the neutralizing capacity of endogenous antioxidants. During ischemia, hypoxanthine is accumulated by catabolizing large amount of ATP. Its normal metabolizing enzyme xanthine dehydrogenase is converted to xanthine oxidase, which could utilize oxygen supplied from following reperfusion to generate superoxide as a byproduct in the metabolism of hypoxanthine and xanthine to uric acid. This reaction does not occur until reperfusion supplies the necessary oxygen (Carden and Granger 2000; Granger et al. 1986). This superoxide anion serves as a starting point to form other free radicals including hydrogen peroxide and hydroxyl radical. Among these reactive oxygen species (ROS), hydroxyl radical is the most reactive and toxic one produced through Fenton reaction and Haber-Weiss reaction with the participation of ischemia-induced iron (Hess and Manson 1984). It can cause severe cellular injury by directly targeting the DNA, protein, and lipids of adjacent cells. The cellular damage by overproduction of ROS suggests antioxidants could be beneficial to ameliorate I/R injury (Cuzzocrea et al. 2001). The event and consequent injury by ischemia following reperfusion along with antioxidant intervention are illustrated in Fig. 11.1.

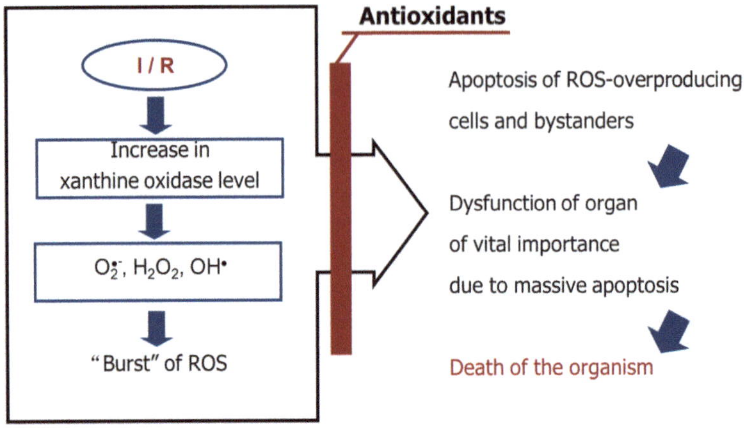

Fig. 11.1 Ischemia/reperfusion induced injury and antioxidant intervention

11.2.1 Antioxidant Effects of Phytochemicals: Role of Heat Shock Proteins

Compounds with capacity to counteract free radicals are categorized into endogenous and exogenous antioxidants. Endogenous antioxidant system includes both enzymatic (superoxide dismutase, catalase, etc.) and non-enzymatic (glutathione, ascorbic acid, etc.) components (Duthie and Crozier 2000; Uttara et al. 2009). A lot of compounds derived from plant sources maintain antioxidant activity and exhibit beneficial effects on oxidative stress – involved I/R injury. Epigallocatechin gallate, a green tea polyphenol, reduces superoxide concentration significantly after ischemia and reperfusion *in vivo*, suggesting it is therapeutically potential to ameliorate I/R injury (Buttemeyer et al. 2003). Procyanidins, a mixture of polyphenolic antioxidants, reduced the sequela of myocardial I/R damage in rats by enhancing plasma antioxidant activity (Facino et al. 1999). Polyphenolic compounds such as curcumin, the intense yellow pigment found in turmeric, and resveratrol, a component of red wine and grapes, exhibited cytoprotective effects against I/R-induced myocardial impairment and blood-brain barrier disruption, possibly through their free radicals scavenging abilities (Jiang et al. 2007; Ray et al. 1999).

Heat shock proteins belong to a highly conserved class of proteins induced by host cells to respond to cellular stress including oxidative stress (Kalmar and Greensmith 2009). These molecular chaperone proteins play an important role in the maintenance of intracellular homeostasis by preventing protein aggregation, facilitating protein folding, and correcting protein misfolding. When cells are subject to oxidative stress, it triggers an array of detrimental events including intracellular redox imbalance, oxidation of vital cellular proteins, aggregation and misfolding of proteins. As a consequence, signal pathways related to inflammation and apoptosis are activated, eventually resulting in the dysfunction of cells and death of cells (Kalmar and Greensmith 2009). HSPs, however, can respond and rescue host cells from oxidative damage at several levels. First, some HSPs such as HSP32 or heme oxygenase-1 can serve as sensors to detect cellular redox changes (Calabrese et al. 2004). Second, upon oxidation of intracellular components, misfolded proteins themselves trigger the expression of HSPs, primarily HSP70 and HSP90, leading to repair and clearance of damaged proteins (Abravaya et al. 1992). In addition, as apoptosis is triggered by prolonged oxidative stress, HSPs such as HSP27, HSP70, and HSP90 can act as anti-apoptotic proteins inhibiting the downstream events of apoptosis protease activating factor-1 (Apaf-1) and block release of mitochondrial cytochrome c (Kalmar and Greensmith 2009). Last but not the least, subsequent inflammatory response following oxidative stress is ameliorated through a negative feedback mechanism between HSP70 and inflammatory mediators including nuclear factor kappa B (NF-kB) and cyclooxygenase-2 (Feinstein et al. 1996; Ialenti et al. 2005).

Despite increasing evidence on the beneficial effects of plant-derived therapeutic agents for the amelioration of oxidative stress-mediated I/R injury, there is limited information on the role of HSPs in the phytochemical-induced cytoprotective effects against oxidative stress.

11.2.1.1 Caffeic Acid Phenethyl Ester (CAPE)

CAPE is first identified from honeybee hive product, propolis, and reported to selectively inhibit the growth of tumor cells (Grunberger et al. 1988). Propolis, a natural resinous material, is collected by bees from various plant sources to seal their hives, reinforce the borders of their combs, and protect against the invasion of pathogenic microorganisms. Propolis has been used as a folk medicine since ancient time in many cultures to treat various diseases. It shows various activities including antibacterial, antifungal, antiviral, antioxidant, and anti-inflammation (Banskota et al. 2001). The chemical compositions and biological activities of propolis, however, vary not only from the local flora and phenology of the host plants but from the geographic regions and season of collection as well. CAPE is reported in relatively high amounts in propolis from European and non-tropic Asian countries such as the Netherlands, China, and Korea (Ahn et al. 2004; Bankova 2005; Nagaoka et al. 2003; Usia et al. 2002). A comprehensive picture of the existence and relative quantity of CAPE in propolis from all over the world is reported recently (Kumazawa et al. 2004). Among all the propolis examined from different countries, CAPE is found most abundant in propolis from China (Kumazawa et al. 2004).

Numerous biological properties of CAPE have been reported since its discovery from bee propolis. These activities include antitumor, antiviral, anti-inflammation, antioxidant, and immunomodulation (Fesen et al. 1994; Huang et al. 1996; Michaluart et al. 1999; Park et al. 2004; Son and Lewis 2002). Recently, it is shown that CAPE is able to ameliorate I/R injury *in vivo* and protect various tissues and organs such as skeletal muscle, myocardium, and kidney (Ozer et al. 2005; Ozyurt et al. 2006; Parlakpinar et al. 2005). In order to elucidate the mechanism of CAPE protection, cytoprotective effects of CAPE are studied using menadione-generated oxidative stress in human umbilical vein endothelial cells (HUVEC) to simulate I/R injury *in vivo* (Wang et al. 2008). The results demonstrate cytoprotective effect of CAPE against menadione cytotoxicity and indicate that this effect is closely regulated by heme oxygenase-1 (HO-1). CAPE highly induced the expression of HO-1 and the cytoprotective effect of CAPE is abolished when the activity of HO-1 is blocked by its pharmacological inhibitor.

HO-1 appears to play an important role in the cytoprotection of CAPE against cellular oxidative stress. This enzyme is considered a member of HSP family because the promoter region of *ho-1* gene contains binding element for heat shock factors, the transcriptional regulatory proteins (Syapin 2008). Interestingly, the transcriptional response of HO-1 (HSP32) to heat shock is found mostly in rodent organs such as rat brain and liver (Ewing et al. 1992; Raju and Maines 1994). The heme metabolizing function of HO-1 is well characterized. During hemorrhage or cell damage, large amount of heme released induces HO-1. HO-1 catalyzes the rate-limiting step in the degradatiogn of heme to free iron, carbon monoxide, and bile pigment biliverdin which can be further converted to bilirubin by biliverdin reductase (Ryter et al. 2006). The induction of HO-1 and consequent production of its heme-metabolizing products is reported to ameliorate oxidative stress-involved I/R injury (Katori et al. 2002). In addition to CAPE, HO-1, the inducible form of heme oxygenase, is reported to be activated in various cell types by many other plant-derived polyphenolic compounds including epigallocatechin-3-gallate (the major active constituent of green tea), resveratrol (a stilbenoid from grapes and berries), carnosol (a phenolic diterpene from herb rosemary), and curcumin (the yellow phenolic compound from turmeric and used as Indian spice) (Juan et al. 2005; Martin et al. 2004; Motterlini et al. 2000; Wu et al. 2006).

Many pharmacological activities of polyphenolic phytochemicals are ascribed to their antioxidant properties to scavenge free radicals and reactive species directly. The structure of CAPE (Fig. 11.2) maintains a 3,4 dihydroxyl configuration of the catechol ring, which is known to deactivate and stabilize ROS. To determine whether the cytoprotective effect of CAPE is due to induction of HO-1 or direct free radical scavenging activity, the activity of CAPE is compared with its structurally modified derivatives (Wang et al. 2006, 2010). Although the 3,4 adjacent hydroxyl groups on the catechol ring are required for the antioxidant activity of CAPE and its derivatives, this functional moiety did not correlate well with their cytoprotection profile. On the other hand, all cytoprotective derivatives of CAPE were found to induce

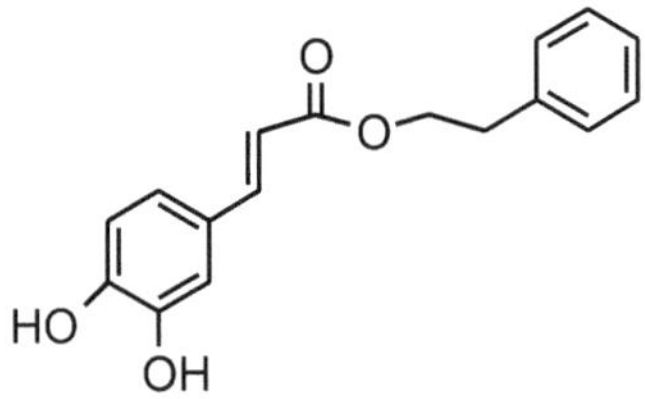

Fig. 11.2 Structure of CAPE

HO-1 confirming that the induction of heat shock protein HO-1 by CAPE and its derivatives, not their direct antioxidant activity, is the main mechanism of cytoprotection against oxidative stress.

11.2.1.2 1-[2-Cyano-3,12-Dioxooleana-1,9(11)-Dien-28-Oyl]Imidazole (CDDO-Im)

CDDO-Im is the imidazole analogue of a synthesized oleanane triterpenoid, 2-cyano-3,12-dioxooleana-1,9-dien-28-oic acid (CDDO). CDDO was synthesized due to an effort to design and develop more potent anti-inflammatory agents from the natural triterpenoids oleanolic acid and ursolic acid (Honda et al. 1998). Triterpenoids can be found in different plant products such as lavender, rosemary, oregano, olives, figs, apples, and cranberries (Bishayee et al. 2011). This group of phytochemicals, biosynthesized by cyclization of squalene, is composed of more than 20,000 members, the largest group of natural products in plants (Liby et al. 2007). Triterpenoids have long been used in many Asian countries because of their multifunctional activities including anti-inflammation, antioxidant, hepatoprotection, cardio-tonicity, antimicrobial, and antiviral (Liu 1995).

A recent study demonstrated that CDDO-Im is a more potent inducer of HO-1 both in cell culture and animal models than its parental CDDO (Liby et al. 2005). Induction of this cytoprotective enzyme was linked to a signaling pathway mediated by a transcription factor, nuclear factor, erythroid 2-like 2 (Nrf2). This transcription factor governs the expression of an array of cytoprotective and detoxifying genes coding for phase 2 enzymes (Dinkova-Kostova et al. 2001). When stimuli are not present, Nrf2 is associated with Kelch-like ECH-associated protein 1 (Keap1) to form a complex inactivating Nrf2 in the cytoplasmic area of cells. Upon activation, Nrf2 is released from the Nrf2-Keap1 complex and translocated into the nucleus, where it binds to the antioxidant response element (ARE) and triggers the up-regulation of phase 2 genes. The structure of CDDO-Im (Fig. 11.3) provides evidence for the activation of the Keap1-Nrf2-ARE pathway. The α, β-unsaturated carbonyl groups on rings A and C of CDDO-Im structure enable the reaction of Michael addition with nucleophilic thiol group on Keap1, therefore dissociating Keap1-Nrf2 complex and releasing Nrf2 for nuclear translocation (Dinkova-Kostova et al. 2005).

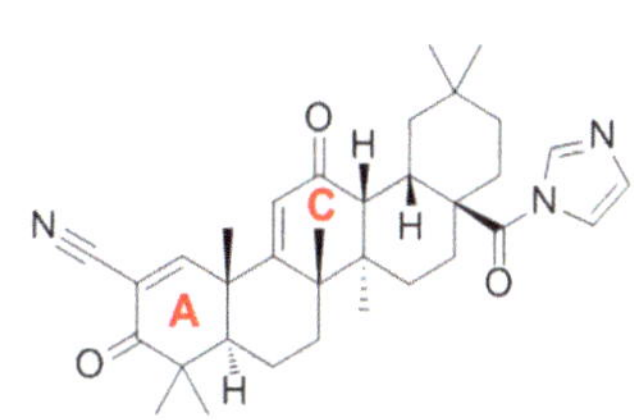

Fig. 11.3 Structure of CDDO-Im

In a continuing effort to identify novel therapeutic agents to ameliorate oxidative stress, cytoprotective effects of CDDO-Im were compared to CAPE (Wang et al. 2014), and found that CDDO-Im at nM level protects HUVEC from menadione-induced oxidative stress even more potent than CAPE at mM level. To explore molecular mechanism of this cytoprotection, microarray analysis is conducted to establish global gene expression profiles in HUVEC when treated with CDDO-Im and CAPE at optimal cytoprotective doses, respectively. The results of microarray analysis identified common genes altered by both compounds, which may suggest a common mechanism of cytoprotection. One of these genes encodes HO-1, the cytoprotective enzymes playing important role for CAPE cytoprotection. Interestingly, CDDO-Im is found to be a much more potent inducer of HO-1 than CAPE at both transcriptional and translational levels, which may partially account for its stronger cytoprotection than CAPE. Noticeably, CDDO-Im also highly upregulated other HSP genes including HSPA1A/HSPA1B and HSPA6 to a greater extent than CAPE. In addition to identify and compare expression of individual genes, Ingenuity Pathway Analysis was used to explore the involvement of transcriptional networks, intracellular functions, and signaling pathways. Based on the list of common genes shared by both compounds, network and functional analysis found that most relevant network and top regulatory functions are related to cellular growth and proliferation, cellular development, and cell death and survival. Pathway analysis of individual lists of genes significantly altered by CDDO-Im and CAPE identified one highly affected canonical pathway, the Nrf2-mediated oxidative stress response. This pathway regulates the expression of genes particularly involved in drug metabolism and transport, oxidant signaling and antioxidant defense, which represents a detoxification strategy that cells utilize in response to stimuli from oxidants, electrophiles, and toxicants (Ma 2013). In comparison to CAPE, this pathway appears to be activated by CDDO-Im at a higher level as evidenced by the upregulation of more genes involved in this pathway to a great extent.

The mechanisms of CAPE and CDDO-lm induced cytoprotection is therefore related to the induction of HO-1, a member of heat shock proteins and phase 2 detoxification system, which is mediated by Keap1-Nrf2-ARE pathway. These novel effects of CDDO-lm on the upregulation of the molecular chaperone HSP family may contribute to its potent cytoprotection against oxidative stress. The induction of HSPs including HO-1 may serve as biomarkers for the identification and development of potential anti-oxidative stress agents from natural sources. The involvement of HSPs and phytochemicals in ischemia/reperfusion injury is still largely unknown. More research efforts are needed to unravel the role of HSPs altered by phytochemicals in the treatment of oxidative stress related disease states.

11.3 Obesity and Related Disorders

Obesity is a global public health crisis affecting over one third of population in the U.S. and about 13 % of the population worldwide. Obesity is a complex disease and increases the risk for other health conditions like cardiovascular disease, type 2 diabetes and stroke to name a few. The economic impact of obesity associated costs in the U.S account for $215 billion annually indicating the severity of the problem (Hammond and Levine 2010). In spite of these staggering numbers, there has been no major progress in the development of successful anti-obesity medications. Currently available FDA approved anti-obesity drugs work by either decreasing appetite or by blocking absorption of fat in the intestines and are associated with potential side effects limiting the usefulness of these drugs (Hanefeld and Sachse 2002; Loke et al. 2002). Thus phytochemicals emerge as an alternative safer approach to address the problem of obesity and related disorders.

11.3.1 Heat Shock Proteins in Obesity and Related Disorders

Obesity is associated with chronic low grade inflammation and altered stress response especially in peripheral tissues like adipose tissue, muscle and liver. Since HSPs are induced in response to stressful conditions like metabolic stress to protect cells, activation of these proteins ameliorates obesity and complications associated with this disease (Sharma et al. 2011). While activation of HSPs protect cells against inflammation and oxidative stress contributing to anti-obesity effects (Morimoto 1993), decreased expression of HSPs impairs the heat shock response which correlates well with the development of insulin resistance (Kurucz et al. 2002). In fact, induction of heat shock proteins provides versatile protection in diabetic patients from vascular complications and majority of these protective effects are mediated through the suppression of NF-kB, a key mediator of the endothelial inflammation (De Martin et al. 2000). Recent studies indicate that enhanced expression of HSP70 protected against obesity-induced insulin resistance in both rodents and humans and rodent models of obesity (Chung et al. 2008). Interestingly, increased expression of HSP70 was also shown to influence the ageing process in round worms and although the extent to which HSPs can boost longevity in mammals is unclear, caloric restriction has been shown to up-regulate HSP induction (Ehrenfried et al. 1996). In contrast, blocking of certain HSPs like HSP90 inhibits adipocyte differentiation indicating the dual regulatory effects of HSPs on adipogenesis.

11.3.2 Role of HSPs in Phytochemical-Induced Anti-obesity Effects

Phytochemicals have gained a lot of public and scientific interest in the past two decades for their beneficial effects on decreasing adiposity and up-regulation of energy expenditure. Although several mechanisms of action have been ascribed to the anti-obesity effects mediated by phytochemicals, the underlying theme for phytochemical-induced anti-obesity effects revolve around inflammation. Since obesity is considered a low-grade systemic inflammatory disease, anti-inflammatory phytochemicals proved to be effective in ameliorating problems associated with obesity.

11.3.2.1 Curcumin

Curcumin is one of the most widely studied phytochemical for its anti-inflammatory effects. It is derived from turmeric, a spice, very popular in Asia. The major anti-inflammatory effects of curcumin are mediated via the down-regulation of tumor necrosis factor – alpha (TNF-α) and NF-kB. In adipocytes, curcumin suppressed differentiation and inhibited adipogenesis (Ejaz et al. 2009) and this is accompanied by the inhibition of stress kinase c-Jun NH2 terminal kinase signaling (Aggarwal 2010). *In vivo*, curcumin significantly improved obesity-associated inflammation and diabetes in diet induced obese mouse model (Weisberg et al. 2008). Anti-inflammatory effects of curcumin are associated with the translocated heat shock factor 1 (HSF-1) into nucleus followed by increased HSP70 expression (Dunsmore et al. 2001). HSF-1 also plays a critical role in HSP90 (Chen et al. 2013; Park and Liu 2001) expression, which in turn increases the expression of peroxisome proliferator-activated receptor gamma (PPAR-γ), a key transcription factor in the regulation of adipocyte differentiation (Desarzens et al. 2014). Recently, HSP90 has been proposed as a potential therapeutic target for obesity. Although HSP90 inhibitory effects of curcumin in adipocytes are not investigated, evidence for these effects in other cell types (Li et al. 2014) indicate a possibility for the anti-obesity effects of curcumin mediated through the inhibition of HSP90. HSPs -72 and -25 on the other hand render cytoprotective effects and evidence from human and rodent studies indicate that decreased HSP72 correlates with the degree of insulin-resistance in type -2 diabetic patients (Bruce et al. 2003; Gupte et al. 2008). Interestingly, curcumin is a strong inducer of HSP72 and HSP25 in response to stress under both in vivo and in vitro conditions (Kato et al. 1998; Lundvig et al. 2015) and these effects may in part contribute to the beneficial effects of curcumin on insulin resistance *in vivo*.

11.3.2.2 Resveratrol

Known to be produced initially as a phytoalexin in the skin of red grapes in response to infection or injury, resveratrol has come a long way in its anti-inflammatory, anti-cancer, anti-aging and anti-obesity effects. Resveratrol is an inducer of hear shock response and increases transcription and translation of HSP70 in a HSF-1 dependent manner (Putics et al. 2008). In a recent study, positive correlation was reported between decreased HSF-1 levels in the adipose tissue of obese patients and the impairment of HSP70. In addition, attenuated anti-inflammatory HSP70 pathway in obese patients contributed to the progression of non-alcoholic fatty liver disease (Di Naso et al. 2015). Furthermore, resveratrol is a sirtuin and AMP-activated protein kinase activator and when coupled with its reciprocal effects on HSF-1 and NF-kB activation, resveratrol emerges as a candidate compound for increasing longevity (Baur and Sinclair 2006).

11.3.2.3 Naringin

As a bioflavonoid found in grapefruit and oranges, naringin is well studied for its anti-oxidant and anti-diabetic effects. Administration of naringin to high fat fed streptozoticin induced diabetic rats resulted in a multitude of beneficial effects including amelioration of insulin resistance, dyslipidemia and prevention of kidney damage (Sharma et al. 2011). These effects are attributed to the upregulation of PPAR-γ and HSP72 and HSP27 by naringenin in these animals. As mentioned earlier, HSP72 protects against obesity – induced insulin resistance and there exists a negative correlation between HSP72 expression and the degree of insulin resistance (Chung et al. 2008). Further, dietary supplementation with naringin attenuated hyperglycemia-induced oxidative stress, reduced inflammatory status, improved glucose intolerance and the structure and function of the heart and liver without decreasing total body weight (Alam et al. 2013; Mahmoud et al. 2012). In myocardial ischemia-reperfusion injury model in rats, naringin significantly reduced infarct size and these cardioprotective effects of are attributed to the increase in HSP70 and HSP27 levels together with suppression of inflammatory pathways and oxidative stress (Rani et al. 2013). Hesperidin, another citrus flavonoid, exhibits similar effects on the activation of HSP70 and 27 contributing to the attenuation of hyperglycemia-mediated oxidative stress in diabetic rodent models (Mahmoud et al. 2012).

11.3.2.4 Epigallocatechin-3-Gallate

Popularly known as green tea, Epigallocatechin-3-Gallate (EGCG), has received enormous attention lately for its anti-cancer effects. Anti-obesity effects of EGCG are mediated through its effects on mitogen activated kinase family. EGCG decreased both adipocyte size and number in vitro and in rodent models, EGCG reduced serum triglycerides and cholesterol, increased energy expenditure as well as reducing body weight (Moon et al. 2007). Human intervention studies indicated that consumption of 25 % EGCG for 3 months reduced body weight and waist circumference in obese patients (Moon et al. 2007). Recently, EGCG has been identified as a novel HSP90 inhibitor and given that HSP90 inhibition deceases adipocyte differentiation, it is possible that anti-obesity effects of EGCG are partly mediated through this mechanism (Yin et al. 2009).

While it is interesting to note that anti-inflammatory and protective effects of anti-obesity phytochemicals like curcumin (Kato et al. 1998), withaferin A (Grogan et al. 2013), resveratrol (Putics et al. 2008) and naringin (Rani et al. 2013) are associated with the upregulation of HSPs like 72, 70 and 27, other anti-obesity phytochemicals like EGCG (Moses et al. 2015) inhibit thc induction of HSPs like HSP90.

11.4 Liver Health and Disease

HSPs are ubiquitous and highly conserved molecules that are induced under conditions of physiological stress. They act as defense mechanisms against various environmental stresses (Morimoto 1993). Over the past few decades, HSPs have been both therapeutic targets as well as therapeutic agents. They play a role in processes like liver repair, resolution of inflammation, and well as liver regeneration. On the other hand, some HSPs have also been implicated in liver disease. Some disease states implicating HSPs are hepatocellular carcinoma, alcoholic steatohepatitis (ASH), fluoride-induced hepatotoxicity, and experimental liver injury induced by experimental chemical carbon tetrachloride (CCl_4). Moreover, some HSPs have gained value as diagnostic/prognostic markers for certain liver diseases. The following section discusses the role of some HSPs as therapeutic agents and others as therapeutic targets. We have also discussed some phytochemicals that have shown to be beneficial in some of these disease states via modulation of HSPs.

11.4.1 Role of HSPs in Liver Repair and Regeneration

HSPs have been shown to play an important role in liver regeneration. Among these are HSPs 27, 60, 70, and 90 (Shi et al. 2007). Early phase of successful liver regeneration requires the presence of HSP70 to induce TNF-α (Wolf et al. 2014). Heat shock pretreatment reduces acute liver injury and accelerates liver repair in mice (Li et al. 2013). HSP27 has been shown to be downregulated in ballooning hepatocytes in nonalcoholic steatohepatitis (NASH) patients (Sookoian et al. 2016). Since HSP27 protects cells against cell death (Fulda et al. 2010), ballooned hepatocytes that lack HSP27 induction fail to mount a robust physiological response to metabolic induced stress (Sookoian et al. 2016). HSP70 has also been shown to aid in liver recovery from I/R injury (Boeri et al. 2003).

11.4.2 Role of HSPs in Liver Disease

Many HSPs have been found to be overexpressed in various liver diseases. Mallory bodies (MBs) are aggresomes, composed of cytokeratin and various other proteins, which form in diseased liver because of disruption in the ubiquitin-proteasome protein degradation pathway. Heat shock proteins are thought to be involved in this process because it was discovered that MB formation is induced by heat shock in drug-primed mice. Liver biopsy sections stained with HSPs 70 and 90 showed increased expression in all liver diseases investigated including primary biliary cirrhosis, NASH, Hepatitis B and C, idiopathic cirrhosis, ASH, and hepatocellular carcinoma (HCC) (Riley et al. 2003). While more studies are required to elucidate the role of overexpressed HSPs in the pathology of liver diseases, some studies outlined below have attempted to investigate this.

Chemoprevention can be defined as an emerging strategy to prevent, reduce, slow, or reverse the occurrence and progression of cancer by the administration of one or more naturally occurring or synthetic compounds (Bishayee et al. 2010; Castrellon and Gluck 2008; Jordan 2007; Sporn and Suh 2000). The following section discusses some plant-derived phytochemicals and their role in liver disease.

11.4.2.1 HSPs in hepatocellular carcinoma (HCC)

HCC is one of the most prevalent life threatening malignancies representing the majority of primary liver cancers (Center and Jemal 2011). The United states has reported more than 70 % increase in the incidence of HCC in the past 25 years (El-Serag 2004). Unresolved inflammation and chronic oxidative stress have been identified as the two major drivers of HCC progression (Kawanishi et al. 2006; Marra et al. 2011). Two major contributors of HCC development and tumor progression include HSP70 and HSP90 (Jolly and Morimoto 2000; Joo et al. 2005). Induction of HSP70 has been shown to promote the proliferation of tumor cells (Jolly and Morimoto 2000; Joo et al. 2005; Mosser and Morimoto 2004). HSP-90 has been shown to regulate expression and function of various hepatocarcinogenic factors, while its inhibition leads to cell cycle arrest and apoptosis in HCC cells (Breinig et al. 2009).

Dietary Pomegranate Emulsion

Pomegranate (*Punica granatum*, Punicaceae) contains a diverse array of phytochemicals, including polyphenolic constituents (anthocyanins), hydrolysable tannins (ellagitannins and gallotannins), and condensed tannins (proanthocyanidins) (Lansky and Newman 2007). Some of these compounds have been shown to exhibit potent antioxidant and anti-inflammatory properties implicated in prevention and treatment of several inflammation driven diseases, including cancer (Faria and Calhau 2011; Lansky and Newman 2007). Its role as a chemoprotective has been tested in a two stage liver cancer model involving treatment with diethylnitrosamine (DENA). Treatment with dietary pomegranate emulsion has demonstrated to suppress upregulated HSP70 and HSP90 in DENA-challenged rats (Bishayee et al. 2013a), indicating its protective effect in HCC.

Black Currant Phytoconstituents

Black currant (Ribes nigrum L.) berries, widely known in Europe, New Zealand, and Northern Asia in cooler weather conditions are particularly rich in the flavonoid antioxidant anthocyanins (Nielsen et al. 2003; Scalzo et al. 2008; Wu et al. 2004) and also known to contain various other antioxidant phenolics, including flavonoids and phenolic acids (Paredes-Lopez et al. 2010). Upon continuous treatment with dietary black currant skin extract, there was a dose dependent decrease in all the elevated inflammatory markers of HCC. Moreover, HSPs 70 and 90 were significantly suppressed indicating attenuation of the inflammation-mediated hepatocarcinogenic stress generated by DENA (Bishayee et al. 2013b) in this two-stage model of HCC.

11.4.2.2 HSPs in alcoholic steatohepatitis (ASH)

HSPs 70 and 85 have been implicated in the pathogenesis of ASH (Omar et al. 1990). This is related to the intracytoplasmic accumulation of HSP85 in alcoholic liver disease. Strong immunoreactivity of hepatocytes from ASH livers with these HSPs also suggests their potential as a “marker” of hepatocellular injury.

11.4.2.3 HSPs in fluoride induced hepatotoxicity

Fluoride is a ground water contaminant in many parts of the world. Liver is one of the vital organs susceptible to fluoride toxicity (Shanthakumari et al. 2004). Some of the mechanisms of toxicity to liver include free radical mediated lipid peroxidation, oxidative stress, DNA damage, inflammation, mitochondrial dysfunction, and necrotic/apoptotic cell death in *in vivo* (Zhan et al. 2006) as well as *in vitro* models (Ghosh et al. 2008). Among the diverse classes of HSPs, HSP60 and HSP27 confer cytoprotection against oxidative damage and hyperthermia caused by xenobiotics (Parcellier et al. 2003; Rezzani et al. 2005). HSP60 has both pro-apoptotic and anti-apoptotic activity and its activity depends on the type of cell and nature of the stimulus (Knowlton and Gupta 2003). Increased levels of HSP60 expression were found in livers of fluoride intoxicated rats (Kanagaraj et al. 2015). In contrast to this, decreased levels of HSP27 have been reported in fluoride induced hepatotoxicity (Kanagaraj et al. 2015). Since HSP27 is shown to regulate glutathione levels and inhibit apoptotic signaling in response to external stress or oxidative stress (Snoeckx et al. 2001), the decrease in HSP27 in fluoride toxicity may be due to oxidative stress mediated apoptosis.

Caffeic Acid

Caffeic acid (CA; 3,4-dihydroxycinnamic acid) is a phenolic compound present in citrus fruits, vegetables and herbs, artichoke, basil, and apple (Clifford et al. 2007). The pharmacological activities of CA include antioxidant (Chen and Ho 1997), anti-inflammatory (Chung et al. 2004), immunomodulatory (Figueiredo-Rinhel et al. 2013), and hepatoprotective (Pari and Prasath 2008). CA treatment has shown to significantly restore the levels of HSP60 and HSP27 expression by maintaining normal chaperone function during fluoride hepatotoxicity (Kanagaraj et al. 2015), thus providing a protective effect in fluoride hepatotoxicity.

11.4.2.4 HSPs in CCl_4-induced hepatotoxicity

CCl_4-induced liver injury is an experimental model of acute liver failure in rats and mice. This model is utilized to study the mechanisms of injury and repair in the liver. HSP25 and HSP70 expression is induced in this model (Liu et al. 1996) implicating its role in causing acute liver failure.

Wei Kang Su and Flavonoids

Wei Kang Su is a commercial herbal product based on the Shengmai San formula composed of Radix Ginseng, Radix Ophiopogonis, and Fructus Schisandrae Chinensis. This formula has been clinically prescribed for energy invigoration and body fluid retention, particularly for the treatment of coronary heart disease in China for more than 800 years (Liang et al. 2002). Wei Kang Su pretreatment causes a reversal of altered HSP25 and HSP70 production in CCl_4-intoxicated rats, indicating an improvement of antioxidant status in these rats (Leong et al. 2010).

11.5 Conclusion

There is limited research on the HSP-mediated beneficial effects of phytochemicals in human health. Since stress response is altered in chronic debilitating conditions like obesity and diabetes, natural compounds that increase the expression of HSPs to aid in the normal functioning of cells may provide beneficial effects in such disease states. Current literature suggests that HSPs can have both, a protective or disruptive effect in liver disease depending on the model and disease condition investigated (Fig. 11.4). In fact, recent findings indicate that novel therapeutic agents identified to ameliorate oxidative stress-induced I/R injury like CAPE and CDDO-Im induce cytoprotective effects through upregulation of HSPs, particularly HO-1 or HSP32. While some HSPs act as prognostic/diagnostic/therapeutic markers in certain diseases, their direct implication in the pathology of these diseases warrants further research to evaluate HSPs as potential diagnostic or therapeutic targets.

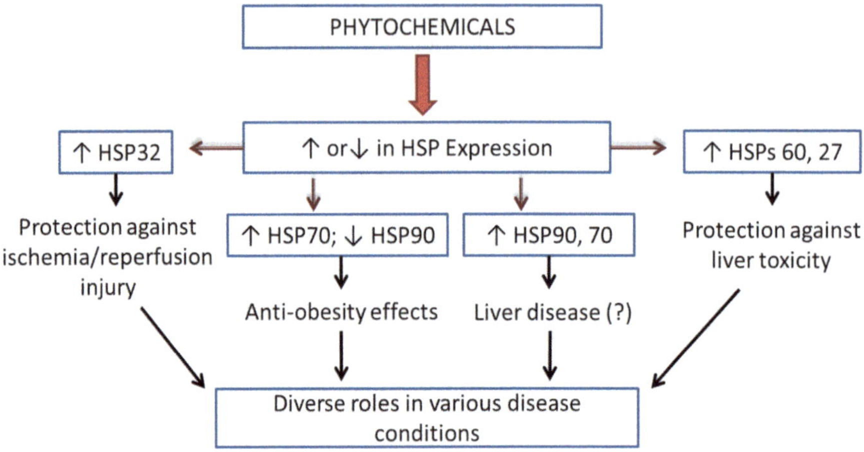

Fig. 11.4 Role of phytochemical-induced expression of HSPs in human health and disease

Acknowledgments Some of the work discussed in this chapter was supported by the grant from the Center for Chronic Disorders of Aging (CCDA) to W. X.

References

Abravaya K, Myers MP, Murphy SP, Morimoto RI (1992) The human heat shock protein hsp70 interacts with HSF, the transcription factor that regulates heat shock gene expression. Genes Dev 6:1153–1164

Aggarwal BB (2010) Targeting inflammation-induced obesity and metabolic diseases by curcumin and other nutraceuticals. Annu Rev Nutr 30:173–199

Ahn MR, Kumazawa S, Hamasaka T, Bang KS, Nakayama T (2004) Antioxidant activity and constituents of propolis collected in various areas of Korea. J Agric Food Chem 52:7286–7292

Alam MA, Kauter K, Brown L (2013) Naringin improves diet-induced cardiovascular dysfunction and obesity in high carbohydrate, high fat diet-fed rats. Nutrients 5:637–650

Arumuggam N, Bhowmick NA, Rupasinghe HP (2015) A review: phytochemicals targeting JAK/STAT signaling and IDO expression in cancer. Phytother Res 29:805–817

Bankova V (2005) Recent trends and important developments in propolis research. Evid Based Complement Alternat Med 2:29–32

Banskota AH, Tezuka Y, Kadota S (2001) Recent progress in pharmacological research of propolis. Phytother Res 15:561–571

Baur JA, Sinclair DA (2006) Therapeutic potential of resveratrol: the in vivo evidence. Nat Rev Drug Discov 5:493–506

Bishayee A, Politis T, Darvesh AS (2010) Resveratrol in the chemoprevention and treatment of hepatocellular carcinoma. Cancer Treat Rev 36:43–53

Bishayee A, Ahmed S, Brankov N, Perloff M (2011) Triterpenoids as potential agents for the chemoprevention and therapy of breast cancer. Front Biosci 16:980–996

Bishayee A, Thoppil RJ, Darvesh AS, Ohanyan V, Meszaros JG, Bhatia D (2013a) Pomegranate phytoconstituents blunt the inflammatory cascade in a chemically induced rodent model of hepatocellular carcinogenesis. J Nutr Biochem 24:178–187
Bishayee A, Thoppil RJ, Mandal A et al (2013b) Black currant phytoconstituents exert chemoprevention of diethylnitrosamine-initiated hepatocarcinogenesis by suppression of the inflammatory response. Mol Carcinog 52:304–317
Boeri D, Dondero F, Storace D, Maiello M, Pasqualini M, Pellicci R (2003) Heat-shock protein 70 favours human liver recovery from ischaemia-reperfusion. Eur J Clin Investig 33:500–504
Breinig M, Caldas-Lopes E, Goeppert B et al (2009) Targeting heat shock protein 90 with non-quinone inhibitors: a novel chemotherapeutic approach in human hepatocellular carcinoma. Hepatology 50:102–112
Bruce CR, Carey AL, Hawley JA, Febbraio MA (2003) Intramuscular heat shock protein 72 and heme oxygenase-1 mRNA are reduced in patients with type 2 diabetes: evidence that insulin resistance is associated with a disturbed antioxidant defense mechanism. Diabetes 52:2338–2345
Buttemeyer R, Philipp AW, Schlenzka L, Mall JW, Beissenhirtz M, Lisdat F (2003) Epigallocatechin gallate can significantly decrease free oxygen radicals in the reperfusion injury in vivo. Transplant Proc 35:3116–3120
Calabrese V, Stella AM, Butterfield DA, Scapagnini G (2004) Redox regulation in neurodegeneration and longevity: role of the heme oxygenase and HSP70 systems in brain stress tolerance. Antioxid Redox Signal 6:895–913
Calabrese V, Cornelius C, Mancuso C et al (2008) Cellular stress response: a novel target for chemoprevention and nutritional neuroprotection in aging, neurodegenerative disorders and longevity. Neurochem Res 33:2444–2471
Carden DL, Granger DN (2000) Pathophysiology of ischaemia-reperfusion injury. J Pathol 190:255–266
Castrellon AB, Gluck S (2008) Chemoprevention of breast cancer. Expert Rev Anticancer Ther 8:443–452
Center MM, Jemal A (2011) International trends in liver cancer incidence rates. Cancer Epidemiol Biomark Prev 20:2362–2368
Chen JH, Ho CT (1997) Antioxidant activities of caffeic acid and its related hydroxycinnamic acid compound. J Agric Food Chem 45:2374–2378
Chen Y, Chen J, Loo A et al (2013) Targeting HSF1 sensitizes cancer cells to HSP90 inhibition. Oncotarget 4:816–829
Chung TW, Moon SK, Chang YC et al (2004) Novel and therapeutic effect of caffeic acid and caffeic acid phenyl ester on hepatocarcinoma cells: complete regression of hepatoma growth and metastasis by dual mechanism. FASEB J 18:1670–1681
Chung J, Nguyen AK, Henstridge DC et al (2008) HSP72 protects against obesity-induced insulin resistance. Proc Natl Acad Sci U S A 105:1739–1744
Clifford MN, Wu W, Kirkpatrick J, Kuhnert N (2007) Profiling the chlorogenic acids and other caffeic acid derivatives of herbal chrysanthemum by LC-MSn. J Agric Food Chem 55:929–936
Cuzzocrea S, Riley DP, Caputi AP, Salvemini D (2001) Antioxidant therapy: a new pharmacological approach in shock, inflammation, and ischemia/reperfusion injury. Pharmacol Rev 53:135–159
De Martin R, Hoeth M, Hofer-Warbinek R, Schmid JA (2000) The transcription factor NF-kappa B and the regulation of vascular cell function. Arterioscler Thromb Vasc Biol 20:E83–E88
Desarzens S, Liao WH, Mammi C, Caprio M, Faresse N (2014) Hsp90 blockers inhibit adipocyte differentiation and fat mass accumulation. PLoS One 9:e94127
Di Naso FC, Porto RR, Fillmann HS et al (2015) Obesity depresses the anti-inflammatory HSP70 pathway, contributing to NAFLD progression. Obesity (Silver Spring) 23:120–129
Dinkova-Kostova AT, Massiah MA, Bozak RE, Hicks RJ, Talalay P (2001) Potency of Michael reaction acceptors as inducers of enzymes that protect against carcinogenesis depends on their reactivity with sulfhydryl groups. Proc Natl Acad Sci U S A 98:3404–3409

Dinkova-Kostova AT, Liby KT, Stephenson KK et al (2005) Extremely potent triterpenoid inducers of the phase 2 response: correlations of protection against oxidant and inflammatory stress. Proc Natl Acad Sci U S A 102:4584–4589

Dunsmore KE, Chen PG, Wong HR (2001) Curcumin, a medicinal herbal compound capable of inducing the heat shock response. Crit Care Med 29:2199–2204

Duthie G, Crozier A (2000) Plant-derived phenolic antioxidants. Curr Opin Clin Nutr Metab Care 3:447–451

Ehrenfried JA, Evers BM, Chu KU, Townsend CM Jr, Thompson JC (1996) Caloric restriction increases the expression of heat shock protein in the gut. Ann Surg 223:592–597; discussion 97–9

Ejaz A, Wu D, Kwan P, Meydani M (2009) Curcumin inhibits adipogenesis in 3T3-L1 adipocytes and angiogenesis and obesity in C57/BL mice. J Nutr 139:919–925

El-Serag HB (2004) Hepatocellular carcinoma: recent trends in the United States. Gastroenterology 127:S27–S34

Ewing JF, Haber SN, Maines MD (1992) Normal and heat-induced patterns of expression of heme oxygenase-1 (HSP32) in rat brain: hyperthermia causes rapid induction of mRNA and protein. J Neurochem 58:1140–1149

Facino RM, Carini M, Aldini G et al (1999) Diet enriched with procyanidins enhances antioxidant activity and reduces myocardial post-ischaemic damage in rats. Life Sci 64:627–642

Faria A, Calhau C (2011) The bioactivity of pomegranate: impact on health and disease. Crit Rev Food Sci Nutr 51:626–634

Feinstein DL, Galea E, Aquino DA, Li GC, Xu H, Reis DJ (1996) Heat shock protein 70 suppresses astroglial-inducible nitric-oxide synthase expression by decreasing NFkappaB activation. J Biol Chem 271:17724–17732

Fesen MR, Pommier Y, Leteurtre F, Hiroguchi S, Yung J, Kohn KW (1994) Inhibition of HIV-1 integrase by flavones, caffeic acid phenethyl ester (CAPE) and related compounds. Biochem Pharmacol 48:595–608

Figueiredo-Rinhel AS, Kabeya LM, Bueno PC et al (2013) Inhibition of the human neutrophil oxidative metabolism by Baccharis dracunculifolia DC (Asteraceae) is influenced by seasonality and the ratio of caffeic acid to other phenolic compounds. J Ethnopharmacol 150:655–664

Fulda S, Gorman AM, Hori O, Samali A (2010) Cellular stress responses: cell survival and cell death. Int J Cell Biol 2010:214074

Garcia JH, Lassen NA, Weiller C, Sperling B, Nakagawara J (1996) Ischemic stroke and incomplete infarction. Stroke 27:761–765

Ghosh J, Das J, Manna P, Sil PC (2008) Cytoprotective effect of arjunolic acid in response to sodium fluoride mediated oxidative stress and cell death via necrotic pathway. Toxicol in Vitro 22:1918–1926

Granger DN, Hollwarth ME, Parks DA (1986) Ischemia-reperfusion injury: role of oxygen-derived free radicals. Acta Physiol Scand Suppl 548:47–63

Grogan PT, Sleder KD, Samadi AK, Zhang H, Timmermann BN, Cohen MS (2013) Cytotoxicity of withaferin A in glioblastomas involves induction of an oxidative stress-mediated heat shock response while altering Akt/mTOR and MAPK signaling pathways. Investig New Drugs 31:545–557

Grunberger D, Banerjee R, Eisinger K et al (1988) Preferential cytotoxicity on tumor cells by caffeic acid phenethyl ester isolated from propolis. Experientia 44:230–232

Gupte AA, Bomhoff GL, Geiger PC (2008) Age-related differences in skeletal muscle insulin signaling: the role of stress kinases and heat shock proteins. J Appl Physiol 105:839–848

Hammond RA, Levine R (2010) The economic impact of obesity in the United States. Diabetes Metab Syndr Obes targets Ther 3:285–295

Hanefeld M, Sachse G (2002) The effects of orlistat on body weight and glycaemic control in overweight patients with type 2 diabetes: a randomized, placebo-controlled trial. Diabetes Obes Metab 4:415–423

Hess ML, Manson NH (1984) Molecular oxygen: friend and foe. The role of the oxygen free radical system in the calcium paradox, the oxygen paradox and ischemia/reperfusion injury. J Mol Cell Cardiol 16:969–985
Honda T, Rounds BV, Gribble GW, Suh N, Wang Y, Sporn MB (1998) Design and synthesis of 2-cyano-3,12-dioxoolean-1,9-dien-28-oic acid, a novel and highly active inhibitor of nitric oxide production in mouse macrophages. Bioorg Med Chem Lett 8:2711–2714
Huang MT, Ma W, Yen P et al (1996) Inhibitory effects of caffeic acid phenethyl ester (CAPE) on 12-O-tetradecanoylphorbol-13-acetate-induced tumor promotion in mouse skin and the synthesis of DNA, RNA and protein in HeLa cells. Carcinogenesis 17:761–765
Ialenti A, Di Meglio P, D'Acquisto F et al (2005) Inhibition of cyclooxygenase-2 gene expression by the heat shock response in J774 murine macrophages. Eur J Pharmacol 509:89–96
Jiang J, Wang W, Sun YJ, Hu M, Li F, Zhu DY (2007) Neuroprotective effect of curcumin on focal cerebral ischemic rats by preventing blood-brain barrier damage. Eur J Pharmacol 561:54–62
Jolly C, Morimoto RI (2000) Role of the heat shock response and molecular chaperones in oncogenesis and cell death. J Natl Cancer Inst 92:1564–1572
Joo M, Chi JG, Lee H (2005) Expressions of HSP70 and HSP27 in hepatocellular carcinoma. J Korean Med Sci 20:829–834
Jordan VC (2007) Chemoprevention of breast cancer with selective oestrogen-receptor modulators. Nat Rev Cancer 7:46–53
Juan SH, Cheng TH, Lin HC, Chu YL, Lee WS (2005) Mechanism of concentration-dependent induction of heme oxygenase-1 by resveratrol in human aortic smooth muscle cells. Biochem Pharmacol 69:41–48
Kalmar B, Greensmith L (2009) Induction of heat shock proteins for protection against oxidative stress. Adv Drug Deliv Rev 61:310–318
Kanagaraj VV, Panneerselvam L, Govindarajan V, Ameeramja J, Perumal E (2015) Caffeic acid, a phyto polyphenol mitigates fluoride induced hepatotoxicity in rats: a possible mechanism. BioFactors 41:90–100
Kato K, Ito H, Kamei K, Iwamoto I (1998) Stimulation of the stress-induced expression of stress proteins by curcumin in cultured cells and in rat tissues in vivo. Cell Stress Chaperones 3:152–160
Katori M, Anselmo DM, Busuttil RW, Kupiec-Weglinski JW (2002) A novel strategy against ischemia and reperfusion injury: cytoprotection with heme oxygenase system. Transpl Immunol 9:227–233
Kauvar DS, Lefering R, Wade CE (2006) Impact of hemorrhage on trauma outcome: an overview of epidemiology, clinical presentations, and therapeutic considerations. J Trauma 60:S3–S11
Kawanishi S, Hiraku Y, Pinlaor S, Ma N (2006) Oxidative and nitrative DNA damage in animals and patients with inflammatory diseases in relation to inflammation-related carcinogenesis. Biol Chem 387:365–372
Knowlton AA, Gupta S (2003) HSP60, bax, and cardiac apoptosis. Cardiovasc Toxicol 3:263–268
Kumazawa S, Hamasaka T, Nakayama T (2004) Antioxidant activity of propolis of various geographic origins. Food Chem 84:329–339
Kurucz I, Morva A, Vaag A et al (2002) Decreased expression of heat shock protein 72 in skeletal muscle of patients with type 2 diabetes correlates with insulin resistance. Diabetes 51:1102–1109
Lansky EP, Newman RA (2007) Punica granatum (pomegranate) and its potential for prevention and treatment of inflammation and cancer. J Ethnopharmacol 109:177–206
Leong PK, Chen N, Chiu PY et al (2010) Long-term treatment with shengmai san-derived herbal supplement (Wei Kang Su) enhances antioxidant response in various tissues of rats with protection against carbon tetrachloride hepatotoxicity. J Med Food 13:427–438
Li SQ, Wang DM, Shu YJ, Wan XD, Xu ZS, Li EZ (2013) Proper heat shock pretreatment reduces acute liver injury induced by carbon tetrachloride and accelerates liver repair in mice. J Toxicol Pathol 26:365–373

Li Z, Jia L, Wang J et al (2014) Design, synthesis and biological evaluation of 17-arylmethylamine-17-demethoxygeldanamycin derivatives as potent Hsp90 inhibitors. Eur J Med Chem 85:359–370
Liang SMC, Chen SY, Liang SQ (2002) Shengmai San-a renowned traditional Chinese medicinal formula. In: Ko KM (ed) Traditional herbal medicines for modern times-Shengmai San. Taylor and Francis, London, pp 1–15
Liby K, Hock T, Yore MM et al (2005) The synthetic triterpenoids, CDDO and CDDO-imidazolide, are potent inducers of heme oxygenase-1 and Nrf2/ARE signaling. Cancer Res 65:4789–4798
Liby KT, Yore MM, Sporn MB (2007) Triterpenoids and rexinoids as multifunctional agents for the prevention and treatment of cancer. Nat Rev Cancer 7:357–369
Liu J (1995) Pharmacology of oleanolic acid and ursolic acid. J Ethnopharmacol 49:57–68
Liu H, Lightfoot R, Stevens JL (1996) Activation of heat shock factor by alkylating agents is triggered by glutathione depletion and oxidation of protein thiols. J Biol Chem 271:4805–4812
Loke YK, Derry S, Pritchard-Copley A (2002) Appetite suppressants and valvular heart disease – a systematic review. BMC Clin Pharmacol 2:6
Lundvig DM, Pennings SW, Brouwer KM et al (2015) Curcumin induces differential expression of cytoprotective enzymes but similar apoptotic responses in fibroblasts and myofibroblasts. Exp Cell Res 330:429–441
Ma Q (2013) Role of nrf2 in oxidative stress and toxicity. Annu Rev Pharmacol Toxicol 53:401–426
Mahmoud AM, Ashour MB, Abdel-Moneim A, Ahmed OM (2012) Hesperidin and naringin attenuate hyperglycemia-mediated oxidative stress and proinflammatory cytokine production in high fat fed/streptozotocin-induced type 2 diabetic rats. J Diabetes Complicat 26:483–490
Maiti P, Manna J, Veleri S, Frautschy S (2014) Molecular chaperone dysfunction in neurodegenerative diseases and effects of curcumin. Biomed Res Int 2014:495091
Marra M, Sordelli IM, Lombardi A et al (2011) Molecular targets and oxidative stress biomarkers in hepatocellular carcinoma: an overview. J Transl Med 9:171
Martin D, Rojo AI, Salinas M et al (2004) Regulation of heme oxygenase-1 expression through the phosphatidylinositol 3-kinase/Akt pathway and the Nrf2 transcription factor in response to the antioxidant phytochemical carnosol. J Biol Chem 279:8919–8929
Michaluart P, Masferrer JL, Carothers AM et al (1999) Inhibitory effects of caffeic acid phenethyl ester on the activity and expression of cyclooxygenase-2 in human oral epithelial cells and in a rat model of inflammation. Cancer Res 59:2347–2352
Moon HS, Lee HG, Choi YJ, Kim TG, Cho CS (2007) Proposed mechanisms of (−)-epigallocatechin-3-gallate for anti-obesity. Chem Biol Interact 167:85–98
Morimoto RI (1993) Cells in stress: transcriptional activation of heat shock genes. Science 259:1409–1410
Moses MA, Henry EC, Ricke WA, Gasiewicz TA (2015) The heat shock protein 90 inhibitor, (−)-epigallocatechin gallate, has anticancer activity in a novel human prostate cancer progression model. Cancer Prev Res (Phila) 8:249–257
Mosser DD, Morimoto RI (2004) Molecular chaperones and the stress of oncogenesis. Oncogene 23:2907–2918
Motterlini R, Foresti R, Bassi R, Green CJ (2000) Curcumin, an antioxidant and anti-inflammatory agent, induces heme oxygenase-1 and protects endothelial cells against oxidative stress. Free Radic Biol Med 28:1303–1312
Nagaoka T, Banskota AH, Tezuka Y, Midorikawa K, Matsushige K, Kadota S (2003) Caffeic acid phenethyl ester (CAPE) analogues: potent nitric oxide inhibitors from the Netherlands propolis. Biol Pharm Bull 26:487–491
Nielsen IL, Haren GR, Magnussen EL, Dragsted LO, Rasmussen SE (2003) Quantification of anthocyanins in commercial black currant juices by simple high-performance liquid chromatography. Investigation of their pH stability and antioxidative potency. J Agric Food Chem 51:5861–5866

Omar R, Pappolla M, Saran B (1990) Immunocytochemical detection of the 70-kd heat shock protein in alcoholic liver disease. Arch Pathol Lab Med 114:589–592

Ozer MK, Parlakpinar H, Vardi N, Cigremis Y, Ucar M, Acet A (2005) Myocardial ischemia/reperfusion-induced oxidative renal damage in rats: protection by caffeic acid phenethyl ester (CAPE). Shock 24:97–100

Ozyurt B, Iraz M, Koca K, Ozyurt H, Sahin S (2006) Protective effects of caffeic acid phenethyl ester on skeletal muscle ischemia-reperfusion injury in rats. Mol Cell Biochem 292:197–203

Parcellier A, Schmitt E, Gurbuxani S et al (2003) HSP27 is a ubiquitin-binding protein involved in I-kappaBalpha proteasomal degradation. Mol Cell Biol 23:5790–5802

Paredes-Lopez O, Cervantes-Ceja ML, Vigna-Perez M, Hernandez-Perez T (2010) Berries: improving human health and healthy aging, and promoting quality life–a review. Plant Foods Hum Nutr 65:299–308

Pari L, Prasath A (2008) Efficacy of caffeic acid in preventing nickel induced oxidative damage in liver of rats. Chem Biol Interact 173:77–83

Park J, Liu AY (2001) JNK phosphorylates the HSF1 transcriptional activation domain: role of JNK in the regulation of the heat shock response. J Cell Biochem 82:326–338

Park JH, Lee JK, Kim HS et al (2004) Immunomodulatory effect of caffeic acid phenethyl ester in Balb/c mice. Int Immunopharmacol 4:429–436

Parlakpinar H, Sahna E, Acet A, Mizrak B, Polat A (2005) Protective effect of caffeic acid phenethyl ester (CAPE) on myocardial ischemia-reperfusion-induced apoptotic cell death. Toxicology 209:1–14

Putics A, Vegh EM, Csermely P, Soti C (2008) Resveratrol induces the heat-shock response and protects human cells from severe heat stress. Antioxid Redox Signal 10:65–75

Raju VS, Maines MD (1994) Coordinated expression and mechanism of induction of HSP32 (heme oxygenase-1) mRNA by hyperthermia in rat organs. Biochim Biophys Acta 1217:273–280

Rani N, Bharti S, Manchanda M et al (2013) Regulation of heat shock proteins 27 and 70, p-Akt/p--eNOS and MAPKs by Naringin Dampens myocardial injury and dysfunction in vivo after ischemia/reperfusion. PLoS One 8:e82577

Ray PS, Maulik G, Cordis GA, Bertelli AA, Bertelli A, Das DK (1999) The red wine antioxidant resveratrol protects isolated rat hearts from ischemia reperfusion injury. Free Radic Biol Med 27:160–169

Rezzani R, Buffoli B, Rodella L, Stacchiotti A, Bianchi R (2005) Protective role of melatonin in cyclosporine A-induced oxidative stress in rat liver. Int Immunopharmacol 5:1397–1405

Riley NE, Li J, McPhaul LW, Bardag-Gorce F, Lue YH, French SW (2003) Heat shock proteins are present in mallory bodies (cytokeratin aggresomes) in human liver biopsy specimens. Exp Mol Pathol 74:168–172

Ryter SW, Alam J, Choi AM (2006) Heme oxygenase-1/carbon monoxide: from basic science to therapeutic applications. Physiol Rev 86:583–650

Scalzo J, Currie A, Stephens J et al (2008) The anthocyanin composition of different Vaccinium, Ribes and Rubus genotypes. BioFactors 34:13–21

Shanthakumari D, Srinivasalu S, Subramanian S (2004) Effect of fluoride intoxication on lipidperoxidation and antioxidant status in experimental rats. Toxicology 204:219–228

Sharma AK, Bharti S, Ojha S et al (2011) Up-regulation of PPARgamma, heat shock protein-27 and -72 by naringin attenuates insulin resistance, beta-cell dysfunction, hepatic steatosis and kidney damage in a rat model of type 2 diabetes. Br J Nutr 106:1713–1723

Shi Q, Dong Z, Wei H (2007) The involvement of heat shock proteins in murine liver regeneration. Cell Mol Immunol 4:53–57

Sirerol JA, Rodriguez ML, Mena S, Asensi MA, Estrela JM, Ortega AL (2016) Role of natural stilbenes in the prevention of cancer. Oxidative Med Cell Longev 2016:3128951

Snoeckx LH, Cornelussen RN, Van Nieuwenhoven FA, Reneman RS, Van Der Vusse GJ (2001) Heat shock proteins and cardiovascular pathophysiology. Physiol Rev 81:1461–1497

Son S, Lewis BA (2002) Free radical scavenging and antioxidative activity of caffeic acid amide and ester analogues: structure-activity relationship. J Agric Food Chem 50:468–472

Sookoian S, Castano GO, Scian R, San Martino J, Pirola CJ (2016) Heat shock protein 27 is down-regulated in ballooned hepatocytes of patients with Nonalcoholic Steatohepatitis (NASH). Sci Rep 6:22528

Sporn MB, Suh N (2000) Chemoprevention of cancer. Carcinogenesis 21:525–530

Syapin PJ (2008) Regulation of haeme oxygenase-1 for treatment of neuroinflammation and brain disorders. Br J Pharmacol 155:623–640

Tapuria N, Kumar Y, Habib MM, Abu Amara M, Seifalian AM, Davidson BR (2008) Remote ischemic preconditioning: a novel protective method from ischemia reperfusion injury–a review. J Surg Res 150:304–330

Turer AT, Hill JA (2010) Pathogenesis of myocardial ischemia-reperfusion injury and rationale for therapy. Am J Cardiol 106:360–368

Usia T, Banskota AH, Tezuka Y, Midorikawa K, Matsushige K, Kadota S (2002) Constituents of Chinese propolis and their antiproliferative activities. J Nat Prod 65:673–676

Uttara B, Singh AV, Zamboni P, Mahajan RT (2009) Oxidative stress and neurodegenerative diseases: a review of upstream and downstream antioxidant therapeutic options. Curr Neuropharmacol 7:65–74

Wang X, Stavchansky S, Bowman PD, Kerwin SM (2006) Cytoprotective effect of caffeic acid phenethyl ester (CAPE) and catechol ring-fluorinated CAPE derivatives against menadione-induced oxidative stress in human endothelial cells. Bioorg Med Chem 14:4879–4887

Wang X, Stavchansky S, Zhao B, Bynum JA, Kerwin SM, Bowman PD (2008) Cytoprotection of human endothelial cells from menadione cytotoxicity by caffeic acid phenethyl ester: the role of heme oxygenase-1. Eur J Pharmacol 591:28–35

Wang X, Stavchansky S, Kerwin SM, Bowman PD (2010) Structure-activity relationships in the cytoprotective effect of caffeic acid phenethyl ester (CAPE) and fluorinated derivatives: effects on heme oxygenase-1 induction and antioxidant activities. Eur J Pharmacol 635:16–22

Wang H, Khor TO, Shu L et al (2012) Plants vs. cancer: a review on natural phytochemicals in preventing and treating cancers and their druggability. Anti Cancer Agents Med Chem 12:1281–1305

Wang X, Bynum JA, Stavchansky S, Bowman PD (2014) Cytoprotection of human endothelial cells against oxidative stress by 1-[2-cyano-3,12-dioxooleana-1,9(11)-dien-28-oyl]imidazole (CDDO-Im): application of systems biology to understand the mechanism of action. Eur J Pharmacol 734:122–131

Weisberg SP, Leibel R, Tortoriello DV (2008) Dietary curcumin significantly improves obesity-associated inflammation and diabetes in mouse models of diabesity. Endocrinology 149:3549–3558

Wolf JH, Bhatti TR, Fouraschen S et al (2014) Heat shock protein 70 is required for optimal liver regeneration after partial hepatectomy in mice. Liver Transpl 20:376–385

Wu X, Gu L, Prior RL, McKay S (2004) Characterization of anthocyanins and proanthocyanidins in some cultivars of Ribes, Aronia, and Sambucus and their antioxidant capacity. J Agric Food Chem 52:7846–7856

Wu CC, Hsu MC, Hsieh CW, Lin JB, Lai PH, Wung BS (2006) Upregulation of heme oxygenase-1 by Epigallocatechin-3-gallate via the phosphatidylinositol 3-kinase/Akt and ERK pathways. Life Sci 78:2889–2897

Yin Z, Henry EC, Gasiewicz TA (2009) (−)-Epigallocatechin-3-gallate is a novel Hsp90 inhibitor. Biochemistry 48:336–345

Zhan XA, Wang M, Xu ZR, Li WF, Li JX (2006) Evaluation of caspase-dependent apoptosis during fluoride-induced liver lesion in pigs. Arch Toxicol 80:74–80

Zhang YJ, Gan RY, Li S et al (2015) Antioxidant phytochemicals for the prevention and treatment of chronic diseases. Molecules 20:21138–21156

Chapter 12
The Induction of *Drosophila* Heat Shock Proteins by Plants That Can Extend Fly Lifespan

Samuel E. Schriner

Abstract The occurrence of most non-infectious diseases increases as we get older. There is now a growing research effort to understand aging, and try to slow it in order to treat a broad range of diseases. The underlying molecular mechanisms of aging are not clear, and there are many theories, each with varying degrees of support. A hallmark of aging is a decreased resistance to environmental stresses such as heat, and the lifespan of many organisms is inversely related to body temperature. One strategy for slowing the aging process is to screen for drugs and botanical extracts that can extend lifespan in model organisms such as the fruit fly, *Drosophila melanogaster*. Several botanical extracts have been recently identified to increase lifespan in flies. Surprisingly, there is no overall correlation between increased lifespan and an increased tolerance to heat. In one case, an extract that increased fly lifespan had a marked decreased in heat tolerance and heat shock protein expression. These results parallel those of the direct study of HSP expression and lifespan in fruit flies. In summary, there is no clear relationship between the induction of HSPs and extension of lifespan in fruit flies through the use of botanical extracts.

Keywords Aging • Botanical extracts • Cinnamon • *Drosophila melanogaster* • Green tea • Heat shock proteins • Lifespan • *Rhodiola rosea* • *Rosa damascena*

Abbreviations

DR	dietary restriction
HSF	heat shock factor
HSP	heat shock protein

S.E. Schriner (✉)
Department of Pharmaceutical Sciences, University of California,
146 Bison Modular, Building 515, Irvine, CA 92697, USA
e-mail: schriner@uci.edu

A.A.A. Asea et al. (eds.), *Heat Shock Proteins and Plants*,
Heat Shock Proteins 10, DOI 10.1007/978-3-319-46340-7_12

12.1 Introduction

Human's age. While this is a seemingly obvious statement, not all animals appear to do so. A notable example is the lobster. As time passes these animals become bigger, stronger, and gain an increasing ability to reproductive (Finch 1990). There are several other examples of animals that exhibit negligible senescence, such as some species of fish, turtles, and clams (Finch 1990). This is unlike many organisms we are familiar with such as fruit flies, mice, and of course, humans. Lobsters do eventually die, often due to molting, but they don't get older in the same way that we do. We become more likely to get sick as time goes by; the occurrence of many diseases, such as heart disease, is strongly associated with our age (Mozaffarian et al. 2015). Muscle mass and our ability to move decreases, as does our ability to reproduce (McCarter 1990; Pellicer et al. 1995; Pal and Santoro 2003; Thompson 2009; Crawford and Steiner 2015). Ultimately, our likelihood of dying *increases* as we age (Gompertz 1825; Finch et al. 1990). This is the fundamental characteristic of aging. Things get *worse* for us as time passes, whereas things seem, for the most part, to be the same and maybe even get better for lobsters. The crucial point here is that while humans age, animals that show negligible senescence demonstrate that there is no intrinsic reason to, and that aging is something that can be potentially slowed, and perhaps cured. The ultimate hope here, though, isn't to allow humans to live hundreds of years, but to cure age-related diseases.

What is it that causes us to age? What is it that could be treated? The evolutionary theory of aging argues that animals are selected to last as long a possible for their particular environment, but no longer (Rose and Graves 1989; Rose et al. 2006). An excellent analogy is a car and its warranty. The optimal situation for the automobile manufacturer is that their car will break down immediately after the warranty expires. In this case, there is no need to honor the warranty, and the buyer is encouraged to acquire a new car. However, cars, like people, can last much longer than their warranties if they are given good care. What then, do we mean, by "good care?" For us, as for our cars, it would mean fixing whatever is going wrong. Hundreds of theories have been proposed to explain the mechanism of aging (Medvedev 1990). Two of the more prominent theories are the free radical theory of aging and the somatic mutation theory of aging (Beckman and Ames 1998; Morley 1998). The former argues that oxygen free radicals, largely produced by our mitochondria, damage our cells. Whereas, the latter argues that aging is due the accumulation of mutations in nuclear DNA causing cellular dysfunction and death. These two ideas, and at least 300 others, are not mutually exclusive. There is evidence for many theories, and each probably contributes some component to the process.

How might we target the potentially hundreds of different genes needed to slow the aging process. One solution might be to devise a cocktail of drugs, with each hitting different molecular pathways. In this scenario, plants may provide provide a potential contribution. A screen was recently undertaken in order to identify drugs, or botanical extracts, that can extend lifespan in the fruit fly, *Drosophila melanogaster*. Out of more than 80 compounds and 10 natural products examined, 6 were found to increase lifespan without any apparent negative effects on fly health (M. Jafari, personal communication). Of these 6, 4 were plant extracts: *Rhodiola rosea*, *Rosa damascena*, cinnamon, and green tea, demonstrating a significant enrichment for natural products (Schriner et al. 2009a, 2012, 2014; Lopez et al. 2014). This shouldn't necessarily be surprising, as these extracts contain hundreds of different molecules that could target multiple age-related pathways.

12.2 Aging, Hormesis, and Heat Shock Proteins

The lifespan of the fruit fly, *Drosophila melanogaster* has been shown to be inversely related to temperature (Loeb and Northrop 1917). As an example, flies commonly live 2–3 months at 25 °C, whereas lifespans of 6 months can be attainable at 18 °C (Mockett and Sohal 2006). Thus, heat is a limiting factor for lifespan in flies, though below a certain optimal temperature, lifespan will be shortened; flies may only live 10–12 days at 4 °C (Mockett and Sohal 2006). A similar phenomenon occurs in mammals where dietary restriction (DR), defined as a decreased caloric intake without malnutrition, increases lifespan while decreasing body temperature (Ferguson et al. 2007), a correlation consistent with that seen in invertebrates.

We are protected against heat by heat shock factors (HSFs) and heat shock proteins (HSFs). Heat shock factors are transcriptional activators that sense heat, and many other stresses such as oxidation, starvation, heavy metals, inflammation, hypoxia, and desiccation, and then drive expression of HSPs (Tower 2011; West et al. 2012). Heat shock proteins are molecular chaperones that bind to and help refold proteins damaged by heat and these other stresses (Tower 2011). Since the expression of these proteins confers a positive effect on survival under conditions of stress, one might ask if their induction could also extend lifespan. In *D. melanogaster*, there is some evidence to support this contention. Over-expression of several of the HSPs have been reported to extend lifespan in flies (Morrow et al. 2004; Wang et al. 2004; Liao et al. 2008). However, there have been some contradictory results where elevated HSP expression actually shortened lifespan (Bhole et al. 2004; Morrow et al. 2004). A meta-analysis of broader range of species, including several other insect species, the worm *Caenorhabditis elegans*, and the yeast, *Saccharomyces cerevisiae*, found no overall benefit on lifespan (Lagisz et al. 2013). However, there did appear to be a general positive effect in non-*Drosophila* insects examined, suggesting that there may be genus-specific responses to HSP activation.

Many plants are thought to provide health benefits through the phenomenon of hormesis (Mattson 2008a, b). This is when a toxic, but sub-lethal, treatment confers an enhanced protection later on to a greater insult (Mattson 2008a, b). One example in flies is the ability of black tea to induce the expression of the antioxidant enzymes superoxide dismutase and catalase and extend lifespan (Li et al. 2007). The tea was unable to extend lifespan when either one of these enzymes was mutated. This suggests that the tea may be acting as a pro-oxidant. The extract induces these enzymes which afford an enhanced protection against oxidative stress. Many botanicals have been reported to exhibit positive health effects in humans. An increasing number these have been found to extend lifespan and improve health in fruit flies, and could be acting by similar molecular mechanisms. Considering that heat has a negative effect of fly lifespan, plant extracts or natural products may confer their benefits in *Drosophila* through the induction of HSPs. In the following sections, specific examples will be discussed.

12.2.1 Rhodiola rosea

The root extract of *Rhodiola rosea*, also known as the golden root, is a succulent that is native to mountainous region throughout the world (Kelly 2001). It has been widely used in traditional medicine, where it has been purported to mediate a variety of beneficial effects, such as improved mood, improved physical and mental stamina, and enhanced protection against high altitude sickness (Kelly 2001). The extract has also been reported to protect against tumor progression in mice, improve endurance in rats, improve blood glucose profiles in diabetic mice, and protect snail eggs against oxidative stress, heat, and heavy metals (Udintsev and Schakhov 1991; Boon-Niermeijer et al. 2000; Abidov et al. 2003; Kim et al. 2006). In addition, *R. rosea* root extract has been found to extend lifespan in *D. melanogaster*, *C. elegans*, *S. cerevisiae*, and more recently, the silkworm, *Bombyx mori* (Jafari et al. 2007; Schriner et al. 2009a, 2013; Wiegant et al. 2009; Bayliak and Lushchak 2011; Chen et al. 2015). The molecular action of *R. rosea* is not known, but has been hypothesized to act as an antioxidant (Kelly 2001). This is consistent with results that the extract can protect flies and human cultured cells against oxidative stress, and decrease the production of H_2O_2 from isolated fly mitochondria (Schriner et al. 2009a, b).

Some have suggested that *R. rosea* may act through hormesis (Mattson 2008a, b). Consistent with this, the extract did have a dose dependent activation of the antioxidant response element (ARE) in cultured cells (Schriner et al. 2009b). However, there was no measurable effect on the major antioxidant enzymes or on the ratio of reduced to oxidized glutathione, a marker of oxidative status (Schriner et al. 2009b). In addition, the minimum dose of extract needed to activate the ARE was 1 μg/mL, whereas 1 ng/mL was sufficient to confer a protective effect against H_2O_2, a concentration 3 orders of magnitude lower. As a result, it is unlikely that *R. rosea* acts as hormetic agent at the protective doses. More likely, compounds within the extract are inhibiting some signaling pathway.

It has also been suggested that *R. rosea* may act through a pathway related to dietary restriction (DR), e.g., as a DR mimetic. Several natural products and drugs have been proposed to act as DR mimetics. Examples include resveratrol, rapapmycin, and metformin (de Cabo et al. 2014). *Rhodiola rosea* could conceivably work in a similar manner. However, in flies, no evidence has been found to support a role of *R. rosea* in three DR-related pathways: the silent information regulator 2 (Sir2) homologs, the target of rapamycin (TOR), and insulin and insulin-like signaling (Schriner et al. 2013). The extract has been found to specifically kill cancerous bladder cells (Liu et al. 2012). In this study, *R. rosea* did appear to inhibit TOR. However, the minimum dose needed was 100 μg/mL, 5 orders of magnitude greater that the minimum need to protect against H_2O_2. The inactivation of TOR may be secondary to killing by *R. rosea* or an artifact due to the high doses used.

While *R. rosea* extended lifespan in flies, it provided no protective effect against starvation or desiccation in either sex, and was unable to protect males against incubation at 37 °C. Though females were protected at 37 °C (Schriner et al. 2013). There was no effect on HSP70 expression levels, and HSP22, a mitochondrial localized HSP, was actually down-regulated (Schriner et al. 2013). Since *R. rosea* extends lifespan in both sexes, protection against heat was not a prerequisite for its action. The decreased expression of HSP22 is curious, and may suggest that *R. rosea* provides some protective effect, possibly specific to the mitochondria, such that normal amounts of HSP22 are no longer required. In any case, there is no clear effect or requirement of HSPs on the action of *R. rosea* root extract.

12.2.2 Cinnamon

Cinnamon is a spice commonly used in various cuisines around the world to flavor desserts, fruits, cereals, breads, and meats. It is derived from the bark of trees in the genus *Cinnamomum*, most commonly, *C. cassia*. The extract has been reported to exhibit numerous health benefits, including anti-inflammatory, antimicrobial, antioxidant, and anti-cancer action (Osawa et al. 1991; Kim et al. 1995, 2007; Friedman et al. 2002; Hong et al. 2002; Huss et al. 2002; Lin et al. 2003; Lopez et al. 2005; Koppikar et al. 2010; Jaganathan and Supriyanto 2012). There has been a recent interest in the ability of cinnamon to have a positive effect on diabetes. In particular, several clinical studies on the anti-diabetic properties of cinnamon may create an important role for this botanical in the management of diabetes and pre-diabetes (Kirkham et al. 2009; Davis and Yokoyama 2011; Lu et al. 2012; Ranasinghe et al. 2012).

Like *R. rosea*, the molecular action of cinnamon is not known. However, cinnamon and one of its putative active compounds, cinnamaldehyde, have been shown to elevate the expression levels of peroxisome proliferator-activated receptor γ (PPARγ) and activate AMP kinase (AMPK) (Sheng et al. 2008; Huang et al. 2011). These activities mimic the action of the thiazolidinediones and metformin; commonly used anti-diabetic drugs (Zhou et al. 2001; Scarsi et al. 2007; Zhang et al. 2007; Lee et al. 2011). Both PPARγ and AMPK elevate metabolism in part by activating mitochondrial biogenesis (Alaynick 2008; Haemmerle et al. 2011; Hardie 2011). Cinnamon has also shown a complex relationship with insulin signaling where it activates insulin-like growth factor 1 (IGF1) signaling in fibroblasts, but down-regulates insulin signaling in adipocytes (Cao et al. 2010; Takasao et al. 2012). Another compound present in cinnamon, β-caryophyllene oxide, has been suggested to activate the mammalian target of rapamycin (mTOR) (Park et al.

2011), which is a complex of proteins that centrally regulates numerous cellular and metabolic processes (Schmelzle and Hall 2000; Wullschleger et al. 2006). In addition, cinnamaldehdye, and eugenol, another constituent compound of cinnamon, have been shown to directly modulate mitochondrial physiology (Usta et al. 2002). Clearly, cinnamon seems to exhibit significant effects on energy metabolism.

Since all of these pathways have been implicated in aging, it is not unexpected that cinnamon can extend fly lifespan (Schriner et al. 2014). Cinnamon also improved fly climbing ability, though its protective effect against oxidative and environmental stresses is quite discordant, showing marked sex-specific effects. Cinnamon protected females, but not males, against paraquat, a superoxide generator, had no effect against H_2O_2 in either sex, and sensitized females to iron. Males were sensitized to starvation and desiccation, while both sexes were sensitized to heat (Schriner et al. 2014). Despite the latter finding, HSP70 was up-regulated, while the levels of HSP22 were unchanged (Schriner et al. 2014). It is somewhat perplexing and difficult to explain how cinnamon can induce HSP70, while yet at the same time sensitize flies to heat stress. A speculative explanation is that cinnamon itself mimics heat stress. Subjecting flies to heat can increase lifespan (Lagisz et al. 2013). Perhaps cinnamon, like heat stress induces HSP70, which confers an extended lifespan at 22 °C, the temperature at which the flies are normally cultured. However, at 37 °C, the combination of cinnamon and the additional heat is too overwhelming for the flies, and their survival is then compromised.

12.2.3 Rosa damascena

Rosa damascena is a hybrid rose species predominantly grown in Turkey and Bulgaria. It is commonly used to produce rose oil and rose water for perfume, and for the cosmetic and food industries. The extract has also been reported to have several medicinal properties. It has an anti-microbial activity; it is able to protect neurons against amyloid β toxicity, a major pathological component of Alzheimer's disease, and to protect rats against seizures (Basim and Basim 2003; Ramezani et al. 2008; Shokouhinejad et al. 2010; Awale et al. 2011). The active components of *R. damascena* are not known. However, its oil is composed of a large number of volatile organic compounds including various terpenes such as citrenellol, heneicosane, and disiloxane (Loghmani-Khouzani et al. 2007). The marc, material left after rose oil is extracted, has significant polyphenol content, including quercetin, myricetin, kaempferol and gallic acid (Kumar et al. 2008). Though, the predominant molecules have been suggested to be glycosides of quercetin and kaempferol (Kumar et al. 2008).

The petals of *R. damascena* have been found to extend lifespan in fruit flies. This correlated with protection against paraquat, H_2O_2, and iron in female flies, but males only had and elevated protection against iron insult (Schriner et al. 2012). No effect was detected on the major antioxidant enzymes, superoxide dismutase and catalase, nor was there any effect on starvation resistance. The most striking feature of *R.*

damascena, was that conferred a marked decrease in tolerance to heat stress (survival at 37 °C), coupled with decreased expression levels of both HSP70 and HSP22 (Schriner et al. 2012). This is a more consistent finding that what was seen in cinnamon. Here the decreased heat tolerance can be correlated to decreased HSP expression levels. What's curious is how these effects relate to the extended lifespan. It's tempting to imagine that the compounds in *R. damascena* are themselves affording an enhanced protective effect of some sort, resulting in the longer lifespan. It these were true; it could be that the HSPs were then down-regulated as they were less needed under normative conditions (22 °C). However, once the flies were subjected to a heat challenge (37 °C), the decreases HSP expression levels resulted in compromised survival. No matter the reason, *R. damascena* is dramatic example of an inverse relationship between stress resistance and lifespan.

12.3 Other Anti-aging Plants and Plant Products

The are several other natural products that have been found to extend lifespan in fruit flies with mixed effects on heat tolerance and HSP expression levels. Green tea polyphenols extend lifespan in male flies, but have no effect on antioxidant defenses and do not afford any protective effect against oxidative stress, starvation, or desiccation (Lopez et al. 2014). These polyphenols do protect male flies against iron toxicity, and it was hypothesized that the restriction of iron entering the organism actually hindered reproduction in males which then conferred the lifespan extension. Expression levels of HSPs were not reported, but green tea polyphenols had a moderate negative effect on heat tolerance in males (Lopez et al. 2014). Cranberries, on the other hand, extend lifespan and exhibit and elevated expression levels of HSF and HSP68 (Wang et al. 2014). Korean mistletoe extended fly lifespan while inducing Sir2 expression and sensitizing flies to starvation (Lee et al. 2014). However, the mistletoe extract did not exhibit any effect on heat tolerance. Extract of the nectarine fruit can extend female fly lifespan and protect flies against a high fat diet (Boyd et al. 2011). In the latter case, they also down-regulate two HSPs, HSP68 and l(2)efl (Boyd et al. 2011). Pectins, a polysaccharide, present in fruits have also been found to extend lifespan in flies, and they induce HSP70 (Shaposhnikov et al. 2014). Similarly, polysaccharides isolated from *Porphyra haitanensis*, a type of seaweed, extend lifespan in flies and protect them against heat stress (Zhao et al. 2008).

12.4 Conclusion

Many diseases are associated with aging, which is underlying a significant research effort to understand and treat the process of aging. One of the strategies is to identify new treatments for aging and age-related is to use short-lived model organisms, such as fruit flies, to screen for drugs or extracts that can extend lifespan. Several

Table 12.1 Botanical extracts that extend lifespan in *Drosophila melanogaster* and their effects on heat tolerance and heat shock protein expression

Extract	Heat tolerance	HSP expression	Reference
Rhodiola rosea	Females protected	HSP22 down	Schriner et al. (2013)
Cinnamon	Both sexes sensitized	HSP70 up	Schriner et al. (2014)
Rosa damascena	Both sexes sensitized	HSP70/22 down	Schriner et al. (2012)
Green tea	Males sensitized	Not reported	Lopez et al. (2014)
Cranberries	Not reported	HSF/HSP68 up	Wang et al. (2014)
Korean mistletoe	No effect in either sex	Not reported	Lee at al. (2014)
Nectarine	Not reported	HSP68/l(2)efl up	Boyd et al. (2011)
Fruit pectins	Not reported	HSP70 up	Shaposhnikov et al. (2014)
Porphyra haitanensis	Males protected	Not reported	Zhao et al. (2008)

such extracts have been described here. While the underlying cause of aging is not known, heat shock proteins, which protect us against heat and many types of environmental stresses, may play an important role. Unexpectedly, there is no obvious relationship between these lifespan-extending extracts and protection against heat and/or the regulation of HSPs (Table 12.1). Some of these plant extracts induce HSPs, while others down-regulate them. Some enhance protection against heat, while others have no effect or actually sensitize flies to heat. The most striking inverse relationship is that of *Rosa damascena*. This extract extends lifespan, but sensitizes both sexes to heat and down-regulates the two HSPs measured (Schriner et al. 2012). The overall relationship between HSP expression and lifespan extension through the use of plant extracts parallels that seen with heat shock alone. Individual experiments sometimes reveal direct, nonexistent, or inverse relationships between heat shock and lifespan (Lagisz et al. 2013). In conclusion, extracts that can extend lifespan in short-lived organism have been identified. However, there appears to be no consistent relationship between heat tolerance, HSP expression and lifespan extension.

Acknowledgments I would like to thank Sally T. Pham and Benjamin V. Nguyen for the critical reading of this manuscript and providing helpful comments.

References

Abidov M, Crendal F, Grachev S, Seifulla R, Ziegenfuss T (2003) Effect of extracts from *Rhodiola rosea* and *Rhodiola crenulata* (Crassulaceae) roots on ATP content in mitochondria of skeletal muscles. Bull Exp Biol Med 136:585–587

Alaynick WA (2008) Nuclear receptors, mitochondria and lipid metabolism. Mitochondrion 8:329–337

Awale S, Tohda C, Tezuka Y, Miyazaki M, Kadota S (2011) Protective effects of *Rosa damascena* and its active constituent on Abeta(25–35)-induced neuritic atrophy. Evid Based Complement Alternat Med 2011:131042
Basim E, Basim H (2003) Antibacterial activity of *Rosa damascena* essential oil. Fitoterapia 74:394–396
Bayliak MM, Lushchak VI (2011) The golden root, *Rhodiola rosea*, prolongs lifespan but decreases oxidative stress resistance in yeast *Saccharomyces cerevisiae*. Phytomedicine 18:1262–1268
Beckman KB, Ames BN (1998) The free radical theory of aging matures. Physiol Rev 78:547–581
Bhole D, Allikian MJ, Tower J (2004) Doxycycline-regulated over-expression of hsp22 has negative effects on stress resistance and life span in adult *Drosophila melanogaster*. Mech Ageing Dev 125:651–663
Boon-Niermeijer EK, van den Berg A, Wikman G, Wiegant FA (2000) Phyto-adaptogens protect against environmental stress-induced death of embryos from the freshwater snail *Lymnaea stagnalis*. Phytomedicine 7:389–399
Boyd O, Weng P, Sun X, Alberico T, Laslo M, Obenland DM, Kern B, Zou S (2011) Nectarine promotes longevity in *Drosophila melanogaster*. Free Radic Biol Med 50:1669–1678
Cao H, Graves DJ, Anderson RA (2010) Cinnamon extract regulates glucose transporter and insulin-signaling gene expression in mouse adipocytes. Phytomedicine 17:1027–1032
Chen C, Song J, Chen M, Li Z, Tong X, Hu H, Xiang Z, Lu C, Dai F (2015) *Rhodiola rosea* extends lifespan and improves stress tolerance in silkworm, *Bombyx mori*. Biogerontology 17:373–381
Crawford NM, Steiner AZ (2015) Age-related infertility. Obstet Gynecol Clin N Am 42:15–25
Davis PA, Yokoyama W (2011) Cinnamon intake lowers fasting blood glucose: meta-analysis. J Med Food 14:884–889
de Cabo R, Carmona-Gutierrez D, Bernier M, Hall MN, Madeo F (2014) The search for antiaging interventions: from elixirs to fasting regimens. Cell 157:1515–1526
Ferguson M, Sohal BH, Forster MJ, Sohal RS (2007) Effect of long-term caloric restriction on oxygen consumption and body temperature in two different strains of mice. Mech Ageing Dev 128:539–545
Finch C (1990) Longevity, senescence, and the genome. The University of Chicago Press, Chicago
Finch CE, Pike MC, Witten M (1990) Slow mortality rate accelerations during aging in some animals approximate that of humans. Science 249:902–905
Friedman M, Henika PR, Mandrell RE (2002) Bactericidal activities of plant essential oils and some of their isolated constituents against *Campylobacter jejuni*, *Escherichia coli*, *Listeria monocytogenes*, and *Salmonella enterica*. J Food Prot 65:1545–1560
Gompertz B (1825) On the nature of the function expressive of the law of human mortality, and on a new mode of determining the value of life contingencies. Philos Trans R Soc Lond 115:513–585
Haemmerle G, Moustafa T, Woelkart G, Buttner S, Schmidt A, van de Weijer T, Hesselink M, Jaeger D, Kienesberger PC, Zierler K, Schreiber R, Eichmann T, Kolb D, Kotzbeck P, Schweiger M, Kumari M, Eder S, Schoiswohl G, Wongsiriroj N, Pollak NM, Radner FP, Preiss-Landl K, Kolbe T, Rulicke T, Pieske B, Trauner M, Lass A, Zimmermann R, Hoefler G, Cinti S, Kershaw EE, Schrauwen P, Madeo F, Mayer B, Zechner R (2011) ATGL-mediated fat catabolism regulates cardiac mitochondrial function via PPAR-alpha and PGC-1. Nat Med 17:1076–1085
Hardie DG (2011) AMP-activated protein kinase: an energy sensor that regulates all aspects of cell function. Genes Dev 25:1895–1908
Hong CH, Hur SK, Oh OJ, Kim SS, Nam KA, Lee SK (2002) Evaluation of natural products on inhibition of inducible cyclooxygenase (COX-2) and nitric oxide synthase (iNOS) in cultured mouse macrophage cells. J Ethnopharmacol 83:153–159

Huang B, Yuan HD, Kim do Y, Quan HY, Chung SH (2011) Cinnamaldehyde prevents adipocyte differentiation and adipogenesis via regulation of peroxisome proliferator-activated receptor-gamma (PPARgamma) and AMP-activated protein kinase (AMPK) pathways. J Agric Food Chem 59:3666–3673

Huss U, Ringbom T, Perera P, Bohlin L, Vasange M (2002) Screening of ubiquitous plant constituents for COX-2 inhibition with a scintillation proximity based assay. J Nat Prod 65:1517–1521

Jafari M, Felgner JS, Bussel II, Hutchili T, Khodayari B, Rose MR, Vince-Cruz C, Mueller LD (2007) Rhodiola: a promising anti-aging Chinese herb. Rejuvenation Res 10:587–602

Jaganathan SK, Supriyanto E (2012) Antiproliferative and molecular mechanism of eugenol-induced apoptosis in cancer cells. Molecules 17:6290–6304

Kelly GS (2001) Rhodiola rosea: a possible plant adaptogen. Altern Med Rev 6:293–302

Kim SJ, Han D, Moon KD, Rhee JS (1995) Measurement of superoxide dismutase-like activity of natural antioxidants. Biosci Biotechnol Biochem 59:822–826

Kim SH, Hyun SH, Choung SY (2006) Antioxidative effects of *Cinnamomi cassiae* and *Rhodiola rosea* extracts in liver of diabetic mice. Biofactors 26:209–219

Kim DH, Kim CH, Kim MS, Kim JY, Jung KJ, Chung JH, An WG, Lee JW, Yu BP, Chung HY (2007) Suppression of age-related inflammatory NF-kappaB activation by cinnamaldehyde. Biogerontology 8:545–554

Kirkham S, Akilen R, Sharma S, Tsiami A (2009) The potential of cinnamon to reduce blood glucose levels in patients with type 2 diabetes and insulin resistance. Diabetes Obes Metab 11:1100–1113

Koppikar SJ, Choudhari AS, Suryavanshi SA, Kumari S, Chattopadhyay S, Kaul-Ghanekar R (2010) Aqueous cinnamon extract (ACE-c) from the bark of *Cinnamomum cassia* causes apoptosis in human cervical cancer cell line (SiHa) through loss of mitochondrial membrane potential. BMC Cancer 10:210

Kumar N, Bhandari P, Singh B, Gupta AP, Kaul VK (2008) Reversed phase-HPLC for rapid determination of polyphenols in flowers of rose species. J Sep Sci 31:262–267

Lagisz M, Hector KL, Nakagawa S (2013) Life extension after heat shock exposure: assessing meta-analytic evidence for hormesis. Ageing Res Rev 12:653–660

Lee KW, Ku YH, Kim M, Ahn BY, Chung SS, Park KS (2011) Effects of sulfonylureas on peroxisome proliferator-activated receptor gamma activity and on glucose uptake by thiazolidinediones. Diabetes Metab J 35:340–347

Lee SH, An HS, Jung YW, Lee EJ, Lee HY, Choi ES, An SW, Son H, Lee SJ, Kim JB, Min KJ (2014) Korean mistletoe (*Viscum album coloratum*) extract extends the lifespan of nematodes and fruit flies. Biogerontology 15:153–164

Li YM, Chan HY, Huang Y, Chen ZY (2007) Green tea catechins upregulate superoxide dismutase and catalase in fruit flies. Mol Nutr Food Res 51:546–554

Liao PC, Lin HY, Yuh CH, Yu LK, Wang HD (2008) The effect of neuronal expression of heat shock proteins 26 and 27 on lifespan, neurodegeneration, and apoptosis in *Drosophila*. Biochem Biophys Res Commun 376:637–641

Lin CC, Wu SJ, Chang CH, Ng LT (2003) Antioxidant activity of *Cinnamomum cassia*. Phytother Res 17:726–730

Liu Z, Li X, Simoneau AR, Jafari M, Zi X (2012) *Rhodiola rosea* extracts and salidroside decrease the growth of bladder cancer cell lines via inhibition of the mTOR pathway and induction of autophagy. Mol Carcinog 51:257–267

Loeb JN, Northrop JH (1917) On the influence of food and temperature upon the duration of life. J Biol Chem 32:103–121

Loghmani-Khouzani H, Sabzi Fini O, Safari J (2007) Essential oil composition of *Rosa damascena* mill cultivated in central Iran. Sci Iran 14:316–319

Lopez P, Sanchez C, Batlle R, Nerin C (2005) Solid- and vapor-phase antimicrobial activities of six essential oils: susceptibility of selected foodborne bacterial and fungal strains. J Agric Food Chem 53:6939–6946

Lopez T, Schriner SE, Okoro M, Lu D, Chiang BT, Huey J, Jafari M (2014) Green tea polyphenols extend the lifespan of male *Drosophila melanogaster* while impairing reproductive fitness. J Med Food 17:1314–1321
Lu T, Sheng H, Wu J, Cheng Y, Zhu J, Chen Y (2012) Cinnamon extract improves fasting blood glucose and glycosylated hemoglobin level in Chinese patients with type 2 diabetes. Nutr Res 32:408–412
Mattson MP (2008a) Dietary factors, hormesis and health. Ageing Res Rev 7:43–48
Mattson MP (2008b) Hormesis defined. Ageing Res Rev 7:1–7
McCarter RJ (1990) Age-related changes in skeletal muscle function. Aging (Milano) 2:27–38
Medvedev ZA (1990) An attempt at a rational classification of theories of ageing. Biol Rev Camb Philos Soc 65:375–398
Mockett RJ, Sohal RS (2006) Temperature-dependent trade-offs between longevity and fertility in the *Drosophila* mutant, methuselah. Exp Gerontol 41:566–573
Morley A (1998) Somatic mutation and aging. Ann N Y Acad Sci 854:20–22
Morrow G, Samson M, Michaud S, Tanguay RM (2004) Overexpression of the small mitochondrial Hsp22 extends *Drosophila* life span and increases resistance to oxidative stress. FASEB J 18:598–599
Mozaffarian D, Benjamin EJ, Go AS, Arnett DK, Blaha MJ, Cushman M, de Ferranti S, Despres JP, Fullerton HJ, Howard VJ, Huffman MD, Judd SE, Kissela BM, Lackland DT, Lichtman JH, Lisabeth LD, Liu S, Mackey RH, Matchar DB, McGuire DK, Mohler ER, Moy CS, Muntner P, Mussolino ME, Nasir K, Neumar RW, Nichol G, Palaniappan L, Pandey DK, Reeves MJ, Rodriguez CJ, Sorlie PD, Stein J, Towfighi A, Turan TN, Virani SS, Willey JZ, Woo D, Yeh RW, Turner MB (2015) Heart disease and stroke statistics–2015 update: a report from the American Heart Association. Circulation 131:e29–e322
Osawa K, Matsumoto T, Yasuda H, Kato T, Naito Y, Okuda K (1991) The inhibitory effect of plant extracts on the collagenolytic activity and cytotoxicity of human gingival fibroblasts by *Porphyromonas gingivalis* crude enzyme. Bull Tokyo Dent Coll 32:1–7
Pal L, Santoro N (2003) Age-related decline in fertility. Endocrinol Metab Clin N Am 32:669–688
Park K, Nam RD, Yun HM, Lee SG, Jang HJ, Sethi G, Cho SK, Ahn KS (2011) beta-Caryophyllene oxide inhibits growth and induces apoptosis through the suppression of PI3K/AKT/mTOR/S6K1 pathways and ROS-mediated MAPKs activation. Cancer Lett 312:178–188
Pellicer A, Simon C, Remohi J (1995) Effects of aging on the female reproductive system. Hum Reprod 10:77–83
Ramezani R, Moghimi A, Rakhshandeh H, Ejtehadi H, Kheirabadi M (2008) The effect of *Rosa damascena* essential oil on the amygdala electrical kindling seizures in rat. Pak J Biol Sci 11:746–751
Ranasinghe P, Jayawardana R, Galappaththy P, Constantine GR, de Vas Gunawardana N, Katulanda P (2012) Efficacy and safety of 'true' cinnamon (*Cinnamomum zeylanicum*) as a pharmaceutical agent in diabetes: a systematic review and meta-analysis. Diabet Med 29:1480–1492
Rose MR, Graves JL Jr (1989) What evolutionary biology can do for gerontology. J Gerontol 44:B27–B29
Rose MR, Rauser CL, Mueller LD, Benford G (2006) A revolution for aging research. Biogerontology 7:269–277
Scarsi M, Podvinec M, Roth A, Hug H, Kersten S, Albrecht H, Schwede T, Meyer UA, Rucker C (2007) Sulfonylureas and glinides exhibit peroxisome proliferator-activated receptor gamma activity: a combined virtual screening and biological assay approach. Mol Pharmacol 71:398–406
Schmelzle T, Hall MN (2000) TOR, a central controller of cell growth. Cell 103:253–262
Schriner SE, Abrahamyan A, Avanessian A, Bussel I, Maler S, Gazarian M, Holmbeck MA, Jafari M (2009a) Decreased mitochondrial superoxide levels and enhanced protection against paraquat in *Drosophila melanogaster* supplemented with *Rhodiola rosea*. Free Radic Res 43:836–843

Schriner SE, Avanesian A, Liu Y, Luesch H, Jafari M (2009b) Protection of human cultured cells against oxidative stress by *Rhodiola rosea* without activation of antioxidant defenses. Free Radic Biol Med 47:577–584
Schriner SE, Katoozi NS, Pham KQ, Gazarian M, Zarban A, Jafari M (2012) Extension of *Drosophila* lifespan by *Rosa damascena* associated with an increased sensitivity to heat. Biogerontology 13:105–117
Schriner SE, Lee K, Truong S, Salvadora KT, Maler S, Nam A, Lee T, Jafari M (2013) Extension of *Drosophila* lifespan by *Rhodiola rosea* through a mechanism independent from dietary restriction. PLoS One 8:e63886
Schriner SE, Kuramada S, Lopez TE, Truong S, Pham A, Jafari M (2014) Extension of *Drosophila* lifespan by cinnamon through a sex-specific dependence on the insulin receptor substrate *chico*. Exp Gerontol 60:220–230
Shaposhnikov M, Latkin D, Plyusnina E, Shilova L, Danilov A, Popov S, Zhavoronkov A, Ovodov Y, Moskalev A (2014) The effects of pectins on life span and stress resistance in *Drosophila melanogaster*. Biogerontology 15:113–127
Sheng X, Zhang Y, Gong Z, Huang C, Zang YQ (2008) Improved insulin resistance and lipid metabolism by cinnamon extract through activation of peroxisome proliferator-activated receptors. PPAR Res 2008:581348
Shokouhinejad N, Emaneini M, Aligholi M, Jabalameli F (2010) Antimicrobial effect of *Rosa damascena* extract on selected endodontic pathogens. J Calif Dent Assoc 38:123–126
Takasao N, Tsuji-Naito K, Ishikura S, Tamura A, Akagawa M (2012) Cinnamon extract promotes type I collagen biosynthesis via activation of IGF-I signaling in human dermal fibroblasts. J Agric Food Chem 60:1193–1200
Thompson LV (2009) Age-related muscle dysfunction. Exp Gerontol 44:106–111
Tower J (2011) Heat shock proteins and *Drosophila* aging. Exp Gerontol 46:355–362
Udintsev SN, Schakhov VP (1991) Decrease of cyclophosphamide haematotoxicity by *Rhodiola rosea* root extract in mice with Ehrlich and Lewis transplantable tumors. Eur J Cancer 27:1182
Usta J, Kreydiyyeh S, Bajakian K, Nakkash-Chmaisse H (2002) *In vitro* effect of eugenol and cinnamaldehyde on membrane potential and respiratory chain complexes in isolated rat liver mitochondria. Food Chem Toxicol 40:935–940
Wang HD, Kazemi-Esfarjani P, Benzer S (2004) Multiple-stress analysis for isolation of *Drosophila* longevity genes. Proc Natl Acad Sci U S A 101:12610–12615
Wang C, Yolitz J, Alberico T, Laslo M, Sun Y, Wheeler CT, Sun X, Zou S (2014) Cranberry interacts with dietary macronutrients to promote healthy aging in *Drosophila*. J Gerontol A Biol Sci Med Sci 69:945–954
West JD, Wang Y, Morano KA (2012) Small molecule activators of the heat shock response: chemical properties, molecular targets, and therapeutic promise. Chem Res Toxicol 25:2036–2053
Wiegant FA, Surinova S, Ytsma E, Langelaar-Makkinje M, Wikman G, Post JA (2009) Plant adaptogens increase lifespan and stress resistance in *C. elegans*. Biogerontology 10:27–42
Wullschleger S, Loewith R, Hall MN (2006) TOR signaling in growth and metabolism. Cell 124:471–484
Zhang L, He H, Balschi JA (2007) Metformin and phenformin activate AMP-activated protein kinase in the heart by increasing cytosolic AMP concentration. Am J Physiol Heart Circ Physiol 293:H457–H466
Zhao T, Zhang Q, Qi H, Liu X, Li Z (2008) Extension of life span and improvement of vitality of *Drosophila melanogaster* by long-term supplementation with different molecular weight polysaccharides from *Porphyra haitanensis*. Pharmacol Res 57:67–72
Zhou G, Myers R, Li Y, Chen Y, Shen X, Fenyk-Melody J, Wu M, Ventre J, Doebber T, Fujii N, Musi N, Hirshman MF, Goodyear LJ, Moller DE (2001) Role of AMP-activated protein kinase in mechanism of metformin action. J Clin Invest 108:1167–1174

Chapter 13
Molecular Chaperones and HSPs in Sugarcane and Eucalyptus

Conrado C. Gonçalves and Carlos H.I. Ramos

Abstract The sessile existence of plants makes these organisms more exposed to unfavorable environmental changes than animals and more likely to have evolved sophisticated ways to combat stresses. Therefore, knowledge about the network of molecular chaperones and heat shock proteins (HSPs) in plants is of great interest not only to improve agricultural production but also to enhance our understanding of the cellular protein-folding process. In this chapter we will review the use of bioinformatics to identify and annotate 5′EST-contigs belonging to molecular chaperones within plant genomes, with an emphasis on sugarcane and eucalyptus. The chapter will show that information concerning the diversity and quantity of expressed mRNAs under diverse developmental and environmental conditions has led to new insights on specific proteins' importance and activities in response to environmental conditions sensed by these organisms. The general findings are as follows: Chaperone and stress-related protein genes are abundantly expressed and have ample diversity. Cytoplasmic chaperones have both higher expression and greater diversity than those from other cellular compartments. Findings regarding cDNA cloning and protein purification and characterization will also be discussed.

Keywords Protein folding • Molecular chaperone • Heat shock protein • HSP90 • HSP70 • Plant

Abbreviations

AUC	analytical ultracentrifugation
CD	circular dichroism
HSP	heat shock protein
SEC-MALS	size exclusion chromatography coupled to multi-angle light scattering

C.C. Gonçalves • C.H.I. Ramos (✉)
Department of Organic Chemistry, Institute of Chemistry, University of Campinas – UNICAMP, Campinas, SP 13083-970, Brazil
e-mail: conradodecampos@gmail.com; cramos@iqm.unicamp.br

A.A.A. Asea et al. (eds.), *Heat Shock Proteins and Plants*,
Heat Shock Proteins 10, DOI 10.1007/978-3-319-46340-7_13

13.1 Introduction

13.1.1 Protein Folding, Structure and Function

Proteins are one of the most important classes of biomolecules, as most of the cellular functions are carried out by these macromolecules (Voet and Voet 2011). The function of a protein is closely related to its structure, which is reached under a specific set of environmental conditions by spontaneous folding into a tridimensional conformation that corresponds to its minimum Gibbs free energy (Anfinsen 1973). Thus, the structure of a protein is determined thermodynamically and may unfold/refold, completely or partially, depending on changes in environmental conditions. Folding is driven by the phenomenon known as hydrophobic collapse, in which proteins have their hydrophobic amino acids buried in their core and polar or charged residues on the surface. Consequently, the isolation of hydrophobic residues from the aqueous environment energetically stabilizes the protein and compensates thermodynamically for the unfavorable entropy caused by the organized structure (Levitt and Warshel 1975; Ramos and Ferreira 2005; Baldwin and Rose 2013; Goloubinoff 2014; Uversky 2015). The aggregation of a polypeptide chain competes with the drive to fold (Zettlmeissl et al. 1979), and during unfolding, proteins expose hydrophobic residues that were buried in the native state and potentiate the formation of aggregates, which could be harmful to the cell (Ferreira and De Felice 2001; Ramos and Ferreira 2005; Goloubinoff 2014; Knowles et al. 2014; Tipping et al. 2015).

Protein folding and misfolding can be modeled into a funnel-shaped schematic energy landscape (Fig. 13.1) that can explain the complexity of the process and also its pathways (Baldwin 1995; Ferreira and De Felice 2001; Ramos and Ferreira 2005). In this scheme, unfolded proteins are found in the upper region where free-energy is maximal. Because of the large number of putative structures that a polypeptide chain is able to assume, protein folding can follow multiple folding pathways without necessarily going through a single pathway. Moreover, the free-energy surface that a polypeptide must follow in order to achieve its native state is very rugged, implying that there is a considerable energetic barrier to be crossed between folding intermediates (Fig. 13.1).

The crowded cellular environment facilitates protein unfolding and the formation of aggregates, a condition that is worsened by environmental forms of stress, such as temperature changes, that disturb cellular homeostasis in all organisms. In such conditions, the hydrophobic collapse tends to lead to the arrangement of amorphous aggregates, which can be very large since there are hydrophobic interactions between polypeptides instead of within a polypeptide. In these cases, if the intermediates (partially folded or misfolded) become numerous, a problematic cellular situation can occur due to the increased probability of aggregate formation (Ferreira and De Felice 2001; Ramos and Ferreira 2005; Goloubinoff 2014; Knowles et al. 2014; Tipping et al. 2015). The intracellular accumulation of misfolded proteins can be a serious threat to the health of an organism. Deposition of large aggregates

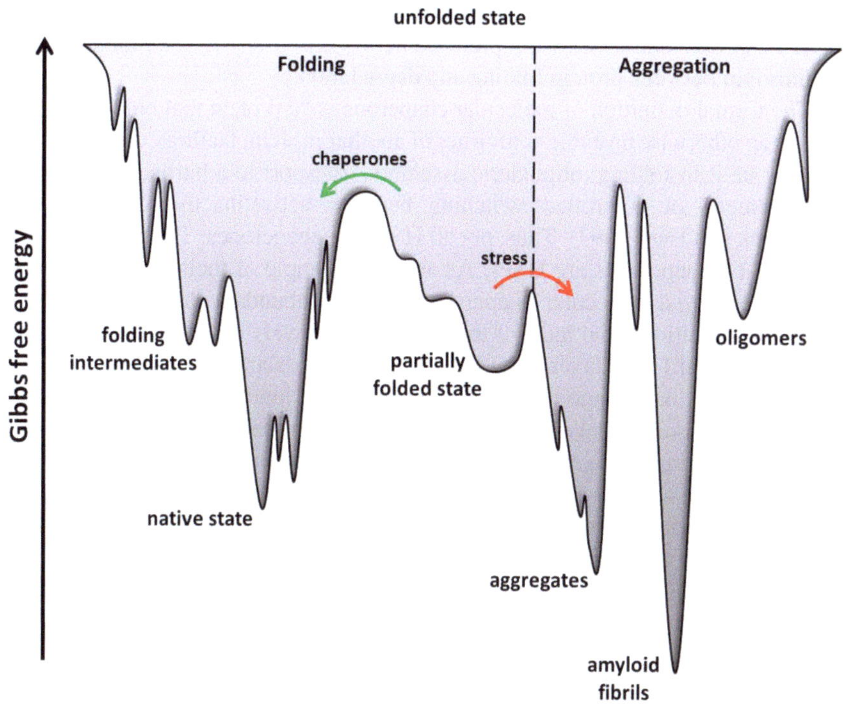

Fig. 13.1 Scheme of the funnel-shaped Gibbs free-energy landscape. Unfolded, recently synthesized proteins are at the upper region, in which free energy is maximal. The free-energy surface is very rugged because protein folding can follow multiple pathways. On such a rugged surface, there are many conformations of folding intermediates that must cross an energetic barrier to achieve the protein's native state. During stress situations, aggregation processes can predominate, creating harmful species such as toxic oligomers, aggregates and amyloid fibrils, but in the cell, the folding process is usually facilitated by molecular chaperones (Adapted from Hartl et al. 2011)

causes loss of function or even gain of toxic function, which generally culminates in diverse disorders such as Parkinson's and Alzheimer's diseases or cystic fibrosis (Ferreira and De Felice 2001; Ramos and Ferreira 2005; Luheshi et al. 2008).

13.1.2 Molecular Chaperones and Heat Shock Proteins (HSPs) Are Part of the Protein Quality Control (PQC) System

To oppose the nonproductive, and sometimes even harmful, effects of misfolding and aggregation, cells evolved a refined system, named Protein Quality Control (PQC), to maintain proteostasis by maximizing proper folding and eliminating damaged polypeptides (Douglas et al. 2009; Morimoto 2011; Tiroli-Cepeda and

Ramos 2011; Kim et al. 2013). Molecular chaperones and heat shock proteins (HSPs), in association with the proteasome system, assist in the maintenance of equilibrium between protein folding and degradation.

The formal definition of molecular chaperone is "a protein that binds to and stabilizes an otherwise unstable conformer of another protein, facilitates its correct fate *in vivo*: be it in folding, oligomeric assembly, transport to a particular subcellular compartment, or controlled switching between active/inactive conformations" (Hendrick and Hartl 1993). Thus, not all HSPs are chaperones. The converse is also true: not all chaperones are HSPs. As another example of their importance to the cell homeostasis, molecular chaperones are also abundant during high-protein-synthesis conditions, and most of them are constitutively expressed (Hendrick and Hartl 1993; Hartl and Hayer-Hartl 2002; Tiroli-Cepeda and Ramos 2011).

Chaperones have important and diverse cellular functions in the life cycle of proteins: (1) ensuring that recently synthesized proteins achieve a folding-competent state; (2) assisting in the assembly and disassembly of multiprotein complexes; (3) membrane translocation of proteins destined for cellular compartments; (4) preventing the formation of protein aggregates; (5) refolding proteins misfolded by stressful conditions; (6) disassembling aggregates that have formed (Ramos 2008; Hartl and Hayer-Hartl 2009; Tiroli-Cepeda and Ramos 2011; Priya et al. 2013; Saibil 2013; Brandvold and Morimoto 2015; Landreh et al. 2015).

The mechanism of interaction with substrates is used to classify molecular chaperones and HSPs, generating three general types, referred to as disaggregases, foldases, and holders (Fig. 13.2 and Table 13.1) (Mayer 2010; Tiroli-Cepeda and Ramos 2011). Disaggregases, as the designation indicates, are involved in rescuing aggregated proteins. They are primarily represented by ClpB/HSP104 chaperones, but recent evidences suggest that there is a HSP70-based system that also has disaggregase functions (Slepenkov and Witt 2002; Zietkiewicz et al. 2004; Shorter 2011; Rampelt et al. 2012; Matoo et al. 2013; Mogk et al. 2015; Mokry et al. 2015; Nillegoda and Bukau 2015; O'Driscoll et al. 2015). A disaggregase can function by itself or cooperate with a foldase. Foldases are directly involved in aiding the folding or refolding of proteins by helping them to adopt their native state in an ATP-dependent manner. Classical examples of foldases are chaperonin, HSP70s and HSP90s (Wegele et al. 2004; Gava and Ramos 2009; Young 2010; Makhnevych and Houry 2012; da Silva and Ramos 2012; Jackson 2013; Batista et al. 2015; Clerico et al. 2015; Duncan et al. 2015; Mayer and Le Breton 2015). Holders prevent protein misfolding or aggregation by binding substrate proteins (or client proteins) in an ATP-independent manner and delivering them to other chaperones, commonly a foldase, for further action. Classical examples of holders are small HSPs and HSP40s (Summers et al. 2009; Tiroli-Cepeda and Ramos 2011; Garrido et al. 2012; Pesce and Blatch 2014; Cyr and Ramos 2015; Haslbeck and Vierling 2015).

Molecular chaperones are also classified into five major families by their molecular masses and assist in protein folding in different ways (Borges and Ramos 2005; Tiroli-Cepeda and Ramos 2011) (Fig. 13.2 and Table 13.1). The HSP70 (70 kDa heat shock protein) family plays a central role in assisting protein folding because it is involved in the folding of nascent proteins, participates in disaggregation systems

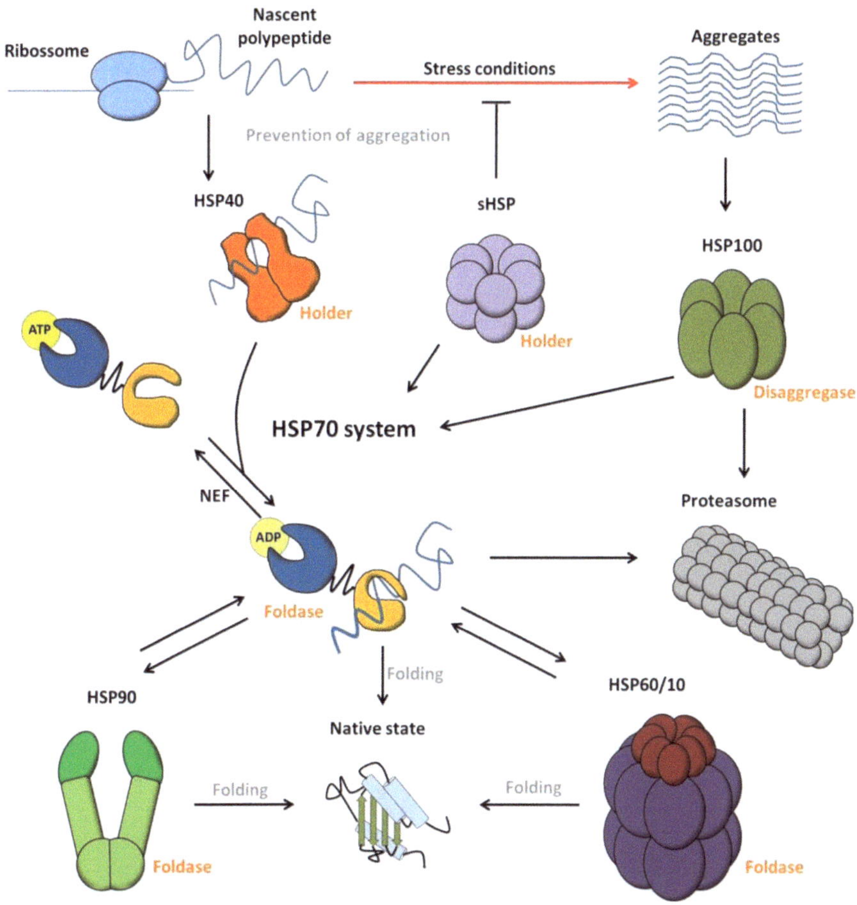

Fig. 13.2 Scheme of main molecular chaperone families and their function in the cell. Unfolded, newly synthesized polypeptides from ribosomes have their hydrophobic domains exposed and interact with holders (mainly HSP40, but also sHSP in stress conditions). Holders deliver client proteins to the HSP70 system that, in an ATP-dependent manner, fold them into the native state, redistribute them to other foldases such as HSP90 or HSP60/10, or else target them for degradation by the proteasome. During stress conditions, if not prevented by sHSP, polypeptides can form aggregates, which can be disassembled by disaggregases such as HSP100. Unless indicated otherwise, *arrows* represent protein transfer between chaperones

and can receive and distribute protein substrates in cooperation with other molecular chaperone families. The HSP100 (100 kDa heat shock protein) family has the ability to recover proteins from aggregates, altering their fate. The HSP90 (90 kDa heat shock protein) family are hub chaperones because they are involved in the maturation of numerous proteins, such as those involved in signal transduction. They act as enablers of phenotypic evolution and have important roles in several diseases, such as cancer and neurodegeneration. The HSP60/HSP10 complex forms a barrel-like structure that encages partially unfolded proteins to provide the proper

Table 13.1 The molecular chaperone families and their main characteristics

Chaperone family	sHSP	HSP60	HSP70	HSP90	HSP100
Mass (kDa)	12–42	~60	~70	~90	~100
Oligomerization (n-mers)	2–32	7 or ~16	1	2	6
Main type	Holder	Foldase	Foldase	Foldase	Disaggregase
Functional characteristics	Prevention of aggregation. Highly expressed in stress conditions (thermotolerance). ATP-independent.	Folding recently synthesized or misfolded client polypeptides. ATP-dependent.	Folding of recently synthesized proteins. Solubilization and refolding of misfolded polypeptides. Disaggregation. Transport across membranes. Distribution of non-native proteins to other chaperones or proteasome. ATP-dependent.	Binding non-native client proteins and keeping them in a refoldable state. ATP-dependent.	Disassembly of aggregates. Prevention of aggregation. ATP-dependent.

Table 13.2 The co-chaperone families and their characteristics

Co-chaperone family	Mass (kDa)	Oligomerization (n-mers)	Functional characteristics
HSP10	~10	7	Acting as a lid for Anfinsen's cage complex.
HSP40	~40	2	Prevention of misfolding and aggregation. Direct client protein to the HSP70 system. Stimulates the ATPase activity of HSP70.
HSP110	~110	1	Nucleotide exchange factor.
TPR proteins	–	–	Binding to EEVD motifs of HSP70 and HSP90

environment for folding and avoiding aggregation. The small HSP (sHSP) family is composed of low-molecular- mass monomers. They lack ATPase activity and work by binding and protecting unfolded and partially folded proteins from aggregation. Additionally, each family has more than one subfamily that is often classified by cellular localization, function and/or expression pattern.

Most chaperones are assisted by co-chaperones, which are greater in number and variety than chaperones (Table 13.2). Surprisingly, single-celled organisms have similar numbers of chaperons as complex ones, but the number and variety of co-chaperones change significantly with organism complexity. This impressive characteristic may be necessary for the quality control system to cope with the more complex intracellular organization characteristic of superior organisms. Co-chaperones interact with client proteins specifically, carrying them to the adequate chaperone and acting to prevent aggregation. Interactions among chaperones, co-chaperones and client proteins generally produce huge complexes that are important to the proper functioning of this system (Bukau and Horwich 1998).

13.1.3 Main Proteins Involved with Stress Response

Below is a brief description of the main proteins involved in the stress response in the cell.

13.1.4 Heat Shock Factor (HSF)

The upregulation of molecular chaperones during stress is caused by the recruitment of heat shock transcription factor 1 (HSF1) to the promoters of chaperone genes, resulting in the induction of their expression (Fujimoto and Nakai 2010; Anckar and-Sistonen 2011; Miozzo et al. 2015). HSF-1 is considered the pivotal regulator of the response to heat stress and is conserved in all eukaryotes. In the cytosol, HSF1 remains inactive, in a monomeric form, bound to the chaperones (mainly HSP90, HSP70 and HSP40) it induces (Abravaya et al. 1992; Shi et al. 1998) (Fig. 13.3).

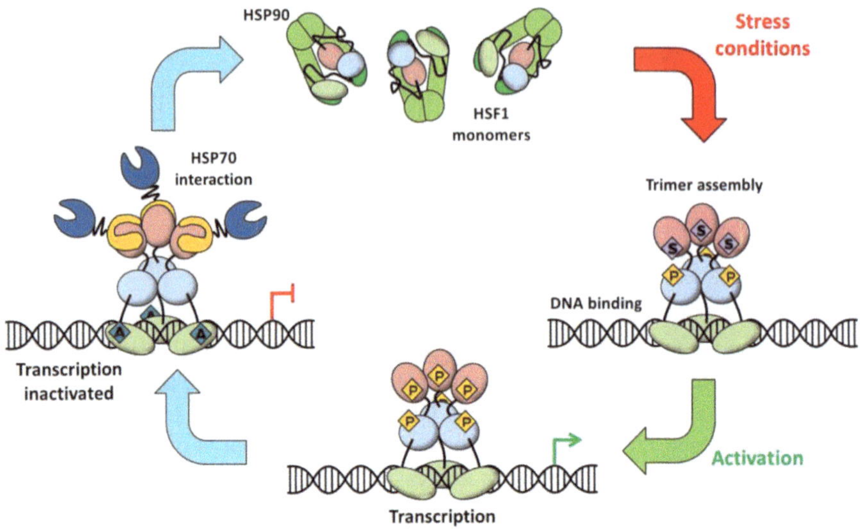

Fig. 13.3 The heat shock factor 1 (HSF1) activation cycle. In the inactive state, HSF1, complexed with HSP90, is found in the cytosol as a monomer. Due to stress signaling, HSF1 dissociates from HSP90, forming trimers and becoming able to interact with DNA, especially in heat shock elements of HSP genes. Phosphorylation and sumoylation are important regulators in this step of the process. After further phosphorylation, HSF1 becomes active, stimulating transcription of HSP genes. Then, attenuation occurs by negative feedback from HSP70 and/or acetylation of the DNA-binding domain of HSF1. Positions of chaperones and HSF1 domains are purely illustrative (Adapted from Akerfelt et al. 2010)

However, upon sufficient induction of the heat shock response pathway, HSF1 is released from the binding of the chaperones and, free of its repressors, translocates to the nucleus in a transcriptionally active DNA-binding trimer (Abravaya et al. 1992; Shi et al. 1998) (Fig. 13.3). Additionally, multiple post-translational regulatory modifications are also involved in the activation of HSF1 (Sorger and Pelham 1998; Hietakangas et al. 2003; Westerheide et al. 2009). Aging is characterized by, among several other characteristics, the reduction of HSP expression, which is caused by defects in the signal transduction pathway that leads to HSF1 activation, (Sóti and Csermely 2000) but since HSF1 is also recruited by many non-chaperones (Mendillo et al. 2012), it may have a broad spectrum of action.

13.1.5 The HSP70/HSP90 System

Among the most important and well studied of the chaperone families are the HSP70 family because of the central role they play in the cellular network of the PQC system (Young 2010; da Silva and Borges 2011; Tiroli-Cepeda and Ramos 2011; Batista et al. 2015; Duncan et al. 2015; Mayer and Kityk 2015) (Fig. 13.2).

These proteins are found in all organisms and are ubiquitous in different cellular compartments, especially in the cytosol, mitochondria and endoplasmic reticulum (Hartl and Mayer-Hartl 2002). HSP70 proteins contain a conserved N-terminal nucleotide- binding domain (NBD), a conserved C-terminal substrate-binding domain (SBD) and a linker between them (Young 2010; da Silva and Borges 2011).

HSP70s are pivot chaperones because they are engaged in a myriad of biological processes, including the folding of recently synthesized proteins to the native state, prevention of aggregation, solubilization and refolding of misfolded or aggregated proteins and transport of proteins across membranes (Bukau et al. 2006). The HSP70 system is mainly composed of HSP70 proteins and the co-chaperones HSP40 and HSP110 (eukaryotes). Together, these proteins are central in the PQC because they receive and distribute non-native proteins to other chaperones and the proteasome. The HSP40 co-chaperones are also known as J-proteins because of the presence of the family-specific, conserved J-domain in the N-terminus (Summers et al. 2009; Kakkar et al. 2012; Cyr and Ramos 2015). Without the complementary work of the HSP40, the functional diversity of HSP70 would be in disagreement with the high sequence identity that different HSP70 orthologs show across species, since not only the amino acid sequence but also the structure and the mechanism by which the HSP70 system folds client proteins seems to be conserved in eukaryotes (Young 2010). Thus, the diversity of HSP40 proteins explains how it is possible that many different substrates can be delivered to HSP70.

In the HSP70 system, consecutive cycles of binding, folding and release of client proteins are connected to the ATPase activity of HSP70 (Young 2010; Tiroli-Cepeda and Ramos 2011). When in the ATP-bound state, the affinity of HSP70 for substrate is low; however, the interaction of HSP70 with J-proteins, through the NBD and J-domain, induces ATP hydrolysis and, consequently, the predominance of the ADP-bound state, increasing the affinity for substrate. In this process, HSP40 proteins deliver the correct client protein and stimulate HSP70 activity. Thus, once again, the diversity of HSP40s is important because they allow that a vast variety of substrates to be selected and delivered to HSP70.

During the cycle of ATPase activity, co-chaperones that act as nucleotide exchange factors (NEFs) are important because they perform the replacement of ATP, promoting the cycling of HSP70 to the ATP-bound state (Borges and Ramos 2005; Young 2010; Tiroli-Cepeda and Ramos 2011). Unlike HSP70 and HSP40, NEF proteins are not conserved between bacteria, where they are known as GrpEs, and eukaryotes, where in human and yeast cytosol they belong to the HSP110/SSE family, which is structurally related to the NBD and SBD domains of HSP70 (Young 2010; Bracher and Verghese 2015).

The HSP70 system also cooperates with molecular chaperones from the Heat-shock 90 kDa protein (HSP90) family and its co-chaperones (Tiroli-Cepeda and Ramos 2011; Batista et al. 2015; Wegele et al. 2004). The interaction of these proteins generates complex chaperone machineries that have various functions during protein folding and refolding processes (Fig. 13.4). Central to this interaction is the fact that HSP70 deliver substrates to HSP90/co-chaperones because HSP90 does not bind recently synthesized polypeptides, although HSP90 proteins also belong to

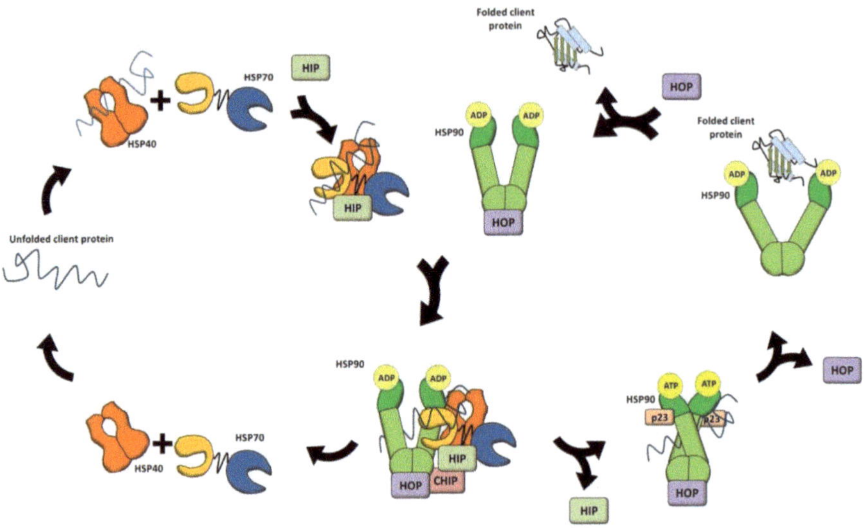

Fig. 13.4 HSP70/HSP90 system. Client proteins first interact with HSP40, which delivers them to HSP70 while promoting the activity of this chaperone. Together with HIP, another co- chaperone, HSP70 interacts with the HSP90 complex, which is formed by HSP90 and the co- chaperones HOP and CHIP. HSP70 then delivers the polypeptide to HSP90, and after the ATPase activity cycle, the protein is folded into its native state

the group of chaperones with ATP-dependent folding activity. Chaperones from the HSP90 family have a highly conserved amino acid sequence and are widely distributed in fungi, animals and plants (Buchner 1999; Young et al. 2001; Wandinger et al. 2008; Gava and Ramos 2009; da Silva and Ramos 2012). Unlike HSP70 proteins, which have promiscuous activity, HSP90 proteins seems to be a more selective protein-folding tool, yet they react with a large number of different client proteins, especially transcription factors and protein kinases. Additionally, HSP90s have important roles during development and growth of eukaryotic cells by binding non-native client proteins and supporting them in a refoldable state.

Structurally, HSP90 is composed of a conserved N-terminal domain responsible for ATPase activity and a C-terminal domain responsible for dimerization and interaction with co-chaperones (Buchner 1999; Young et al. 2001; Wandinger et al. 2008; Gava and Ramos 2009; da Silva and Ramos 2012). These domains are connected by a flexible domain. The nucleotide binding in the N-terminus affects the interaction of HSP90 with the substrate. Its interaction with co-chaperones is indispensable for proper functioning of the system. Many co-chaperones bind HSP90, mainly at the motif MEEVD in the C- terminus of HSP90. This binding happens through the degenerate motif named the tetratricopeptide repeat (TPR), which is conserved in these co-chaperones (Scheufler et al. 2001).

13.1.6 HSP60/HSP10

Another molecular chaperone family with ATP-dependent folding activity is the HSP60/HSP10 family (Tiroli-Cepeda and Ramos 2011; Boshoff 2015). This family is commonly named Anfinsen's cage because of the cavity formed within the rings in the cylindrical tridimensional conformation. Like the HSP70 system, HSP60 and HSP10 are present in several cellular compartments (cytosol, plastids and mitochondria, for example) of eukaryotes and eubacteria (Fig. 13.2 and Table 13.1). HSP60/HSP10 can be classified in two groups. Type I consists of the mitochondrial and chloroplast HSP60/HSP10, which are found in huge homo-oligomeric, tetradecameric complexes (two heptameric rings in a cage shape) together with a lid formed by a single heptameric ring of HSP10. Type II are cytoplasmic and can form hetero-oligomeric complexes of ~16 members (eight or nine members per ring), but without co- chaperones (Hill and Hemmingsen 2001; Boshoff 2015). Type I chaperones were first known as Rubisco-binding protein since they were first identified as helpers of the folding of this photosynthesis protein. Because of their action with other protein of the chloroplast, these chaperones are essential in plants, and knockout experiments of their genes are lethal (Barraclough and Ellis 1980).

Aside from the differences between the type I and II HSP60/HSP10 chaperones, all of them have similar subunit organization and structure (Borges and Ramos 2005; Hartl and Hayer-Hartl 2002; Boshoff 2015). Their mechanism of action is also comparable between types, and both are called Anfinsen's cage. In the center of the cavity occurs the interaction and the encaging of the substrate, which leads to a productive environment of folding recently synthesized and misfolded client polypeptides. HSP60/HSP10 not only facilitate folding but also prevent aggregation by binding unfolded substrate. Additionally, the HSP70 system has an important role in cooperating with the HSP60/HSP10 family to fold nascent proteins: HSP70 is the main protein responsible for selecting and then delivering client proteins to the HSP60/HSP10 family (Hartl et al. 1992; Borges and Ramos 2005).

13.1.7 HSP100 and sHSP

Heat-shock 100 kDa proteins (HSP100), also known as the HSP104/ClpB family, have intrinsic ATPase activity. They belong to the superfamily of ATPases named AAA+ proteins (ATPases Associated with cellular Activities; Sauer et al. 2004; Snider and Houry 2008) and are very important because of the disaggregase activity that some members present (Hodson et al. 2012; Mokry et al. 2015) (Fig. 13.2 and Table 13.1). Particularly during some kinds of stress, several proteins can aggregate through inter- polypeptide interactions formed by exposed hydrophobic residues, even in the presence of chaperones from HSP70, HSP90 and HSP60/HSP10 families (Winkler et al. 2012; Zolkiewski et al. 2012; Doyle et al. 2013). For this reason, in plants, fungi and eubacteria, proteins from the HSP100 family are responsible for disassembling these aggregates and for allowing client substrates to be refolded or targeted for degradation. A collaborative action with the HSP70 machinery is once again crucial, since HSP100 cannot refold or redirect the client proteins that were recovered from aggregates. (Glover and Lindquist 1998; Mokry et al. 2015).

In addition to their disaggregase activity, HSP100 proteins are engaged in a myriad of other functions in the cell, such as stress tolerance, intracellular trafficking, cell cycle regulation, gene regulation and DNA transposition (Schirmer et al. 1996; Tiroli-Cepeda and Ramos 2011; Mokry et al. 2015). HSP100/Clp proteins have varied structures but share some conserved domains, such as the N-terminal domain (NTD) that is essential for substrate interaction, generally aggregates. Moreover, they have one or two nucleotide-binding domains (NBDs) that have affinity for ATP or ADP and are responsible for the ATPase activity. These proteins are frequently classified by the number of NBDs: class I have two NBDs and class II have one NBD (Barnett et al. 2005; Tiroli-Cepeda and Ramos 2011; Mokry et al. 2015). Class I HSP100 proteins function as hexamers in the presence of ATP, while class II are smaller; however, all HSP100 proteins present a ring-shaped structure, where the binding to ATP influences conformational changes (Zolkiewski 2006; Cagliari et al. 2011; Tiroli-Cepeda and Ramos 2011; Mokry et al. 2015).

Small heat-shock proteins (sHSP) are a family of chaperones of 12–42 kDa that, despite being present in prokaryotes and eukaryotes, are poorly conserved; they present less than 30 % amino acid similarity between family members (Boston et al. 1996; Bakthisaran et al. 2015; Haslbeck and Vierling 2015; Morrow et al. 2015). When these chaperones are expressed, they are targeted to specific cellular compartments, such as the cytosol, mitochondria, chloroplast and endoplasmic reticulum. The sHSPs are classified by their compartment because sHSPs within each compartment have higher sequence identity than sHSPs from different compartments.

Basically, sHSP monomers consist of a conserved domain known as the α- crystallin domain (ACD). The ACD is located between the variable N-terminal domain, responsible for substrate binding, and the C-terminus (Caspers et al. 1995) (Fig. 13.2 and Table 13.1). Although they are smaller monomers compared with other chaperones, sHSPs can form huge oligomeric complexes, from dimers to larger than 32-mers, arranged in two flattened discs with a cavity in the center. The oligomers are quite dynamic, and the subunit equilibrium can change according to the cellular environment (van Montfort et al. 2001a, b; Tiroli and Ramos 2007; Basha et al. 2012; Basha et al. 2013). sHSP function does not depend on ATPase activity. These polypeptides supply thermotolerance by binding exposed hydrophobic residues of substrates through the sHSPs' N-termini within the cavity of the disc (Mani et al. 2015). Like other chaperones, sHSPs are highly expressed during different types of cellular stress (Haslbeck et al. 1999), acting as holders for non-native but refoldable client proteins, thereby preventing irreversible aggregation. In plants, sHSPs are important and abundant, mainly during heat stress. More and more sHSPs are being characterized in these organisms because plants have more of these proteins than other organisms (Waters 2013).

13.2 Stress Response and Chaperones in Plants

13.2.1 Stress Response

Throughout the process of evolution, plants have developed specific adaptations to survive sudden changes in environmental conditions and minimize potential damage to their cellular homeostasis (Vierling 1991; Nover and Miernyk 2001; Baniwal et al. 2004; Wang et al. 2004; Kotak et al. 2007; Cramer et al. 2011; Qu et al. 2013). Because they are sessile organisms, most plants, including economically important crops, have to respond quickly to abiotic stresses, such as temperature fluctuations, quantity of water supply, light source, salinity, pH, and the presence of heavy metals (Vierling 1991; Baniwal et al. 2004; Wang et al. 2004; Kotak et al. 2007; Cramer et al. 2011; Qu et al. 2013). The efficiency of this response system is essential for plant growth and yield, and its regulation can be complex and controlled by cascades of molecular networks.

Plant responses to abiotic stresses appear to be genetically more complex then biotic ones because of their multigenic character, while biotic responses are mostly monogenic (Vinocur and Altman 2005). During abiotic stress responses, heat shock factors (HSFs) stimulate the expression of HSPs, allowing for the acquisition of thermotolerance (Baniwal et al. 2004; Kotak et al. 2007). A testimony to the complexity of the plant stress response is the fact that plants have more than 20 HSF family members, more than other organisms such as yeast and Drosophila, which have only one member, and even vertebrates, which usually have just 3 members (Baniwal et al. 2004; Anckar and Sistonen 2011).

13.2.2 Chaperone Genome Analysis and Gene Expression Pattern in Plants

Large-scale DNA sequencing has revolutionized the study of biology. The sequencing of whole genomes, or even of their expressed genes, has generated a large amount of data that is used to reveal the uniqueness of each species. In the genome sequencing era, systematic analysis of genome organization, gene structure and gene expression patterns has become much more accessible, generating a colossal amount of data that allows for the identification of extensive gene and protein families and their diversity in genomes of different organisms (Stein 2001). Much of what is known about the genomic diversification of molecular chaperones in different living beings is available thanks to molecular biology tools developed in the 2000s.

Regarding genomic information on molecular chaperone genes, the first organisms that had their genome completely sequenced, *E. coli*, yeast and human, delivered a large amount of information (Miernyk 2001). *Arabidopsis thaliana*, which has been the main model system for plants, was the first flowering plant whose

Table 13.3 Number of annotated genes from different molecular chaperone families in plant databases

Chaperone family	*A. thaliana*	*O. sativa*	*P. trichocarpa*[a]	*Saccharum* spp.[j]	*Eucalyptus* spp.[k]
HSF	21[a]	25[a]	28	22	–
sHSP	27[a]	29[g]	37	24	28
HSP60	18[b]	20[h]	28	24	15
HSP10	5[b]	–	–	6	4
HSP70	14[c]	26[h]	20	19	27
HSP40	89[d]	104[i]	–	87	91
HSP110	4[c]	–	–	4	5
HSP90	7[e]	9[h]	10	9	12
HSP100	4[f]	5[h]	5	4	4

[a]Zhang et al. (2015), [b]Hill and Hemmingsen (2001), [c]Lin et al. (2001), [d]Miernyk (2001), [e]Krishna and Gloor (2001), [f]Agarwal et al. (2001), [g]Sarkar et al. (2009), [h]Hu et al. (2009), [i]Sarkar et al. (2013), [j]Borges et al. (2007), [k]Cagliari et al. (2005)

genome was completely sequenced (Arabidopsis Genome Initiative 2000). Due to its economic importance in world alimentation, the rice genome (*Oryza sativa*) is also well studied and was the first monocotyledon whose genome was sequenced (International Rice Genome Sequencing Project 2005). Subsequently, many other plants have been investigated by large-scale DNA sequencing tools, generating unprecedented amounts of information and many studies aiming to identify new genes and proteins in plants (Table 13.3). Molecular chaperones and HSPs are among the main families of genes in plants analyzed by large-scale DNA sequencing tools, especially because these proteins arose as promises of the biotechnological revolution in the fight against environmental stresses.

The diversity of each molecular chaperone family in *A. thaliana* is evaluated by analyses of *A. thaliana*'s genome sequence (Table 13.3). From this analysis, 14 members from the HSP70 family were found (Lin et al. 2001), 18 from HSP60 (Hill and Hemmingsen 2001), 7 from HSP90 (Krishna and Gloor 2001) and 4 from HSP100 (Agarwal et al. 2001). The most abundant chaperone family is the sHSPs with 27 members characterized (Zhang et al. 2015). The diversity is even more impressive when HSP40s (J-proteins), the HSP70 co-chaperones, are analyzed, as they include 89 members so far (Miernyk 2001). This high number of genes from HSP40s supports the hypothesis that many of these co-chaperones evolved to satisfy the need for more specific affinity for a large number of different substrates in order to bind and deliver them to the HSP70 system (Miernyk 2001). On the other hand, the HSP10 family, which has more general activity, has only 5 genes in the genome of *A. thaliana* (Hill and Hemmingsen 2001).

The rice genome has a similar diversity of genes coding for chaperone proteins (Table 13.3). HSP60, HSP70, HSP90 and HSP100 have respectively 20, 26, 9 and 5 members (Hu et al. 2009; Zhang et al. 2015). Once again, the sHSP family is the largest chaperone family, with 29 genes, highlighting the importance of these proteins for thermotolerance acquisition in plants (Sarkar et al. 2009). One hundred four J-proteins have been described in rice (Sarkar et al. 2013), indicating that the high diversity of HSP40 is an indicative of their scanning function, selecting client proteins for HSP70 activities.

More recently, the plant species *Populus* spp., sugarcane (*Saccharum* spp.), sorghum and *Eucalyptus* spp., have gained interest from some research groups. Populus is the most important model for perennial woody species of plants, having different characteristics from *Arabidopsis*. The genome analysis of *Populus trichocarpa* has identified 28 genes coding HSP60 proteins, 20 genes coding HSP70 proteins and 5 genes coding HSP100 proteins (Zhang et al. 2015). Moreover, it found 10 candidates for HSP90 sequences in this genome (Zhang et al. 2013), although there is no information about other HSPs, such as HSP40 or HSP10.

13.2.3 Chaperone Genome Analysis and Gene Expression Pattern in Sugarcane and Eucalyptus

As one of the many developments from large-scale DNA sequencing, the interest in information about chaperones in sugarcane and eucalyptus had become significant. Gene annotation followed by *in vivo* protein characterization have been performed in these species. Sugarcane is one of the most important crops in tropical countries such as Brazil because of its economic importance, not only for sugar production but also for ethanol production as an alternative fuel (Vettore et al. 2003). In these cases, when genomes are still under investigation or when there is no sequenced genome (for instance, in sugarcane), different strategies can provide valuable information about molecular chaperones.

The Sugar Cane EST (SUCEST) consortium has been important to many groups interested in cataloguing the abundance and distribution of different protein families of sugarcane. SUCEST was designed to implement a large scale sequencing strategy of sugarcane ESTs (Expression Sequenced Tags; Fig. 13.5) from a database made of libraries built from different sugarcane tissues in different conditions (Vettore et al. 2003). Using EST sequencing to replace the lack of genomic information is a powerful complementary tool since it is possible to gather information on the sequences of important orthologs, to estimate the abundance and distribution of gene families and also to predict gene expression profiles by the quantity of annotated ESTs sequences (Fig. 13.5). Molecular chaperone transcripts were identified in SUCEST by homological similarity between known sequences and those in the database (Borges et al. 2001).

As commercially significant as sugarcane, eucalyptus has also been a target for genome investigation. Eucalyptus is the most cultivated perennial woody plant in the world thanks to its fast development and its excellent wood characteristics. In addition, these trees attenuate human exploitation of forests while supplying renewable resources for paper, biomaterial and bioenergy industries. The genome of *Eucalyptus grandis* was newly sequenced in mid-2014 (Myburg et al. 2014) and

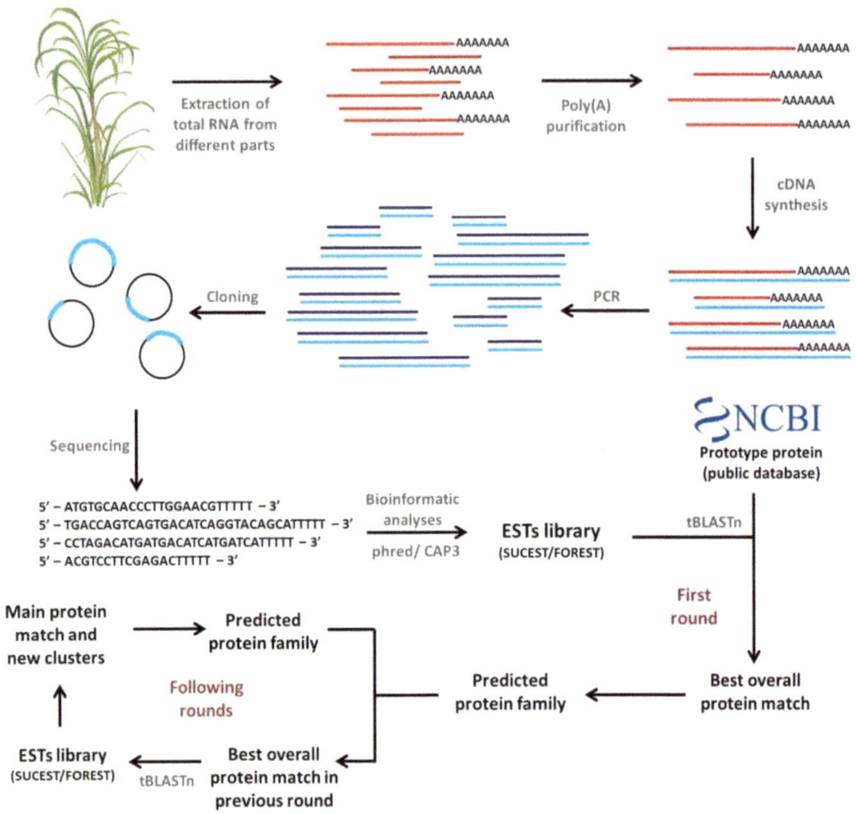

Fig. 13.5 EST library constructs and gene annotation strategies. The construction of EST libraries and the strategies used to annotate genes related to molecular chaperones and HSPs are summarized. Firstly, total RNA is isolated from different tissues. Posteriorly, mRNA molecules are purified, by their poly(A), and used as templates to cDNA synthesis, which are amplified and then cloned into bacterial plasmids for sequencing. Softwares, such as phred and CAP3, are used to estimate error probability, for trimming process, to read quality parameters and to cluster trimmed reads, thereby generating ESTs libraries (Vettore et al. 2003). Data mining and annotation follows: translated amino acid sequences of specific mRNAs of known chaperones in public database are used for comparison with first level cluster consensus generated by SUCEST or ForESTs, using the algorithm tBLASTn. Only sequences with E-value lower than $1e^{-5}$ continue to next steps. Finally, matches found in the first round proceed to a second or more rounds of mining in order to improve the prediction accuracy (Borges et al. 2001; Cagliari et al. 2005)

deep analysis of molecular chaperones and heat shock proteins are underway. However, similar to what has been done with SUCEST, there is also a database of ESTs sequenced and annotated for Eucalyptus, ForESTs (Eucalyptus Genome Sequencing Project Consortium), architected by a Brazilian Consortium (https://forests.esalq.usp.br/). These libraries allow for the searching and identification of molecular chaperone transcripts by sequence similarity between known sequences and those in the database (Cagliari et al. 2005).

Similar mining strategies were used to search for and recognize molecular chaperones in both the SUCEST and ForESTs consortia (Borges et al. 2001; Cagliari et al. 2005). Initially, a choice of known translated amino acid sequences of mRNAs coding for chaperones is done in public databases. This step is important in order to use these deposited sequences as queries for comparing them with first-level cluster consensus obtained from SUCEST or ForESTs through an alignment tool. After this first round, a second phase of mining is done to increase the exactitude of the identification, using the matches found in the first step (Borges et al. 2001; Cagliari et al. 2005). Applying this approach in SUCEST and ForESTs databases, sequences coding for molecular chaperones from sugarcane and eucalyptus have been identified, annotated and classified (Fig. 13.5).

SUCEST analyses have allowed the annotation of 4702 sequences of ESTs as molecular chaperones, which corresponds to 2.2 % of all ESTs present in the database (approximately 230,000 sequences). This high percentage of chaperones indicates the functional importance of these proteins to proteostasis in plant cells. Analysis after removing redundant singleton expressed sequences shows that there are 425 clusters related to chaperone proteins, of which 142 are predicted to have an entire coding region and 283 incomplete sequences. After the annotation endeavor, it was shown that SUCEST contains approximately 300 cDNAs of molecular chaperones (Borges et al. 2007).

The results from the analysis using the SUCEST database are summarized in Table 13.3. There are at least 19 members of the HSP70 family, 9 of the HSP90 family, 4 of the HSP100 family and 6 of the HSP10 family. The HSP60 and sHSP families were the most abundant, with 24 members each. Eighty-seven sequences were found for co- chaperone HSP40 and 4 for co-chaperone HSP110. These numbers are similar to what was observed in *Arabidopsis*, rice and Populus, demonstrating their high conservation (Borges et al., 2007). Furthermore, 22 sequences corresponded to the HSF family were found.

For eucalyptus, ForESTs analyses have identified almost 2000 ESTs belonging to molecular chaperones. This means that 1.6 % of the ESTs sequenced are associated with chaperones, since the whole database is composed of approximately 124,000 sequences. After annotation, 232 clusters of different chaperone proteins were identified (Table 13.3). Of these, the majority of sequences are related to the HSP70 system: 27 sequences were identified as belonging to HSP70 proteins and 91 as belonging to HSP40. From other families, the ForESTs database includes 15 genes belonging to HSP60 proteins, 6 to HSP10, 12 to HSP90 and 4 to HSP100. Once more, sHSPs are the most diverse family, with 28 members (Cagliari et al. 2005).

Because EST databases are usually in accordance with real mRNA expression patterns, we can assume that the quantity of EST clones is closely associated with the protein expression level in each studied tissue (Cagliari et al. 2005). Thus, using digital RNA blot analyses with SUCEST and ForESTs, it is possible to generate a "molecular chaperone expressome" in the different tissues. In sugarcane, the major tissues analyzed were apical meristem (AM), callus (CL), flower (FL), lateral bud (LB), leaf roll (LR), leaves (LV), root (RT), leaf-root transition zone (RZ), stem bark (SB), stem (ST) and seeds (SD) (Vettore et al. 2003). Tissue samples with chaperone expression above the average were LB, LV and CL, possibly because they are young organs that are still differentiating (Fig. 13.6). However, even though it had a high level of cell division and differentiation, AM showed expression levels below average. Other tissues, such as RT and SD, also had low expression of HSPs, probably because they are less exposed to stress conditions (Borges et al. 2007).

The most expressed of the chaperone families in sugarcane tissues was HSP70 and its co-chaperone HSP40 and HSP110, with 41 % of annotated ESTs sequences (Table 13.4), demonstrating their importance for protein folding not only in stress situations but also in high-protein-synthesis conditions, mainly cell division. Meanwhile, families more involved in the stress response, such as HSP100 or sHSP, had low overall expression levels, approximately 3.3 % and 4.7 %, respectively (Borges et al. 2001). Nevertheless, several studies have demonstrate that the expression of these chaperones increases under different kinds of stress in plants and other organisms (Haslbeck et al. 1999, 2004; Löw et al. 2000; Swindell et al. 2007). HSP60 chaperones comprised 3.1 % of all expressed chaperones, and although they were present in all tissues, their expression was higher in AM and CL. HSP90 expression was also confirmed in all libraries/tissues, corresponding to 9.7 % of the total expression level, and it was the highest in leaves (LV) (Borges et al. 2007).

In terms of tissues and conditions, ForESTs database is made of several libraries of different *Eucalyptus* species tissues, such as *E. grandis*, *E. globulus*, *E. urophylla*, *E. saligna* and *E. camaldulensis*, in diverse conditions: cultivated in the dark or light or in water deficit, just to mention a few examples. Digital RNA blot analyses in ForESTs libraries revealed similar results compared to sugarcane. Once again, HSP70 and HSP40 transcripts were found in all libraries as the most expressed chaperones, with 50 % of all annotated chaperone-related transcripts (Table 13.4). HSP70 and its co- chaperones were found in all libraries investigated, always representing a high percentage of chaperone expression. The highest expression was in *E. grandis* seedlings cultivated in the dark (Cagliari et al. 2005). HSP90 proteins were the second most expressed chaperone, and they were also found in all libraries examined, which is in accordance with its high abundance among cytosolic proteins in several organisms. Particularly, root and wood libraries revealed low expression of HSP90, while callus and bark showed high expression. On the other hand, proteins belonging to the HSP60 system were found in all tissues, but when compared to the HSP70/HSP90 system, they had a small number of transcripts. Root and seedling libraries exhibited the highest expression level of HSP60. Nonetheless, in comparison with sugarcane, the expression level of the HSP60 system and HSP90 were higher, approximately 15 % and 18 %, respectively (Cagliari et al. 2005). Moreover, sHSP and HSP100 proteins presented low expression compared to other chaperones, approximately 4 % and 8 %, respectively. However, the low expression of sHSP and HSP100 is expected since they are chaperones related to stress and aggregate formation, not directly facilitating (re)folding of polypeptides, and thus not expected to be highly expressed in non-stress conditions (Cagliari et al. 2005).

It is noteworthy that, even though the expression profiles described here for sugarcane and eucalyptus originated from EST library analyses, they corroborate well-

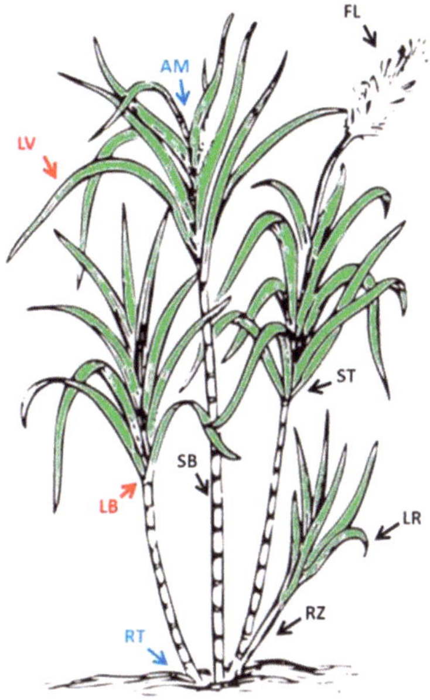

Fig. 13.6 Scheme of a sugarcane plant highlighting tissues analyzed by SUCEST. The tissues analyzed were (*AM*) apical meristema, (*CL*) callus, (*FL*) flower, (*LB*) lateral bud, (*LR*) leaf roll, (*LV*) leaves, (*RT*) root, (*RZ*) leaf-root transition zone, (*SB*) stem bark, (*ST*) stem and (*SD*) seeds. (*CL*) callus and (*SD*) seeds are not shown. *Red*, highest chaperone expression; *blue*, lowest chaperone expression (Borges et al. 2007)

Table 13.4 Comparison of expression levels of molecular chaperones from different families, described as the percentage of annotated EST sequences, in sugarcane and eucalyptus

Chaperone family	*Saccharum* spp.[a]	*Eucalyptus* spp.[b]
HSF	2.4 %	Not done
sHSP	4.7 %	4.1 %
HSP60	3.1 %	13.6 %
HSP10	1.6 %	1.4 %
HSP70	18.5 %	18.0 %
HSP40	21. 9 %	28.9 %
HSP110	0.8 %	2.7 %
HSP90	9.7 %	18.4 %
HSP100	3.3 %	7.7 %

[a]Borges et al. (2001), [b]Cagliari et al. (2005)

established results from the analysis using the *Arabidopsis* genome. This is important because the understanding of chaperones' functional roles in *Arabidopsis* under stress and non-stress conditions is much better than for other plants due to the possibility of genome-wide datasets generated by the DNA microarray approach (Swindell et al. 2007). Thanks to this technology, it was possible to outline HSPs' expression under different types of stress conditions: non-stress, high or low temperatures, salt, osmotic, dry, oxidative, intensity of light, presence of pathogen, injury, etc.

High temperature treatments generate a kind of stress that affects mostly the expression of sHSP, HSP70 and HSP90, while HSP100 expression is more influenced by osmotic stress in *Arabidopsis* roots and shoots. sHSP was the family most affected by these different kind of stresses. Heat, osmotic, salt, injury, and cold stresses significantly increased its expression (Swindell et al. 2007). These results are in agreement with those collected for sugarcane and eucalyptus: HSP70, HSP90 and HSP60 have high expression in normal conditions, while sHSP and HSP100 are low- expressed (Cagliari et al. 2005). In stress conditions, by contrast, sHSP has a strong response to general environmental stress situations, while HSP70, HSP90 and HSP100 responded but less strongly (Cagliari et al. 2005).

13.3 Functional and Structural Advances on Sugarcane Chaperones

Equally as important as the knowledge about the genomic distribution and expression pattern of molecular chaperones in plants is obtaining data on their structure and function to understand where they act in the cell, what they do and what their mechanism of action is. The first experiments on HSPs in plants were in the 1980s, when seedlings cultivated in temperatures above optimal showed differential gene expression patterns, with some mRNAs decreasing and others increasing, along with their encoded proteins. The latter were named "heat shock proteins" (Vierling 1991). Currently, the combination of the information obtained from the genome with molecular biology tools allows for the production of recombinant proteins to validate their functions. Structural studies on these proteins allow for structure-function relationship studies that further the knowledge about proteins in general. In this sense, studies on the structure of sugarcane molecular chaperones has been important to understand the general function of these proteins, especially in plants. Bellow, we give a brief description of the results of structure-function studies of molecular chaperones from sugarcane.

13.3.1 HSP70

The cloning and purification of the first sugarcane HSP70 (SsHSP70) was possible due to the identification of chaperones in the analyses of SUCEST (Borges et al. 2001). As expected, the sequencing of *Saccharum* sp. HSP70 (SsHSP70) revealed that the protein has a high identity with HSP70s from other plants, for instance 92 % with the HSP70 from *Arabidopsis*. The identity is particularly high with the HSP70 genes from other monocotyledons, such as sorghum (99 %), corn (99 %) and rice (96 %) (Tiroli-Cepeda et al. 2014).

The high molecular masses of HSPs complicate the study of these proteins by high-resolution structural techniques such as NMR or crystallography. However, much effort has been taken to structurally characterize the separate domains, which are much smaller and usually soluble, of these proteins by these techniques. Otherwise, biophysical spectroscopic techniques can supply conformational information of entire proteins and make it possible to indirectly obtain helpful insights into their structure.

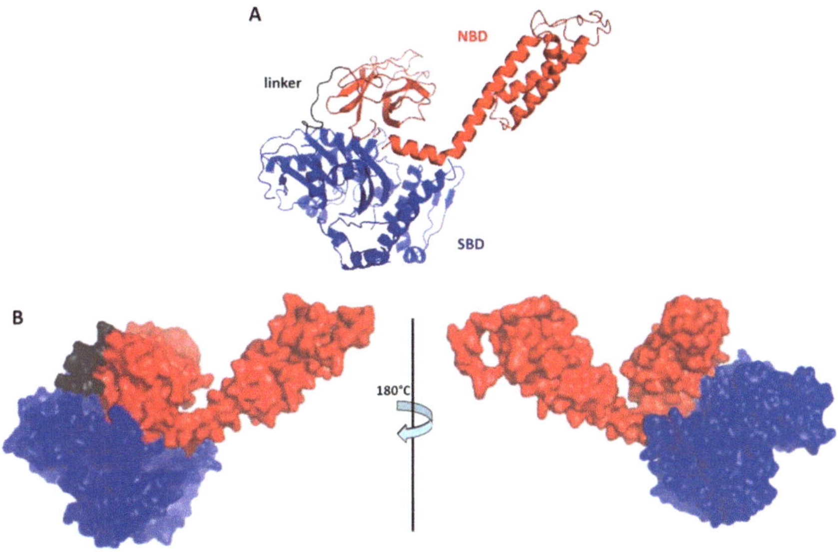

Fig. 13.7 SsHSP70 structure based on cross-linking experiments and comparative modeling. (**a**) SsHSP70 model structure highlighting the positions of the domains and secondary structure content. (**b**) Model surface of monomeric SsHsp70 under different views (Tiroli-Cepeda et al. 2014). The images were prepared using PyMOL

To use these biophysical tools to study protein structure, they must be purified in a folded and functional form (Batista et al. 2015). Circular Dichroism (CD), which probes secondary structure (Correa and Ramos 2009; Batista et al. 2015) and intrinsic fluorescence, which probes the Tryptophan residue environment (Batista et al. 2015), are efficient and fast techniques to evaluate the folding state of pure proteins in solution. Through these techniques, SsHSP70 was purified in a well-folded condition and consisted of 33 % α-helices (Tiroli-Cepeda et al. 2014). According to crystallographic findings of domain structures (Flaherty et al. 1990; Zhu et al. 1996) and to CD experiments on multiple orthologs (Borges and Ramos 2006), HSP70 is predominantly composed of α-helices, since most of its NBD consists of two α-helices forming two lobes (Fig. 13.7), in which the interaction with nucleotides (ATP and ADP) takes place, and its SBD is composed of a β-sheet base and an α-helical lid (Flaherty et al. 1990; Young 2010). Size-exclusion chromatography combined with multi-angle light scattering (SEC-MALS) is a reliable biophysical technique to evaluate molecular mass precisely and, consequently, the oligomeric state of pure proteins in solution. Corroborating the well-established knowledge about these chaperones, the molecular mass of SsHSP70 measured by SEC-MALS is 78 ± 2 kDa (Table 13.5), which means that this protein is a monomer in solution (Tiroli-Cepeda et al. 2014).

Through CD experiments, thermal stability of SsHSP70 was also determined to be similar to that of the *E. coli* DnaK (Montgomery et al. 1999), consisting of two defined transitions at 42 and 74 °C that probably correspond to the unfolding of the

Table 13.5 Hydrodynamic and structural parameters studied from sugarcane and orange chaperones

Protein name	SsHSP70	SsHSP90	CsHSP90	SHSP101	SsHsp17.2	SsHsp17.9
Mass (kDa)	78 ± 2[a]	172 ± 2[a]	150 ± 10[a]	619 ± 6[a, b]	206 ± 10[b]	214 ± 11[b]
Oligomer	monomer	dimer	dimer	hexamer	dodecamer	dodecamer
Rs (Å)	–	66[c]	62 ± 2[d]	–	48.0 ± 1.5[d]	48.5 ± 1.5[d]
D ($cm^2 \cdot s^{-1}$)[e]	–	3.2 ± 0.2 × 10^{-7}	3.3 ± 0.2 × 10^{-7}	2.4 ± 0.1 × 10^{-7}	4.6 ± 0.3 × 10^{-7}	4.4 ± 0.1 × 10^{-7}
% α-helix[f]	33	34	35	40	20	15

[a]Measured by SEC-MALS; [b]measured by AUC; [c]calculated using Stokes-Einstein equation; [d]measured by AGF; [e]measured by DLS; [f]measured by CD spectra

SBD and NBD, respectively. These experiments also revealed that the interaction with either ATP or ADP increases the stability of SsHSP70, as observed for human HSP70 (Tiroli-Cepeda et al. 2014; Borges and Ramos 2006).

Human HSP70 and SsHSP70 have 77 % identity, which is in accordance with the high conservation of this chaperone family. This identity is not only recognized in the sequence alignment but also by comparing the results of functional essays. ATPase activity essays with SsHSP70 together with human J proteins class I (DJA1) and class II (DJB4) revealed that these co-chaperones are able to stimulate the hydrolysis of the ATP of SsHSP70 (Tiroli-Cepeda et al. 2014). This result shows that SsHSP70 is stimulated by HSP40 proteins, as expected, although not as efficiently as the stimulation of yeast HSP70 by yeast HSP40 because the latter pair were from the same organism (Lu and Cyr 1998).

As discussed above, structural characterization of HSP70 is difficult because of the techniques used and because this protein can be in at least two different conformations (ATP- or ADP-bound). Actually, it is more complicated than that, considering that the interactions with co-chaperones or other chaperones can also change its conformation. However, cross-linking reaction of proteins followed by digestion and LC-MS/MS provide a reliable structural model for proteins, and further structural investigations of SsHSP70 was made by this strategy (Tiroli-Cepeda et al. 2014). Using the cross-linker disuccinimidyl suberate (DSS), lysine residues of the protein can react if they are interacting in a distance less than 22.4 Å (Tiroli-Cepeda et al. 2014). The structural model of SsHSP70 was made based on homology and on the map of lysines that interacted after digestion with trypsin and mass spectroscopy analysis (Fig. 13.7).

13.3.2 HSP90

A detailed characterization is also available for sugarcane HSP90 (SsHSP90) (da Silva et al. 2013). As done for HSP70, using the nucleotide sequence annotated in SUCEST, SsHSP90 was cloned and the protein was purified monodispersed in solution. HSP90s are evolutionarily conserved. Analyses of amino acid sequence identity revealed high similarity: 64 % identity with *S. cerevisiae* HSP90, 92 % with *Arabidopsis* HSP90 and even higher with HSP90 from monocotyledons – 97 % with corn, 97 % with rice and 99 % with sorghum (da Silva et al. 2013).

Circular Dichroism experiments, which investigate the secondary structure, showed that α-helices constitute the majority of the secondary structure of SsHSP90, with 34 % of content (Table 13.5), which is in line with crystallographic information found in the literature for HSP90 orthologs (Jackson 2013; Mayer and Le Breton 2015). According to these further structural experiments conducted on human and *S. cerevisiae* HSP90, all domains (N-terminal, middle and C-terminal domains) are essentially formed by α-helices (Obermann et al. 1998; Ali et al. 2006; Jackson 2013; Mayer and Le Breton 2015). SsHSP90's molecular mass of 172 ± 2 kDa (Table 13.5) was determined using SEC-MALS (Size Exclusion Chromatograph coupled to a Multi Angle Light Scattering; da Silva et al. 2013), which corroborates the previous finding that HSP90 is found as a homodimer in solution and that homodimerization is essential for this protein's function (Chadli et al. 2000). Other hydrodynamic features (Table 13.5) obtained for SsHSP90 are its diffusion coefficient (D), measured through Dynamic Light Scattering (DLS), and its Stokes radius (Rs), calculated using D values and the Stokes- Einstein equation (Edward 1970; Borges and Ramos 2011; Batista et al. 2015). The values measured for D and calculated for Rs are $3.2 \pm 0.2 \times 10^{-7}$ $cm^2.s^{-1}$ and 66 Å, respectively (Table 13.5), which is in good agreement with the same parameters studied in orange HSP90 (Mendonça and Ramos 2012).

Functional essays were also performed, and they showed the chaperone was able to protect client proteins from thermal aggregation in the absence of ATP. The model protein citrate synthase (CS) aggregates at 47 °C, but the presence of SsHSP90 at the ratio of 2:1 (dimer of HSP90:monomer of CS) at the same temperature completely protects CS from aggregation (da Silva et al. 2013). This activity of preventing aggregation corroborates what was observed for HSP90s from other organisms (Jakob et al. 1995), as well as the activity of aiding the folding of unfolded or partially folded client proteins in an ATP-dependent way (Freeman and Morimoto 1996; Yonehara et al. 1996; Picard 2002; Table 13.6).

Similarly to SsHSP70, a dimeric model of SsHSP90 was generated using DSS cross-linking, mass spectrometry essays and homology modeling (Fig. 13.8). The model resembles those of other cytosolic HSP90 ortholog structures (Obermann et al. 1998; Jackson 2012) and is reliable since it agrees with fluorescence investigation in exposition of tryptophans (da Silva et al. 2013). Moreover, hydrodynamic parameters predicted for the dimeric model ($D = 3.8 \times 10^{-7}$ $cm^2 \cdot s^{-1}$ and Rs = 55 Å) are quite similar to those mentioned above for SsHSP90 in solution (Table 13.5), which also gives credibility to the model.

Table 13.6 Functional parameters studied from sugarcane and orange chaperones

Protein name	SsHSP70	SsHSP90	CsHSP90	SHSP101	SsHsp17.2	SsHsp17.9
Chaperone activity	Human DJA1 and DJB4 stimulate ATP hydrolysis. Refolding activity of β-galactosidase in association with SsHsp17.2	Protect citrate synthase from aggregation in absence of ATP	Protect citrate synthase from aggregation in absence of ATP	–	Protect luciferase, CS, MDH and a *E. coli* extract from aggregation.	Protect luciferase, CS, MDH and a *E. coli* extract from aggregation.
Nucleotide interaction	ATP and ADP	–	–	ATP and ADP	–	–
Stability (T°C)[a]	42 and 74	>90	>90	–	60	55

[a]Maximum temperature without losing secondary structure conformation as measured by circular dichroism

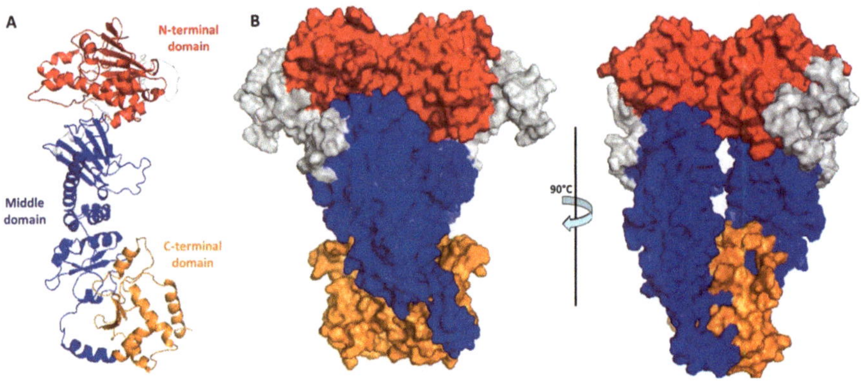

Fig. 13.8 SsHSP90 structure based on cross-linking experiments and comparative modeling. (**a**) SsHSP90 model structure highlighting the positions of the domains and secondary structure content. (**b**) Model surface of dimeric SsHSP90 under different views (The images were prepared using PyMOL (da Silva et al. 2013))

13.3.3 HSP100

The HSP100 disaggregase protein from sugarcane (SHSP101) also has been studied to better understand the function of this type of chaperone and the acquisition of thermotolerance in plants. Through EST sequences annotated from SUCEST, SHSP101 was cloned and the protein purified to enable further characterization. The CD spectrum of SHSP101 showed that the protein has approximately 40 % α-helical content (Table 13.5); in other words, the protein was purified in its folded native state (Cagliari et al. 2011). The predominance of α-helices agrees with the structures of ClpB from *T. thermophilus* and *E. coli* (Lee et al. 2003; Carroni et al. 2014). The similarity of SHSP101 and ClpB from *T. thermophilus* is 50 %. This high similarity allowed for the generation of a model for SHSP101 based on the crystal structure of its *T. thermophilus* counterpart (Fig. 13.9). The secondary structure of all domains (N-terminal, middle, NBD1 and NBD2) is primarily composed of α-helices.

As discussed above, HSP100s are functional only after oligomerization, when they form a hexamer of a ring-shaped quaternary structure with a central channel (Zolkiewski et al. 1999; Lee et al. 2004; Mackay et al. 2008). The oligomeric state of SHSP101 was characterized by SEC-MALS and analytical ultracentrifugation (AUC) (Cagliari et al. 2011). The results of these experiments reveling molecular masses of 619 ± 6 and 620 ± 20 kDa (Table 13.5), respectively, which indicates the presence of a hexamer in solution, since the monomer of SHSP101 has a theoretical molecular mass of 103.1 kDa. The presence of ATP or ADP does not change the molecular mass obtained by AUC, and nor does the environment of the tryptophan residues in intrinsic fluorescence experiments, suggesting that the quaternary structure of SHSP101 does not dependent on nucleotide binding. Despite that, saturation transfer difference (STD) NMR experiments revealed that these nucleotides bind to SHSP101 (Cagliari et al. 2011) (Table 13.6). Moreover, several studies point out that ATP or ADP interacts with NBD domains and are crucial for HSP100's function (Zolkiewski et al. 2012; Doyle et al. 2013).

13.3.4 sHSP

Two cytosolic sHSPs from sugarcane, SsHSP17.2 and SsHSP17.9, have also been cloned based on EST sequences from SUCEST (Tiroli and Ramos 2007). Nucleotide sequence analyses revealed high identity between them (75 %) and also with a sHSP from wheat, TaHSP16.9. The high similarity is essentially confined to the conserved α-crystallin domain, while the N-terminus is less conserved. Both sugarcane sHSPs were expressed and purified to characterize their structure and function (Tiroli and Ramos 2007).

As a standard for protein characterization, CD spectra have been used to identify the folded state and define the secondary structure of pure proteins. As expected from the contribution of the α-crystallin domain (van Montfort et al. 2001a, b; Gusev et al. 2002), the secondary structures of SsHSP17.2 and SsHSP17.9 consist predominantly of β- sheets: SsHSP17.2 has 40 % β-sheets and 20 % α-helices, and SsHSP17.9 has 40 % β-sheets and 15 % α-helices (Table 13.5). Data about chemical stability was obtained by measuring the CD signals of these proteins in different concentrations of the denaturing molecule urea, resulting in the observation that SsHSP17.2 is more stable since it unfolds at higher concentrations than SsHSP17.9 (Tiroli and Ramos 2007). Assays using bis-ANS, a probe that only emits fluorescence when it interacts with unfolded structures (Stryer 1965), corroborates CD experiments because more bis-ANS was bound to SsHSP17.9 than SsHSP17.2,

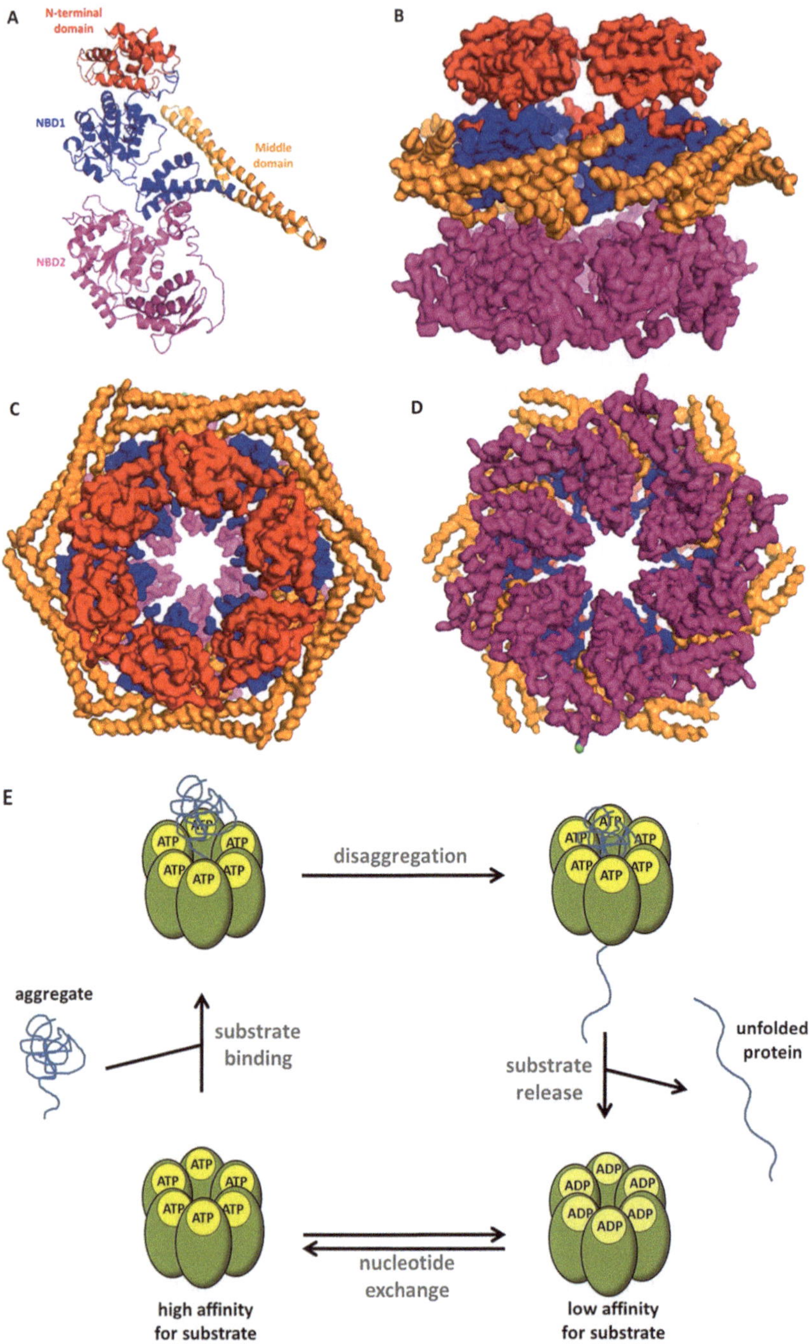

Fig. 13.9 Structure of HSP100 chaperones. (**a**) SHsp101 structure based on comparative modeling predicted by I-TASSER (Yang et al. 2015), highlighting the positions of the domains and secondary structure content (Cagliari et al. 2011). (**b–d**) Model surface of hexameric *E. coli* ClpB mutant E432A (PDB ID: 4D2Q) at different positions (Carroni et al. 2014). The images were prepared using PyMOL. (**e**) A model for the disaggregase action of Hsp100 chaperones (Adapted from Bösl et al. 2006)

indicating that more hydrophobic regions of the protein in solutions were exposed to the solvent (Tiroli and Ramos 2007).

Previous studies have described functional sHSPs as large oligomeric complexes that can be formed from 2 to 32 subunits (Haley et al. 1998; Kim et al. 1998; Basha et al. 2012). Therefore, it is important to determine the hydrodynamic parameters of SsHSP17.2 and SsHSP17.9, such as their molecular mass and oligomeric state by AUC, diffusion coefficient (D) by DLS and Stokes radius (Rs) by analytical gel filtration (AGF) chromatography (Table 13.5) (Tiroli and Ramos 2007). Additionally, those data were used to obtain structural information by comparing them with the data of TaHSP16.9, which has an available high-resolution structure. Both sugarcane sHSPs are dodecameric species in solution, presenting molecular masses of 206.8 kDa (SsHSP17.2) and 214.9 kDa (SsHSP17.9) (Table 13.5). Their Rs values are 48.0 ± 1.5 Å and 48.5 ± 1.5 Å, respectively, in good agreement with the fact that both chaperones consist of the same number of monomers. Comparing the measured Rs values of molecules in solution with hypothetical Rs values based on a spherical shape, it is possible to calculate the Perrin factor (*f*/*fo*), which indicates the shape of the protein (the closer to 1.0, more globular/spherical it is) (Borges and Ramos 2011; Batista et al. 2015). SsHSP17.2 and SsHSP17.9 have *f*/*fo* values of 1.2, suggesting that the complex has a spherical shape (Tiroli and Ramos 2007). Most of the crystal structures pf sHSP complexes in the literature corroborate this spherical shape, since the complexes are arranged in flattened discs with a cavity in the center (van Montfort et al. 2001a, b). The diffusion coefficient D measured by DLS for SsHSP17.2 and SsHSP17.9 was $4.6 \pm 0.3 \times 10^{-7}\ cm^2 \cdot s^{-1}$ and $4.4 \pm 0.1 \times 10^{-7}\ cm^2 \cdot s^{-1}$, respectively (Table 13.5). All these parameters agree with theoretical parameters calculated for TaHSP16.9, which has a solved crystal structure and whose complex is formed by 12 monomers (Tiroli and Ramos 2007). Due to the similarity of the secondary structures, hydrodynamic parameters and nucleotide sequences between SsHSP and TaHSP16.9, it seems reasonable to assume that both sugarcane sHSPs share a similar quaternary structure organization: two disks formed by three dimers with a central cavity where the N-terminal domains are positioned (Fig. 13.10).

Functionally, sHSPs are the most highly expressed chaperones under stress conditions in plants, and for this reason they are frequently associated with the acquisition of thermotolerance in these organisms. sHSPs prevent aggregation of different substrates under thermal stress by keeping them in a refoldable state (Lee et al. 1997; Lee and Vierling 2000). Chaperone functional essays performed with SsHSP17.2 and SsHSP17.9 have shown their capacity to protect model client proteins, such as luciferase, citrate synthase (CS) and malate dehydrogenase (MDH), from heat- induced aggregation. At a sHSP:client protein ratio of 1:20, SsHSP17.2 protects approximately 50 % against the aggregation of luciferase and SsHSP17.9 protects approximately 30 %, which indicates that the first is more efficient at thermoprotecting this substrate. However, for the other model clients (CS and MDH), SsHSP17.9 showed more efficiency at a ratio of 1:1, reaching almost 100 % prevention against aggregation (Tiroli and Ramos 2007) (Table 13.6). These differences in protection activity of SsHSPs may suggest that, although they share similar confor-

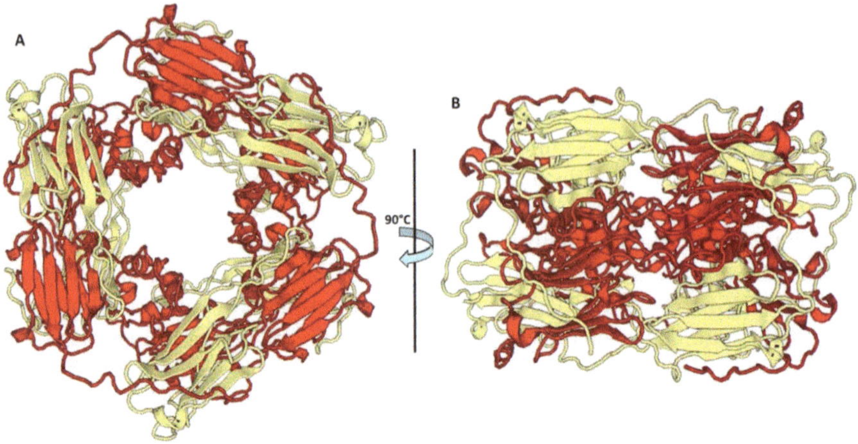

Fig. 13.10 Crystal structure and assembly of wheat sHSP (TaHSP16.9) (PDB ID: 1GME). (**a**) *Top* and (**b**) *side* views (Images are from PDB)

```
                                     N-terminal domain
SsHSP17.2     MSLIRRSNVFDPFSLDLWDPFEGFPFGSGSSSSSLFPSFPRISSETAAFAGARIDWKETP
SsHSP17.9     MSLVRRSNVFDPFSMDLWDPFDNM-------FRSIVPSASSTDSETAAFANARIDWKETP
TaHSP16.9     MSIVRRTNVFDPFAD-LWADPFDT-------FRSIVPAISGGGSETAAFANARMDWKETP
              **::**:******:  **               *:.*:     ******* **:******
                                     α-crystallin domain
SsHSP17.2     EAHVFKADVPGLKKEEVKVEIEDGNILQISGERNKEQEEKTDTWHRVERSSGRFLRRFRL
SsHSP17.9     EAHVFKADLPGVKKEEVKVEVEDGNVLVISGQRSREKEDKNDKWHRVERSSGQFMRRFRL
TaHSP16.9     EAHVFKADLPGVKKEEVKVEVEDGNVLVVSGERTKEKEDKNDKWHRVERSSGKFVRRFRL
              ********:**:********:****:* :**:*.:*:*:*.*.*********:*:*****
                                               C-terminal domain
SsHSP17.2     PENAKTEQIRAAMENGVLTVTVPKEDVKKPEVKSIQISG
SsHSP17.9     PENAKTEEVKAALENGVLTVTVPKAEVKKPEVKSIQISG
TaHSP16.9     LEDAKVEEVKAGLENGVLTVTVPKAEVKKPEVKAIQISG
               *:**.*:::*.:*********** :*******:*****
```

Fig. 13.11 Amino acid sequence comparison of two sHSP from sugarcane. The software online Clustal Omega was used to perform the aligments (Sievers et al. 2011). The identity between SsHSP17.2 and SsHSP17.9 is 77 %; between SsHSP17.2 and TaHSP16.9 is 82 %; and between SsHSP17.9 and TaHSP16.9 is 69 %. (*) Identical residues; (:) conserved substitutions, (.) semi-conserved substitutions

mation, they have distinct specificities for substrates. This is probably because of divergence in their N-terminal domains, which is the most variable domain of sHSPs and is responsible for interacting with client proteins (Basha et al. 2006; Patel et al. 2014; Mani et al. 2015; Fig. 13.11).

Each sHSP may have specificity for certain substrates, but they are promiscuous in that they protect many client proteins from aggregation under heat stress conditions. In an experiment in which proteins from an *E. coli* extract were heated at 40 °C in the presence or absence of SsHSP17.2 and SsHSP17.9, SDS-PAGE showed

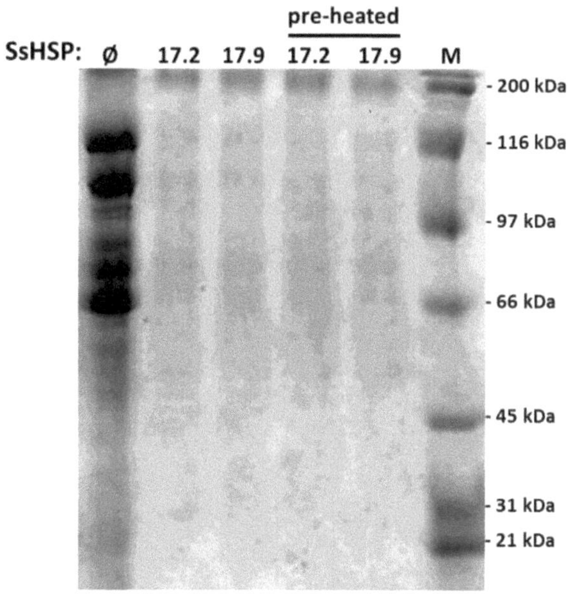

Fig. 13.12 SDS-PAGE showing that sugarcane sHSPs solubilized a wide range of *E. coli* proteins, preventing their aggregation and precipitation. *M*, molecular mass marker. The other lanes are the unsoluble fraction of *E. coli* extract incubated for 15 min at 40 °C. A large fraction of proteins precipitated in the absence of sHSPs (ϕ) but not when sHSPs are present. Pre-heated sHSPs (90 °C for 10 min) are still active

that both chaperones solubilized a wide range of *E. coli* proteins, preventing their aggregation and precipitation (Tiroli-Cepeda and Ramos 2010; Fig. 13.12).

The assembly of sHSPs into huge oligomers is dynamic and depends on environmental conditions (Liu et al. 2015; Fleckenstein et al. 2015). These traits help fine-tune their function to different types of stress. For example, the Rs values of SsHSP17.2 and SsHSP17.9 were measured by AGF chromatography at different temperatures. For SsHSP17.2, there was no change in the Rs (48 Å) between 25 and 30 °C, but at 35 °C there were two peaks, at 48 Å and 24 Å, and at 40–55 °C there was just one peak at 24 Å. The profile for SsHSP17.9 was almost the same, but the change in Rs started at 45 °C. Calculating the molecular masses from the measured Rs data suggests that sugarcane sHSPs disassemble from dodecamers to dimers under high temperatures (Tiroli-Cepeda and Ramos 2010; Fig. 13.13).

The action of sHSPs as holders has been well stablished. Interaction of HSP70 proteins with sHSPs is important under stress conditions since, acting as holders, sHSPs prevent substrates from aggregating but can only (re)fold them in conjunction with the HSP70 system (Haslbeck et al. 2005; Torrente and Shorter 2013; Nillegoda et al. 2015). Refolding experiments using SsHSP70 and SsHSP17.2 have revealed that alone, none of these proteins can refold the unfolded model substrate β-galactosidase at 50 °C, whereas approximately 30 % of the activity of β-galactosidase is restored when the substrate is first incubated with SsHSP17.2 and then heated with SsHSP70 and ATP, indicating that the protein is refolded by this system and establishing SsHSP70 and SsHSPs as foldases (Tiroli-Cepeda et al. 2014; Fig. 13.14).

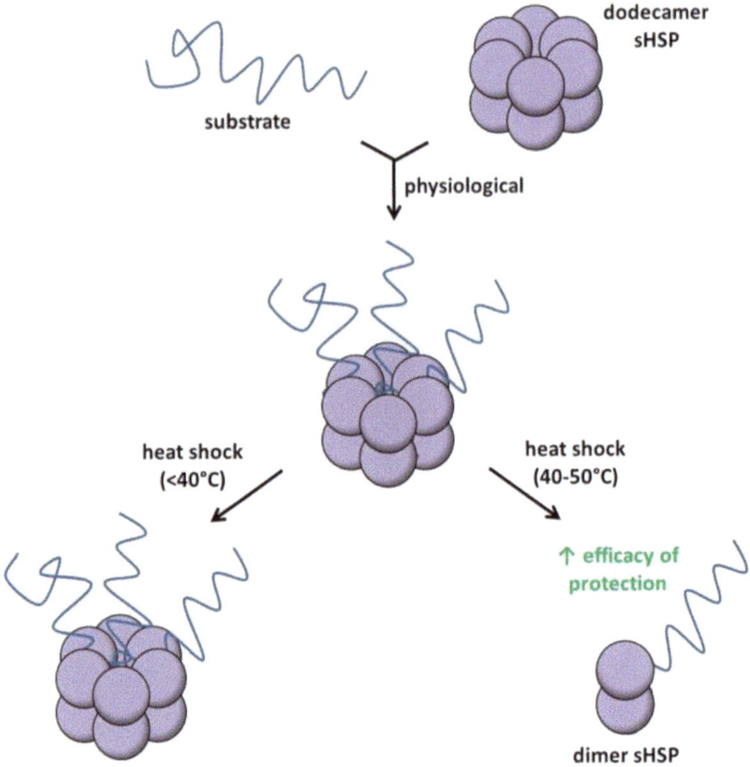

Fig. 13.13 Functional model of sugarcane sHSP under thermal stress. Under physiological conditions, dodecameric sHSP binds substrates. Under mild thermal stress, from 25 to 40 °C, sHSP starts to deoligomerize into dimers, increasing its ability to prevent from aggregation. Under over thermal stress, from 40 to 55 °C, sHSP completely disassemble into dimers, increasing even more its protection efficacy (Tiroli-Cepeda and Ramos 2010)

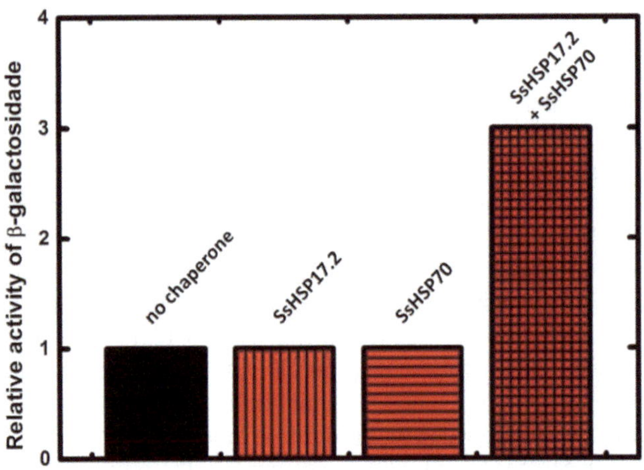

Fig. 13.14 Sugarcane HSP70 cooperates with sHSP to refold thermally unfolded proteins. Together, SsHSP70 and SsHSP17.2 refold the unfolded model substrate β-galactosidase at 50 °C, establishing SsHSP70 as a foldase and SsHSP as a holder (Tiroli-Cepeda and Ramos 2010). The activity of thermally unfolded β-galactosidase is 1 and refolded (first incubated with SsHSP17.2 and then heated with SsHSP70 and ATP) is 3× higher

13.4 Final Remarks

The aim of this chapter was to summarize the available information regarding the expression of molecular chaperones and HSPs in the presence and absence of stress and also about the relationship between the structure and function of these proteins in plants. Sugarcane was the main source of the information described in this chapter, but chaperones and HSPs from other organisms, such as eucalyptus, rice, *Arabidopsis* and orange, were also discussed. The general findings are as follows. Chaperone and stress-related protein genes are abundantly expressed and have ample diversity. Protein classes present in the cytoplasm have both higher expression and diversity than the classes present in other cellular compartments. Cloning, purification and characterization of HSP genes from sugarcane has been effective in the classification of sugarcane chaperones as foldases, holders and disaggregases.

Acknowledgments The authors wish to apologize to the authors of all the studies that we did not have space to include. Research in the laboratory of CHIR is supported by grants from Fundação de Amparo a Pesquisa do Estado de São Paulo (FAPESP, 2012/50161-8) and Ministério da Ciência e Tecnologia/Conselho Nacional de Pesquisa e Desenvolvimento (MCT/CNPq). CCG thanks FAPESP for fellowship.

References

Abravaya K, Myers MP, Murphy SP, Morimoto RI (1992) The human heat shock protein hsp70 interacts with HSF, the transcription factor that regulates heat shock gene expression. Genes Dev 6:1153–1164

Agarwal M, Katiyar-Agarwal S, Sahi C, Gallie DR, Grover A (2001) Arabidopsis thaliana Hsp100 proteins: kith and kin. Cell Stress Chaperones 6:219–224

Akerfelt M, Morimoto RI, Sistonen L (2010) Heat shock factors: integrators of cell stress, development and lifespan. Nat Rev Mol Cell Biol 11:545–555

Ali MM, Roe SM, Vaughan CK, Meyer P, Panaretou B, Piper PW, Prodromou C, Pearl LH (2006) Crystal structure of an Hsp90-nucleotide-p23/Sba1 closed chaperone complex. Nature 440:1013–1017

Anckar J, Sistonen L (2011) Regulation of HSF1 function in the heat stress response: implications in aging and disease. Annu Rev Biochem 80:1089–1115

Anfinsen CB (1973) Principles that govern the folding of protein chains. Science 181:223–230

Arabidopsis Genome Initiative (2000) Analysis of the genome sequence of the flowering plant *Arabidopsis thaliana*. Nature 408:796–815

Bakthisaran R, Tangirala R, Rao Ch M (2015) Small heat shock proteins: role in cellular functions and pathology. Biochim Biophys Acta 1854:291–319

Baldwin RL (1995) The nature of protein-folding pathways – the classical versus the new view. J Biomol NMR 5:103–109

Baldwin RL, Rose GD (2013) Molten globules, entropy-driven conformational change and protein folding. Curr Opin Struct Biol 23:4–10

Baniwal SK, Bharti K, Chan KY, Fauth M, Ganguli A, Kotak S, Mishra SK, Nover L, Port M, Scharf KD, Tripp J, Weber C, Zielinski D, von Koskull-Döring P (2004) Heat stress response in plants: a complex game with chaperones and more than twenty heat stress transcriptions factors. J Biosci 29:471–487

Barnett ME, Nagy M, Kedzierska S, Zolkiewski M (2005) The amino-terminal domain of ClpB supports binding to strongly aggregated proteins. J Biol Chem 280:34940–34945
Barraclough R, Ellis RJ (1980) Protein synthesis in chloroplasts. IX. Assembly of newly- synthesized large subunits into ribulose bisphosphate carboxylase in isolated intact pea chloroplasts. Biochim Biophys Acta 608:19–31
Basha GJ, Friedrich KL, Vierling E (2006) The N-terminal arm of small heat shock proteins is important for both chaperone activity and substrate specificity. J Biol Chem 281:39943–39952
Basha E, O'Neill H, Vierling E (2012) Small heat shock proteins and alpha-crystallins: dynamic proteins with flexible functions. Trends Biochem Sci 37:106–117
Basha E, Jones C, Blackwell AE, Cheng G, Waters ER, Samsel KA, Siddique M, Pett V, Wysocki V, Vierling E (2013) An unusual dimeric small heat shock protein provides insight into the mechanism of this class of chaperones. J Mol Biol 425:1683–1696
Batista FA, Gava LM, Pinheiro GM, Ramos CH, Borges JC (2015) From conformation to interaction: techniques to explore the Hsp70/Hsp90 network. Curr Protein Pept Sci 16:735–753
Borges JC, Ramos CH (2005) Protein folding assisted by chaperones. Protein Pept Lett 12(3):257–261
Borges JC, Ramos CH (2006) Spectroscopic and thermodynamic measurements of nucleotide-induced changes in the human 70-kDa heat shock cognate protein. Arch Biochem Biophys 452:46–54
Borges JC, Ramos CH (2011) Analysis of molecular targets of *Mycobacterium tuberculosis* by analytical ultracentrifugation. Curr Med Chem 18:1276–1285
Borges JC, Peroto MC, Ramos CH (2001) Molecular chaperone genes in the sugarcane expressed sequence database (SUCEST). Genet Mol Biol 24:85–92
Borges JC, Cagliari TC, Ramos CH (2007) Expression and variability of molecular chaperones in the sugarcane expressome. J Plant Physiol 164:505–513
Boshoff A (2015) Chaperonin-co-chaperonin interactions. Subcell Biochem 78:153–178
Bösl B, Grimminger V, Walter S (2006) The molecular chaperone Hsp104 – a molecular machine for protein disaggregation. J Struct Biol 156:139–148
Boston RS, Viitanen PV, Vierling E (1996) Molecular chaperones and protein folding in plants. Plant Mol Biol 32:191–222
Bracher A, Verghese J (2015) GrpE, Hsp110/Grp170, HspBP1/Sil1 and BAG domain proteins: nucleotide exchange factors for Hsp70 molecular chaperones. Subcell Biochem 78:1–33
Brandvold KR, Morimoto RI (2015) The chemical biology of molecular chaperones–implications for modulation of proteostasis. J Mol Biol 427:2931–2947
Buchner J (1999) Hsp90 and co. – a holding for folding. Trends Biochem Sci 24:136–141
Bukau B, Horwich AL (1998) The Hsp70 and Hsp60 chaperone machines. Cell 92:351–366
Bukau B, Weissman J, Horwich A (2006) Molecular chaperones and protein quality control. Cell 125:443–451
Cagliari TC, Tiroli AO, Borges JC, Ramos CH (2005) Identification and *in silico* pattern analysis of *Eucalyptus* expressed sequencing tags (ESTs) enconding molecular chaperones. Genet Mol Biol 28:520–528
Cagliari TC, da Silva VC, Borges JC, Prando A, Tasic L, Ramos CH (2011) Sugarcane Hsp101 is a hexameric chaperone that binds nucleotides. Int J Biol Macromol 49:1022–1030
Carroni M, Kummer E, Oguchi Y, Wendler P, Clare DK, Sinning I, Kopp J, Mogk A, Bukau B, Saibil HR (2014) Head-to-tail interactions of the coiled-coil domains regulate ClpB activity and cooperation with Hsp70 in protein disaggregation. Elife 3:e02481
Caspers GJ, Leunissen JA, de Jong WW (1995) The expanding small heat shock protein family and structure predictions of the conserved "α-crystallin-domain". J Mol Evol 40:238–248
Chadli A, Bouhouche I, Sullivan W, Stensgard B, McMahon N, Catelli MG, Toft DO (2000) Dimerization and N-terminal domain proximity underlie the function of molecular chaperone heat shock protein 90. Proc Natl Acad Sci U S A 97:12524–12529
Clerico EM, Tilitsky JM, Meng W, Gierasch LM (2015) How HSP70 molecular machines interact with their substrates to mediate diverse physiological functions. J Mol Biol 427:1575–1588

Correa DHA, Ramos CH (2009) The use of circular dichroism spectroscopy to study protein folding, form and function. Afr J Biochem Res 3:164–173
Cramer GR, Urano K, Delrot S, Pezzotti M, Shinozaki K (2011) Effects of abiotic stress on plants: a systems biology perspective. BMC Plant Biol 11:163
Cyr DM, Ramos CH (2015) Specification of Hsp70 function by type I and type II Hsp40. Subcell Biochem 78:91–102
da Silva KP, Borges JC (2011) The molecular chaperone Hsp70 family members function by a bidirectional heterotrophic allosteric mechanism. Protein Pept Lett 18:132–142
da Silva VC, Ramos CH (2012) The network interaction of human 90 kDa heat shock protein Hsp90: A target for cancer therapeutics. J Proteome 75:2790–2802
da Silva VC, Cagliari TC, Lima TB, Gozzo FC, Ramos CH (2013) Conformational and functional studies of a cytosolic 90 kDa heat shock protein Hsp90 from sugarcane. Plant Physiol Biochem 68:16–22
Douglas PM, Summers DW, Cyr DM (2009) Molecular chaperones antagonize proteotoxicity by differentially modulating protein aggregation pathways. Prion 3:51–58
Doyle SM, Genest O, Wickner S (2013) Protein rescue from aggregates by powerful molecular chaperone machines. Nat Rev Mol Cell Biol 14:617–629
Duncan EJ, Cheetham ME, Chapple JP, van der Spuy J (2015) The role of HSP70 and its co- chaperones in protein misfolding, aggregation and disease. Subcell Biochem 78:243–273
Edward JT (1970) Molecular volumes and the Stokes-Einstein Equation. J Chem Educ 47:261–270
Ferreira ST, De Felice FG (2001) PABMB lecture. Protein dynamics, folding and misfolding: from basic physical chemistry to human conformational diseases. FEBS Lett 498:129–134
Flaherty KM, DeLuca-Flaherty C, McKay DB (1990) Three-dimensional structure of the ATPase fragment of a 70K heat-shock cognate protein. Nature 346:623–628
Fleckenstein T, Kastenmüller A, Stein ML, Peters C, Daake M, Krause M, Weinfurtner D, Haslbeck M, Weinkauf S, Groll M, Buchner J (2015) The chaperone activity of the development small heat shock protein Sip1 is regulated by pH-dependent conformational changes. Mol Cell 58:1067–1078
Freeman BC, Morimoto RI (1996) The human cytosolic molecular chaperones HSP90, HSP70 (HSC70) and hdj-1 have distinct roles in recognition of a non-native protein and protein refolding. EMBO J 15:2969–2979
Fujimoto M, Nakai A (2010) The heat shock factor family and adaptation to proteotoxic stress. FEBS J 277:4112–4125
Garrido C, Paul C, Seigneuric R, Kampinga HH (2012) The small heat shock proteins family: the long forgotten chaperones. Int J Biochem Cell Biol 44:1588–1592
Gava L, Ramos CH (2009) Human 90 kDa heat shock protein Hsp90 as a target for cancer therapeutics. Curr Chem Biol 3:330–341
Glover JR, Lindquist S (1998) Hsp104, Hsp70, and Hsp40: a novel chaperone system that rescues previously aggregated proteins. Cell 94:73–82
Goloubinoff P (2014) Recent and future grand challenges in protein folding, misfolding, and degradation. Front Mol Biosci 1:1
Gusev NB, Bogatcheva NV, Marston SB (2002) Structure and properties of small heat shock proteins (sHsp) and their interaction with cytoskeleton proteins. Biochemistry (Mosc) 67:511–519
Haley DA, Horwitz J, Stewart PL (1998) The small heat-shock protein, alphaB-crystallin, has a variable quaternary structure. J Mol Biol 277:27–35
Hartl FU, Hayer-Hartl M (2002) Molecular chaperones in the cytosol: from nascent chain to folded protein. Science 295:1852–1858
Hartl FU, Hayer-Hartl M (2009) Converging concepts of protein folding in vitro and in vivo. Nat Struct Mol Biol 16:574–581
Hartl FU, Martin J, Neupert W (1992) Protein folding in the cell: the role of molecular chaperones Hsp70 and Hsp60. Annu Rev Biophys Biomol Struct 21:293–322
Hartl FU, Bracher A, Hayer-Hartl M (2011) Molecular chaperones in protein folding and proteostasis. Nature 475:324–332

Haslbeck M, Vierling E (2015) A first of stress defense: small heat shock proteins and their function in protein homeostasis. J Mol Biol 427:1537–1548
Haslbeck M, Walke S, Stromer T, Ehrnsperger M, White HE, Chen S, Saibil HR, Buchner J (1999) Hsp26: a temperature-regulated chaperone. EMBO J 18:6744–6751
Haslbeck M, Braun N, Stromer T, Richter B, Model N, Weinkauf S, Buchner J (2004) Hsp42 is the general small heat shock protein in the cytosol of *Saccharomyces cerevisiae*. EMBO J 23:638–649
Haslbeck M, Miess A, Stromer T, Walter S, Buchner J (2005) Disassembling protein aggregates in the yeast cytosol. The cooperation of Hsp26 with Ssa1 and Hsp104. J Biol Chem 280:23861–23868
Hendrick JP, Hartl FU (1993) Molecular chaperone functions of heat-shock proteins. Annu Rev Biochem 62:49–384
Hietakangas V, Ahlskog JK, Jakobsson AM, Hellesuo M, Sahlberg NM, Holmberg CI, Mikhailov A, Palvimo JJ, Pirkkala L, Sistonen L (2003) Phosphorylation of serine 303 is a prerequisite for the stress-inducible SUMO modification of heat shock factor 1. Mol Cell Biol 23:2953–2968
Hill JE, Hemmingsen SM (2001) *Arabidopsis thaliana* type I and II chaperonins. Cell Stress Chaperones 6:190–200
Hodson S, Marshall JJ, Burston SG (2012) Mapping the road to recovery: the ClpB/Hsp104 molecular chaperone. J Struct Biol 179:161–171
Hu W, Hu G, Han B (2009) Genome-wide survey and expression profiling of heat shock proteins and heat shock factors revealed overlapped and stress specific response under abiotic stresses in rice. Plant Sci 176:583–590
International Rice Genome Sequencing Project (2005) The map-based sequence of the rice genome. Nature 436:793–800
Jackson SE (2013) Hsp90: structure and function. Top Curr Chem 328:155–240
Jakob U, Lilie H, Meyer I, Buchner J (1995) Transient interaction of Hsp90 with early unfolding intermediates of citrate synthaseeimplications for heat shock in vivo. J Biol Chem 270:7288–7294
Kakkar VV, Prins LC, Kampinga HH (2012) DNAJ proteins and protein aggregation diseases. Curr Top Med Chem 12:2479–2490
Kim KK, Kim R, Kim SH (1998) Crystal structure of a small heat-shock protein. Nature 394:595–599
Kim YE, Hipp MS, Bracher A, Hayer-Hartl M, Hartl FU (2013) Molecular chaperone functions in protein folding and proteostasis. Annu Rev Biochem 82:323–355
Knowles TP, Vendruscolo M, Dobson CM (2014) The amyloid state and its association with protein misfolding diseases. Nat Rev Mol Cell Biol 15:384–396
Kotak S, Larkindale J, Lee U, von Koskull-Döring P, Vierling E, Scharf KD (2007) Complexity of the heat stress response in plants. Curr Opin Plant Biol 10:310–316
Krishna P, Gloor G (2001) The Hsp90 family of proteins in *Arabidopsis thaliana*. Cell Stress Chaperones 6:238–246
Landreh M, Rising A, Presto J, Jörnvall H, Johansson J (2015) Specific chaperones and regulatory domains in control of amyloid formation. J Biol Chem 290:26430–26436
Lee GJ, Vierling E (2000) A small heat shock protein cooperates with heat shock protein 70 systems to reactivate a heat-denatured protein. Plant Physiol 122:189–198
Lee GJ, Roseman AM, Saibil HR, Vierling E (1997) A small heat shock protein stably binds heat-denatured model substrates and can maintain a substrate in a folding-competent state. EMBO J 16:659–671
Lee S, Sowa ME, Watanabe YH, Sigler PB, Chiu W, Yoshida M, Tsai FT (2003) The structure of ClpB: a molecular chaperone that rescues proteins from an aggregated state. Cell 115:229–240
Lee S, Sowa ME, Choi JM, Tsai FT (2004) The ClpB/Hsp104 molecular chaperone – a protein disaggregation machine. J Struct Biol 146:99–105
Levitt M, Warshel A (1975) Computer simulation of protein folding. Nature 253:694–698
Lin BL, Wang JS, Liu HC, Chen RW, Meyer Y, Barakat A, Delseny M (2001) Genomic analysis of the Hsp70 superfamily in *Arabidopsis thaliana*. Cell Stress Chaperones 3:201–208

Liu L, Chen J, Yang B, Wang Y (2015) Oligomer-dependent and –independent chaperone activity of sHsps in different stressed conditions. FEBS Open Bio 5:155–162
Löw D, Brändle K, Nover L, Forreiter C (2000) Cytosolic heat-stress proteins Hsp17.1 class I and Hsp17.3 class II of tomato act as molecular chaperones in vivo. Planta 211:575–582
Lu Z, Cyr DM (1998) Protein folding activity of Hsp70 is modified differentially by the hsp40 co-chaperones Sis1 and Ydj1. J Biol Chem 273:27824–27830
Luheshi LM, Crowther DC, Dobson CM (2008) Protein misfolding and disease: from the test tube to the organism. Curr Opin Chem Biol 12:25–31
Mackay RG, Helsen CW, Tkach JM, Glover JR (2008) The C-terminal extension of *Saccharomyces cerevisiae* Hsp104 plays a role in oligomer assembly. Biochemistry 47:1918–1927
Makhnevych T, Houry WA (2012) The role of Hsp90 in protein complex assembly. Biochim Biophys Acta 1823:674–682
Mani N, Ramakrishna K, Suguna K (2015) Characterization of rice small heat shock proteins targeted to different cellular organelles. Cell Stress Chaperones 20:451–460
Matoo RU, Sharma SK, Priya S, Finka A, Goloubinoff P (2013) Hsp110 is a bona fide chaperone using ATP to unfold stable misfolded polypeptides and reciprocally collaborate with Hsp70 to solubilize protein aggregates. J Biol Chem 288:21399–21411
Mayer MP (2010) Gymnastics of molecular chaperones. Mol Cell 39:321–331
Mayer MP, Kityk R (2015) Insights into the molecular mechanism of allostery in Hsp70s. Front Mol Biosci 20:58
Mayer MP, Le Breton L (2015) Hsp90: breaking the symmetry. Mol Cell 58:8–20
Mendillo ML, Santagata S, Koeva M, Bell GW, Hu R, Tamimi RM, Fraenkel E, Ince TA, Whitesell L, Lindquist S (2012) HSF1 drives a transcriptional program distinct from heat shock to support highly malignant human cancers. Cell 150:549–562
Mendonça YA, Ramos CH (2012) Cloning, purification and characterization of a 90 kDa heat shock protein from Citrus sinensis (sweet orange). Plant Physiol Biochem 50:87–94
Miernyk JA (2001) The J-domain proteins of *Arabidopsis thaliana*: an unexpectedly large and diverse family of chaperones. Cell Stress Chaperones 6:209–218
Miozzo F, Sabéran-Djoneidi D, Mezger V (2015) HSFs, stress sensors and sculptors of transcription compartments and epigenetic landscapes. J Mol Biol 427:3793–3816
Mogk A, Kummer E, Bukau B (2015) Cooperation of Hsp70 and Hsp100 chaperone machines in protein disaggregation. Front Mol Biosci 19:22
Mokry DZ, Abrahao J, Ramos CH (2015) Disaggregases, molecular chaperones that resolubilize protein aggregates. An Acad Bras Cienc 87:1273–1292
Montgomery DL, Morimoto RI, Giesrasch LM (1999) Mutations in the substrate binding domain of the Escherichia coli 70 kDa affinity or interdomain coupling. J Mol Biol 286:915–932
Morimoto RI (2011) The heat shock response: systems biology of proteotoxic stress in aging and disease. Cold Spring Harb Symp Quant Biol 76:91–99
Morrow G, Hightower LE, Tanguay RM (2015) Small heat shock proteins: big folding machines. Cell Stress Chaperones 20:207–212
Myburg AA, Grattapaglia D, Tuskan GA, Hellsten U, Hayes RD, Grimwood J, Jenkins J, Lindquist E, Tice H, Bauer D, Goodstein DM, Dubchak I, Poliakov A, Mizrachi E, Kullan AR, Hussey SG, Pinard D, van der Merwe K, Singh P, van Jaarsveld I, Silva-Junior OB, Togawa RC, Pappas MR, Faria DA, Sansaloni CP, Petroli CD, Yang X, Ranjan P, Tschaplinski TJ, Ye CY, Li T, Sterck L, Vanneste K, Murat F, Soler M, Clemente HS, Saidi N, Cassan-Wang H, Dunand C, Hefer CA, Bornberg-Bauer E, Kersting AR, Vining K, Amarasinghe V, Ranik M, Naithani S, Elser J, Boyd AE, Liston A, Spatafora JW, Dharmwardhana P, Raja R, Sullivan C, Romanel E, Alves-Ferreira M, Külheim C, Foley W, Carocha V, Paiva J, Kudrna D, Brommonschenkel SH, Pasquali G, Byrne M, Rigault P, Tibbits J, Spokevicius A, Jones RC, Steane DA, Vaillancourt RE, Potts BM, Joubert F, Barry K, Pappas GJ, Strauss SH, Jaiswal P, Grima-Pettenati J, Salse J, Van de Peer Y, Rokhsar DS, Schmutz J (2014) The genome of Eucalyptus grandis. Nature 510:356–362
Nillegoda NB, Bukau B (2015) Metazoan Hsp70-based protein disaggregases: emergence and mechanisms. Front Mol Biosci 2:57

Nillegoda NB, Kirstein J, Szlachcic A, Berynskyy M, Stank A, Stengel F, Arnsburg K, Gao X, Scior A, Aebersold R, Guilbride DL, Wade RC, Morimoto RI, Mayer MP, Bukau B (2015) Crucial Hsp70 co-chaperone complex unlocks metazoan protein disaggregation. Nature 524:247–251
Nover L, Miernyk JA (2001) A genomics approach to the chaperone network of *Arabidopsis thaliana*. Cell Stress Chaperones 6:175–176
Obermann WM, Sondermann H, Russo AA, Pavletich NP, Hartl FU (1998) In vivo function of Hsp90 is dependent on ATP binding and ATP hydrolysis. J Cell Biol 143:901–910
O'Driscoll J, Clare D, Saibil H (2015) Prion aggregate structure in yeast cells is determined by the Hsp104-Hsp110 disaggregase machinery. J Cell Biol 211:145–158
Patel S, Vierling E, Tama F (2014) Replica exchange molecular dynamics simulations provide insight into substrate recognition by small heat shock proteins. Biophys J 106:2644–2655
Pesce ER, Blatch GL (2014) Plasmodial Hsp40 and Hsp70 chaperones: current and future perspectives. Parasitology 141:1167–1176
Picard D (2002) Heat-shock protein 90, a chaperone for folding and regulation. Cell Mol Life Sci 59:1640–1648
Priya S, Sharma SK, Goloubinoff P (2013) Molecular chaperones as enzymes that catalytically unfold misfolded polypeptides. FEBS Lett 587:1981–1987
Qu AL, Ding YF, Jiang Q, Zhu C (2013) Molecular mechanisms of the plant heat stress response. Biochem Biophys Res Commun 432:203–207
Ramos CH (2008) In: O'Doherty CB, Byrne, AC (eds) Protein misfolding. Nova Science Publishers, New York. ISBN: 978-1-60456-881-3
Ramos CH, Ferreira ST (2005) Protein folding, misfolding and aggregation: evolving concepts and conformational diseases. Protein Pept Lett 12:213–222
Rampelt H, Kirstein-Miles J, Nillegoda NB, Chi K, Scholz SR, Morimoto RI, Bukau B (2012) Metazoan Hsp70 machines use Hsp110 to power protein disaggregation. EMBO J 31:4221–4235
Saibil H (2013) Chaperone machines for protein folding, unfolding and disaggregation. Nat Rev Mol Cell Biol 14:630–642
Sarkar NK, Kim YK, Grover A (2009) Rice sHsp genes: genomic organization and expression profiling under stress and development. BMC Genomics 24:393
Sarkar NK, Thapar U, Kundnani P, Panwar P, Grover A (2013) Functional importance of J- protein family of rice (*Oryza sativa*). Cell Stress Chaperones 18:321–331
Sauer RT, Bolon DN, Burton BM, Burton RE, Flynn JM, Grant RA, Hersch GL, Joshi SA, Kenniston JA, Levchenko I, Neher SB, Oakes ES, Siddiqui SM, Wah DA, Baker TA (2004) Sculpting the proteome with AAA(+) proteases and disassembly machines. Cell 119:9–18
Scheufler C, Brinker A, Bourenkov G, Pegoraro S, Moroder L, Bartunik H, Hartl FU, Moarefi I (2001) Structure of TPR domainpeptide complexes: critical elements in the assembly of the Hsp70-Hsp90 multichaperone machine. Cell 101:199–210
Schirmer EC, Glover JR, Lindquist S (1996) Hsp100/Clp proteins: a common mechanism explains diverse functions. Trends Biochem Sci 21:289–296
Shi Y, Mosser DD, Morimoto RI (1998) Molecular chaperones as HSF1-specific transcriptional repressors. Genes Dev 12:654–666
Shorter J (2011) The mammalian disaggregase machinery: Hsp110 synergizes with Hsp70 and Hsp40 to catalyze protein disaggregation and reactivation in a cell-free system. PLoS One 6:e26319
Sievers F, Wilm A, Dineen D, Gibson TJ, Karplus K, Li W, Lopez R, McWilliam H, Remmert M, Söding J, Thompson JD, Higgins DG (2011) Fast, scalable generation of high-quality protein multiple aligments using Clustal Omega. Mol Syst Biol 11:7–539
Slepenkov SV, Witt SN (2002) The unfolding story of the Escherichia coli Hsp70 DnaK: is DnaK a holdase or an unfoldase? Mol Microbiol 45:1197–1206
Snider J, Houry WA (2008) AAA+ proteins: diversity in function, similarity in structure. Biochem Soc Trans 36:72–77
Sorger PK, Pelham HR (1998) Yeast heat shock factor is an essential DNA-binding protein that exhibits temperature-dependent phosphorylation. Cell 54:855–864

Sóti C, Csermely P (2000) Molecular chaperones and the aging process. Biogerontology 1:225–233
Stein L (2001) Genome annotation: from sequence to biology. Nat Rev Genet 2:493–503
Stryer L (1965) The interaction of a naphthalene dye with apomyoglobin and apohemoglobin. A fluorescent probe of nonpolar binding sites. J Mol Biol 13:482–495
Summers DW, Douglas PM, Ramos CH, Cyr DM (2009) Polypeptide transfer from Hsp40 to Hsp70 molecular chaperones. Trends Biochem Sci 34:230–233
Swindell WR, Huebner M, Weber AP (2007) Transcriptional profiling of *Arabidopsis* heat shock proteins and trancription factors reveals extensive overlap between heat and non-heat stress response pathways. BMC Genomics 8:125
Tipping KW, van Oosten-Hawle P, Hewitt EW, Radford SE (2015) Amyloid fibres: inert end- stage aggregates or key players in disease? Trends Biochem Sci 40:719–727
Tiroli AO, Ramos CH (2007) Biochemical and biophysical characterization of small heat shock proteins from sugarcane. Involvement of a specific region located at the N-terminus with substrate specificity. Int J Biochem Cell Biol 39:818–831
Tiroli-Cepeda AO, Ramos CH (2010) Heat causes oligomeric disassembly and increases the chaperone activity of small heat shock proteins from sugarcane. Plant Physiol Biochem 48:108–116
Tiroli-Cepeda AO, Ramos CH (2011) An overview of the role of molecular chaperones in protein homeostasis. Protein Pept Lett 18:101–109
Tiroli-Cepeda AO, Lima TB, Balbuena TS, Gozzo FC, Ramos CH (2014) Structural and functional characterization of the chaperone Hsp70 from sugarcane. Insights into conformational changes during cycling from cross-linking/mass spectrometry assays. J Proteome 104:48–56
Torrente MP, Shorter J (2013) The metazoan protein disaggregase and amyloid depolymerase system: Hsp110, Hsp70, Hsp40 and small heat shock proteins. Prion 7:457–463
Uversky VN (2015) Functional roles of transiently and intrinsically disordered regions within proteins. FEBS J 282:1182–1189
van Montfort R, Slingsby C, Vierling E (2001a) Structure and function of the small heat shock protein/alpha-crystallin family of molecular chaperones. Adv Protein Chem 59:105–156
van Montfort RL, Basha E, Friedrich KL, Slingsby C, Vierling E (2001b) Crystal structure and assembly of a eukaryotic small heat shock protein. Nat Struct Biol 8:1025–1030
Vettore AL, da Silva FR, Kemper EL, Souza GM, da Silva AM, Ferro MI, Henrique-Silva F, Giglioti EA, Lemos MV, Coutinho LL, Nobrega MP, Carrer H, França SC, Bacci JM, Goldman MH, Gomes SL, Nunes LR, Camargo LE, Siqueira WJ, Van Sluys MA, Thiemann OH, Kuramae EE, Santelli RV, Marino CL, Targon ML, Ferro JA, Silveira HC, Marini DC, Lemos EG, Monteiro-Vitorello CB, Tambor JH, Carraro DM, Roberto PG, Martins VG, Goldman GH, de Oliveira RC, Truffi D, Colombo CA, Rossi M, de Araujo PG, Sculaccio SA, Angella A, Lima MM, de Rosa JVE, Siviero F, Coscrato VE, Machado MA, Grivet L, Di Mauro SM, Nobrega FG, Menck CF, Braga MD, Telles GP, Cara FA, Pedrosa G, Meidanis J, Arruda P (2003) Analysis and functional annotation of an expressed sequence tag collection for tropical crop sugarcane. Genome Res 13:2725–2735
Vierling E (1991) The roles of heat shock proteins in plants. Annu Rev Plant Physiol Plant Mol Biol 42:579–620
Vinocur B, Altman A (2005) Recent advances in engineering plant tolerance to abiotic stress: achievements and limitations. Curr Opin Biotechnol 16:123–132
Voet D, Voet J (2011) Biochemistry, 4th edn. Wiley, Hoboken
Wandinger SK, Richter K, Buchner J (2008) The Hsp90 chaperone machinery. J Biol Chem 283:18473–18477
Wang W, Vinocur B, Shoseyov O, Altman A (2004) Role of plant heat-shock proteins and molecular chaperones in the abiotic stress response. Trends Plant Sci 9:244–252
Waters ER (2013) The evolution, function, structure and expression of the plant sHSPs. J Exp Bot 64:391–403
Wegele H, Müller L, Buchner J (2004) Hsp70 and Hsp90 – a relay team for protein folding. Rev Physiol Biochem Pharmacol 151:1–44

Westerheide SD, Anckar J, Stevens SMJ, Sistonen L, Morimoto RI (2009) Stress-inducible regulation of heat shock factor 1 by the deacetylase SIRT1. Science 323:1063–1066
Winkler J, Tyedmers J, Bukau B, Mogk A (2012) Chaperone networks in protein disaggregation and prion propagation. J Struct Biol 179:152–160
Yang J, Yan R, Roy A, Xu D, Poisson J, Zhang Y (2015) The I-TASSER suite: protein structure and function prediction. Nat Methods 12:7–8
Yonehara M, Minami Y, Kawata Y, Nagai J, Yahara I (1996) Heat-induced chaperone activity of HSP90. J Biol Chem 271:2641–2645
Young JC (2010) Mechanisms of the Hsp70 chaperone system. Biochem Cell Biol 88:291–300
Young JC, Moarefi I, Hartl FU (2001) Hsp90: a specialized but essential protein-folding tool. J Cell Biol 154:267–273
Zettlmeissl G, Rudolph R, Jaenicke R (1979) Reconstitution of lactic dehydrogenase. Noncovalent aggregation vs. reactivation 1. Physical properties and kinetics of aggregation. Biochemistry 18:5567–5571
Zhang J, Li J, Liu B, Zhang L, Chen J, Lu M (2013) Genome-wide analysis of the Populus Hsp90 gene family reveals differential expression patterns, localization and heat stress responses. BMC Genomics 14:532
Zhang J, Liu B, Li J, Zhang L, Wang Y, Zheng H, Lu M, Chen J (2015) Hsf and Hsp gene families in Populus: genome-wide identification, organization and correlated expression during development and in stress responses. BMC Genomics 16:181
Zhu X, Zhao X, Burkholder WF, Gragerov A, Ogata CM, Gottesman ME, Hendrickson WA (1996) Structural analysis of substrate binding by the molecular chaperone DnaK. Science 272:1606–1614
Zietkiewicz S, Krzewska J, Liberek K (2004) Successive and synergistic action of the Hsp70 and Hsp100 chaperones in protein disaggregation. J Biol Chem 279:44376–44383
Zolkiewski M (2006) A camel passes through the eye of a needle: protein unfolding activity of Clp ATPases. Mol Microbiol 61:1094–1100
Zolkiewski M, Kessel M, Ginsburg A, Maurizi MR (1999) Nucleotide-dependent oligomerization of ClpB from *Escherichia coli*. Protein Sci 9:1899–1903
Zolkiewski M, Zhang T, Nagy M (2012) Aggregate reactivation mediated by the Hsp100 chaperones. Arch Biochem Biophys 520:1–6

Chapter 14
Role of Heat Shock Proteins in Improving Heat Stress Tolerance in Crop Plants

Palakolanu Sudhakar Reddy, Thammineni Chakradhar, Ramesha A. Reddy, Rahul B. Nitnavare, Srikrishna Mahanty, and Malireddy K. Reddy

Abstract High temperature response (HTR) or heat stress response (HSR) is a highly conserved phenomenon, which involves complex networks among different crop species. Heat stress usually results in protein dysfunction by improper folding of its linear amino acid chains to non-native proteins. This leads to unfavourable interactions and subsequent protein aggregation. To tackle this, plants have developed molecular chaperone machinery to maintain high quality proteins in the cell. This is governed by increasing the level of pre-existing molecular chaperones and by expressing additional chaperones through signalling mechanism. Dissecting the molecular mechanism by which plants counter heat stress and identification of important molecules involved are of high priority. This could help in the development of plants with improved heat stress tolerance through advanced genomics and genetic engineering approaches. Owing to this reason molecular chaperones/Heat shock proteins (Hsps) are considered as potential candidates to address the issue of heat stress. In this chapter, recent progress on systematic analyses of heat shock proteins, their classification and role in plant response to heat stress along with an overview of genomic and transgenic approaches to overcome the issue, are summarized.

Keywords Heat shock element • Heat shock factors • Heat shock proteins • Heat shock response • Heat stress

---

P.S. Reddy (✉)
Plant Molecular Biology Group, International Centre for Genetic Engineering and Biotechnology (ICGEB), Aruna Asaf Ali Marg, New Delhi 110 067, India

Crop Physiology, Dryland Cereals, International Crops Research Institute for the Semi-Arid Tropics (ICRISAT), Patancheru, Hyderabad 502 324, Telangana, India
e-mail: p.sudhakarreddy@cgiar.org; palakolanusreddy@gmail.com

T. Chakradhar • R.A. Reddy • S. Mahanty • M.K. Reddy
Plant Molecular Biology Group, International Centre for Genetic Engineering and Biotechnology (ICGEB), Aruna Asaf Ali Marg, New Delhi 110 067, India

R.B. Nitnavare
Institute of Bioinformatics and Biotechnology, Savitribai Phule Pune University, Pune, Maharashtra 411 027, India

A.A.A. Asea et al. (eds.), *Heat Shock Proteins and Plants*,
Heat Shock Proteins 10, DOI 10.1007/978-3-319-46340-7_14

Abbreviations

HSE	heat-shock element
HSF	heat shock factor
HSPs	heat shock proteins
HSR	heat stress response
HTR	high temperature response

14.1 Introduction

Global warming, along with the inevitable climatic changes is estimated to affect the global temperatures by an average of 3–5 °C increase in near future (Kerr 2007). With this predicted rise in temperature, heat stress is gaining as the trait of importance to breed for climate resilient crops. Prolonged incidents of heat waves caused by frequent fluctuations in daily and seasonal temperatures pose a serious challenge for agricultural production worldwide, affecting plant growth and yield with annual loss estimated up to billions of dollars (Mittler et al. 2012). Hence increasing crop productivity in view of escalating population and diminishing arable land and natural resources has become a matter of urgency than merely a research theme. To overcome such heat stress conditions, plants have developed several tolerance mechanisms. To understand the molecular basis of the tolerance mechanisms, knowledge of modern tools in molecular and genetic engineering is essential. Many abiotic stress-inducible genes were dissected and their functions are precisely characterized using functional genomics approaches. Another significant progress made in understanding this complex trait of heat tolerance is completion of the genome sequence information in major crop species including rice, maize, sorghum etc. This information has allowed identification and monitoring of transcript profiling for all the predicted genes at a single shot by either microarray or RNA sequencing approaches. The availability of vast amount of genome data has also enabled the identification of potential *cis*-regulatory elements and *trans*-factors.

Heat stress usually effects in protein dysfunction by improper folding of its linear peptide chains to non-native proteins leading to unfavourable interactions and subsequent protein aggregations (Moriwaki et al. 1999). Under stress conditions not only the nascent polypeptides face error-prone folding but also a large portion of the folded proteins gets partially or completely denatured and re-enter the protein quality control machinery assisted by molecular chaperones (Hebert and Molinari 2007). Nature has developed efficient molecular chaperon machinery in plants to maintain high quality proteins in the cells by increasing the level of pre-existing molecular chaperones and by expressing additional chaperones through signalling mechanism (Buchberger et al. 2010). Many proteins in a living cell will not fold properly without the assistance of molecular chaperones (Buchberger et al. 2010). Heat shock proteins (Hsps) are class of molecular chaperones that play an essential role in preserving cellular functions under stressful conditions. All living organisms are equipped with evolutionarily conserved Hsps to encounter sudden climate changes of nature. Hsps have broad range of functions ranging from the prevention of protein aggregation, refolding of misfolded proteins, and degradation of unstable proteins and dissolution of protein complexes, besides some act as transcription factors. Based on their differences in molecular weight, Hsps are classified into five sub-classes: Hsp100, Hsp90, Hsp70, Hsp60 and low molecular weight Hsps or small sHsps (Wang et al. 2004). Various members of Hsps have been cloned and functionally characterized and some of these have resulted in developing transgenic plants showing tolerance to various abiotic stresses (Lavania et al. 2015). Hsps and heat shock transcription factors (Hsfs) play a crucial role in heat stress tolerance during flowering and grain filling stages as evident in several examples (Waters 2003; Bita and Gerats 2013). However, detailed characterization and the role of plant Hsps as chaperones have been investigated only in a few model plants. The mechanisms of Hsps underlying abiotic stress adaptation in plants and the pivotal role of molecular chaperons will be discussed in the light of recent developments in genomics and genetic engineering approaches. The information and list of the transgenic plants developed for heat stress tolerance are discussed under the following sections.

14.2 Heat Shock Proteins (Hsps)

Heat stress disturbs cellular homeostasis, causes severe growth retardation effecting plant development, and become more vulnerable if occurs during flowering. Higher plants are unable to cope up with the extended exposure to temperatures above 45 °C (Herrenkohl and Politch 1978). The loss of biological activity of proteins upon high temperature stress may be due to aggregation and/or protein misfolding (Grover et al. 2013). The stress-induced accumulation of aggregated and mis-folded proteins is irreversible and deleterious to the cell functioning. To balance the homeostasis of cellular proteins under heat stress, plant cell upregulates several heat inducible genes, commonly referred as "heat shock genes" (HSGs), which encode Hsps that makes plants survival under high temperature (Chang et al. 2007a, b). A wide range of proteins have been reported to possess chaperone activity (Lindquist and Craig 1988). These are also called as molecular chaperones because with the help of several other proteins, commonly called as co-chaperones, they bind to partially folded or denatured proteins and prevent them from self-aggregation or promote their proper folding both in ATP dependent and independent manner. However, during their function they neither covalently bind to the substrate proteins nor form the

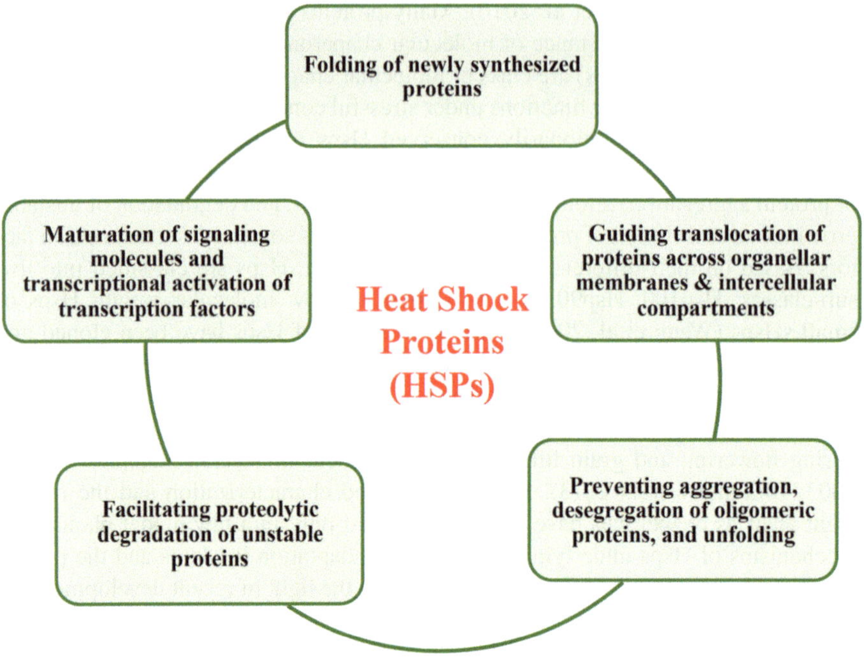

Fig. 14.1 Diverse functions of Heat shock proteins (Hsps)

part of the final product. These Hsps are broadly divided into two major families i.e., low and large molecular weight Hsps which again subdivided into five major classes based on the sizes of the corresponding proteins such as Hsp100/Clp, Hsp90, Hsp70, Hsp60/chaperonin and sHsps (Wang et al. 2004). Under normal conditions they perform many cellular functions such as (1) folding or assisting folding of newly synthesized proteins (Hsp70, Hsp60), (2) guiding translocation of proteins across organellar membranes and between intercellular compartments (Hsp70) (3) preventing aggregation, desegregation of oligomeric proteins, and unfolding (Hsp70, Hsp100, Hsp90, small Hsps) (4) facilitating proteolytic degradation of unstable proteins (Hsp70, Hsp100), (5) maturation of signaling molecules, signal transduction and transcriptional activation of transcription factors (Hsp70, Hsp90) (Driedonks et al. 2015) (Fig. 14.1). Many plant biotechnologists characterized the transcription and translation of Hsps in response to heat stress in different plant species (*Arabidopsis*, rice, wheat, tomato and maize) and their involvement in regulating thermotolerance has been established through forward and reverse genetic approaches (Lavania et al. 2015; Driedonks et al. 2015; Usman et al. 2014).

14.3 Small Heat Shock Proteins (sHsps)

Among five conserved families of Hsps, the sHsps are found to be most prevalent in plants and their expression can be increased up to 200 folds under heat stress (Wang et al. 2004). sHsps range in size from 10 to 42 kDa and share a conserved C-terminal domain that is common to all eukaryotic organisms (Waters et al. 1996). sHsps family shows diversity with respect to sequence similarity, cellular location and functions (Reddy et al. 2014; Reddy et al. 2015). In plants six different multi gene families that encode for sHsp proteins are localized in compartments like cytosol, endoplasmic reticulum (ER), mitochondria and chloroplast (Reddy et al. 2014). sHsps do not actively participate in refolding of non-native proteins (Veinger et al. 1998; Lee and Vierling 2000). They possess a high capacity of binding to non-native proteins, through hydrophobic interaction (Reddy et al. 2000). sHsps perhaps prevent non-native aggregation, thereby facilitating subsequent refolding through ATP-dependent chaperones such as the DnaK system or ClpB/DnaK.

The abundance of sHsps in plants and their functional characteristics of binding and stabilizing denatured proteins suggest that sHsps play an important role in plant-acquired stress tolerance (Sun et al. 2002). To support this, transgenic carrot cell lines with *Hsp17.7* gene under the control of CaMV35s promoter were developed, which resulted in enhanced survival of cell lines and plants at high temperature (48 °C) (Sun et al. 2002). The transformed seedlings with Class I sHsps showed higher cotyledon opening rate in tobacco plant (Park et al. 2002). In contrary, seedlings raised with the antisense construct in this experiment showed increased sensitivity to heat shock indicating the role of sHsps in seed germination at high temperatures. Transgenic rice plants over expressing with *OsHsp17.7* gene showed increased thermo tolerance as well as increased resistance to UV-B irradiation (Murakami et al. 2004). Tomato *mtLeHsp* gene when over expressed in tobacco-conferred thermotolerance up to 48 °C compared to their counter transgenics developed through antisense construct of the same gene (Sanmiya et al. 2004). Transgenic *Arabidopsis* plants over expressing with *NnHsp17.5, RcHsp17.8, ZmHsp22, ScHsp26* and *LdHsp16.45* showed heat tolerance to varied extents (Rhoads et al. 2005; Jiang et al. 2009; Sun et al. 2012; Zhou et al. 2012). Transgenic *Arabidopsis* plants over expressed with *WsHsp26* was tolerant under continuous high temperature and produced bold seeds under high temperature, having higher germination rate than wild type (Mu et al. 2013). In *Arabidopsis*, over expression of *RcHsp17.8* enhanced SOD activity (Jiang et al. 2009) whereas over expression studies of *ZmHsp16.9* in tobacco enhanced POD, CAT and SOD activity indicating the role of sHsps in oxidative stress tolerance (Chauhan et al. 2012). Altogether, it may be hypothesized that the sHsp proteins positively affect thermotolerance by maintaining the threshold levels of ROS scavenging enzymes, that could initiate the signaling pathway of thermotolerance (Driedonks et al. 2015). The updated list of the transgenic plants developed for sHsps is listed in Table 14.1.

Table 14.1 Transgenic plants made by means of dissimilar Hsp genes for heat stress tolerance

S. No	Gene	Source	Transgenic	Promoter	Reference
1	*Hsp17.7*	*D. carota*	*D. carota*	35s	Malik et al. (1999)
2	*sHsp17.7*	*O. sativa*	*O. sativa*	35s	Murakami et al. (2004)
3	*sHsp17.7*	*O. sativa*	*O. sativa*	35s	Sato and Yokoya (2008)
4	*Hsp17.5*	*N. nucifera*	*A. thaliana*	35s	Zhou et al. (2012)
5	*Hsp17.8*	*R. chinensis*	*A. thaliana*	35s	Jiang et al. (2009)
6	*Hsp17.8*	*A. thaliana*	*L. sativa*	35s	Kim et al. (2013)
7	*Hsp17/Hsp23*	*O. sativa*	*O. sativa*	35s	Zou et al. (2012)
8	*Hsp17.9*	*P. mume*	*A. thaliana*	35s	Wang et al. (2016)
9	*Tlhs1*	*N. tabacum*	*N. tabacum*	35s	Park and Hong (2002)
10	*mtsHsp*	*S. lycopersicon*	*N. tabacum*	35s	Sanmiya et al. (2004)
11	*Hsp21*	*S. lycopersicon*	*S. lycopersicum*	35s	Neta-Sharir et al. (2005)
12	*Hsp16.9*	*Z. mays*	*N. tabacum*	35s	Sun et al. (2012)
13	*Hsp16.45*	*L. davidii*	*A. thaliana*	35s	Mu et al. (2013)
14	*Hsp18*	*O. streptacantha*	*A. thaliana*	35s	Salas-Munoz et al. (2012)
15	*Hsp22*	*Z. mays*	*A. thaliana*	35s	Rhoads et al. (2005)
16	*Hsp23*	*M. sativa*	*A. stolonifera*	35s	Lee et al. (2015)
17	*Hsp23*	*M. sativa*	*F. arundinacea*	35s	Lee et al. (2012)
18	*Hsp24.4*	*M. acuminata*	*S. lycopersicum*	35s	Mahesh et al. (2013)
19	*Hsp26*	*O. sativa*	*F. arundinacea*	35S	Kim et al. (2012)
20	*Hsp26*	*S. cerevisiae*	*A. thaliana*	35s	Xue et al. (2010)
21	*Hsp26*	*T. aestivum*	*A. thaliana*	35s	Chauhan et al. (2012)
22	*ChlDnaJ/Hsp40*	*L. esculentum*	*L. esculentum*	35s	Kong et al. (2014)

(continued)

Table 14.1 (continued)

S. No	Gene	Source	Transgenic	Promoter	Reference
23	*DnaK/Hsp70*	*A. halophytica*	*N. tabacum*	35s	Ono et al. (2001)
24	*DnaK/Hsp70*	*A. halophytica*	*N. tabacum, O. sativa*	35s	Uchida et al. (2008)
25	*Hsp70*	*N. tabacum*	*N. tabacum*	35s	Cho and Choi (2009)
26	*Hsp70*	*T. harzianum*	*A. thaliana*	35s	Montero-Barrientos et al. (2010)
27	*mtHsp70*	*O. sativa*	*O. sativa*	35s	Qi et al. (2011)
28	*Hsp70*	*C. morifolium*	*A. thaliana*	35s	Song et al. (2014)
29	*Hsp70*	*B. campestris*	*N. tabacum*	35s	Wang et al. (2015)
30	*Hsp70*	*E. arundinaceus*	*Saccharum spp.*	Ubi2.3	Augustine et al. (2015b)
31	*Hsp70*	*A. thaliana*	*M. sativa*	35s	Ferradini et al. (2015)
32	*Hsp70*	*M. uniflorum*	*A. thaliana*	35s	Masand and Yadav (2016)
33	*Hsp90*	*G. max*	*A. thaliana*	35s	Xu et al. (2013)
34	*Hsp90.7*	*A. thaliana*	*A. thaliana*	35s	Chong et al. (2015)
35	*Hsp101*	*A. thaliana*	*A. thaliana*	35s	Queitsch et al. (2000)
36	*Hsp101*	*A. thaliana*	*O. sativa*	ZmUbi	Katiyar-Agarwal et al. (2003)
37	*Hsp101*	*O. sativa*	*N. tabacum*	35s	Chang et al. (2007a)

14.4 Heat Shock Protein 70 (Hsp70)

The second most evolutionarily conserved Hsp family in diverse organisms is Hsp70 (Boorstein et al. 1994). Hsp70's have two major functional domains, an ATPase domain of 44 kDa at the N-terminus and a 25 kDa peptide-binding domain at C-terminus and, are separated by small linker region (Reddy et al. 2010). The substrate-binding domain comprises of a sandwich of 2-four-stranded β-sheets, where the peptide-binding cleft resides. Another feature of plant Hsp70's is the presence of identifiable unique amino acid signature motif at the C-terminus that can be used to distinguish the protein's sub-cellular location. The EEVD motif indicates the cytosol-specific, HDEL for endoplasmic reticulum-specific;

PEGDVIDADFTDSK for plastid-specific and PEAEYEEAKK for mitochondrion-specific location of Hsp70 proteins (Reddy et al. 2010; Guy and Li 1998). Hsp70 class of proteins involved in many functions like controlling the biological activity of folded regulatory proteins, negative repressors of heat-shock factor (Hsf) mediated transcription. Some Hsp70's also exists in symbiosome membrane, which is known to play an important role in nodule development (He et al. 2008). The activity of Hsp70's can also be regulated by post-translational modifications (Napolitano et al. 1987) and by interaction with other co-chaperones (Santacruz et al. 1997). Hsp70's are also involved in protein import and translocation processes, and in facilitating the proteolytic degradation of unstable proteins by targeting the proteins to lysosomes or proteasomes (Hartl 1996). In addition to its general chaperone functions, Hsp70 also displays a regulatory role in other stress-associated gene expression (Lee and Schoffl 1996). Unfortunately, the role of Hsp70's in the modulation of signal transduction has not been studied in plants.

Hsp70's have been reported to be involved in ABA responses, redox signalling, chloroplast development, and protein translocation into chloroplasts and mitochondria and hence over expression of this class of chaperons leads to increased resistance against drought, high salt and heat stresses in plants (Lee et al. 2012). A halotolerant cynobacterial Hsp70/DnaK gene, when over expressed in tobacco and rice exhibited increased levels of anti-oxidant enzymes and enzymes involved in Calvin cycle conferring temperature and drought stress tolerance particularly during reproductive stage (Uchida et al. 2008). The over expression of Hsp70 from fungus *Trichoderma harzianum* in *A. thaliana* resulted in increased level of Na/H transporter (SOS1) and APX1 with decreased levels of Hsf and Hsp transcripts (Montero-Barrientos et al. 2010). Over expression of rice *mtHsp70* in rice resulted in lesser production of heat induced ROS, higher mitochondrial membrane potential and suppressed programmed cell death (Qi et al. 2011). Constitutive expression of a chrysanthemum Hsp70 in *A. thaliana* enhanced the tolerance against heat, drought and salinity stresses (Song et al. 2014). Hsp 70 from *E. arundinaceus* in sugarcane was shown tolerance to drought and salt stresses (Augustine et al. 2015a). Over expression of *B. campestris* Hsp70 in transgenic tobacco plant had shown heat stress tolerance by enhancing superoxide dismutase (SOD) and peroxidase (POD) activity, soluble sugar content and reduced electrical conductivity than control plant (Masand and Yadav 2016). Transgenic *A. thaliana* over expressing Hsp70 of *M. uniflorum* confers tolerance to multiple abiotic stresses and further shown the reduced levels of malondialdehyde (MDA), H_2O_2 and proteolytic activity. The transgenics have maintained the better shoot biomass, root length, relative water content and chlorophyll content during exposure to stresses relative to wild type plant (Chen et al. 2006). Other studies have found similar effects of Hsps on ROS scavenging proteins up on heat stress. The current status and updated list of the Hsp70 transgenic plants is given in Table 14.1.

14.5 Heat Shock Protein 90 (Hsp90)

Hsp90 family, which is highly, conserved molecular chaperones that are ubiquitously present in a wide range of organisms from prokaryotes to eukaryotes, except Archaea (Johnson and Brown 2009). In eukaryotic organisms, the cytosolic Hsp90 exists in two isoforms, inducible α-form and constitutive ß-form and at least one of these isoforms is functionally essential for the survival of the organism (Reddy et al. 2011). Due to slight variations in their relative molecular masses, these protein homologs have been represented by different names in literature (e.g. Hsp80, Hsp81, Hsp82, Hsp83, Hsp84, Hsp90 etc.). Amino acid sequence analysis of Hsp90 gene family can reveal their subcellular localization. This is possible due to presence of distinguishable amino acid signature motifs either at the N- or C-terminus region i.e. C-terminus MEEVD penta-peptide motif for cytosol-specific Hsp90 isoforms and C-terminus HDEL motif for endoplasmic reticulum-specific Hsp90 isoforms, whereas a characteristic N-terminus extension of signal peptide sequence for chloroplast and mitochondrion-specific Hsp90's (Pearl and Prodromou 2006). Hsp90 family predominantly occurs as a homodimer with three modular structural domains (Sangster and Queitsch 2005). The N-terminal domain contained the ATP-binding site responsible for the weak intrinsic ATPase activity of Hsp90. The middle domain, deliberated as a major site for client protein interaction, was connected to the N-terminal domain through a highly charged linker region. The C-terminal domain confined with the dimerization interface and a conserved C-terminal MEEVD motif, which was responsible for interaction with tetratricopeptide repeat (TPR) domain-containing co-chaperones. Hsp90's are constitutively present up to 1–2 % in cellular proteins; however, their expression is increased further by several folds on exposure to abiotic stresses mainly heat stress. Hsp90's are also considered as marker for morphological evolution (Sangster and Queitsch 2005). This suggests that Hsp90 functions as regulatory housekeeping protein as well as a molecular chaperone (Liu et al. 2006; Xu et al. 2013). Similar results were obtained during over expression of five Hsp90 genes of *Glycine max* in *A. thaliana.* Results obtained showed involvement of Hsp90 in different plant functions like higher biomass production, pod setting, reduction in lipid peroxidation and loss of chlorophyll under heat stress (Neuwald et al. 1999). The updated list of the transgenic plants developed for Hsp90 is presented in the Table 14.1.

14.6 Heat Shock Protein 100 (Hsp100)

The Hsp100/Clp are hexameric rings belonging to the large AAA ATPase super family with a broad spectrum of diverse functional properties (Agarwal et al. 2001; Keeler et al. 2000). Hsp100 was first described as components of the two-subunit bacterial Clp protease system, which consists of regulatory ATPase/chaperones (such as ClpA and ClpX) and proteolytic (ClpP) subunits. So far, Hsp100/Clp proteins have been reported in many plant species, such as *Arabidopsis*, soybean, tobacco, rice, maize, lima bean (*Phaseolus lunatus*) and wheat (Keeler et al. 2000; Adam et al. 2001; Schirmer et al. 1996). Hsp100 family is divided into two major classes and eight distinct subfamilies. Members of the first class (A-D) contain two nucleotide-binding domains (also called ATP-binding domains), whereas those in the second class (M, N, X, Y) have only one nucleotide-binding domain (Schirmer et al. 1994). In lima bean, Hsp100's are revealed to have expression in cytosol and chloroplasts when exposed to heat stress (Adam et al. 2001). Genetic evidence indicates a role for this family of proteins in thermo protection (Lee et al. 1994; Glover and Lindquist 1998). Contrasting to the regular chaperone function of preventing protein aggregation and misfolding, the Hsp100/Clp family has a functional role in protein disaggregation and/or protein degradation. The removal of non-functional but potentially harmful polypeptides arising from misfolding, denaturation or aggregation is important for the maintenance of cellular homeostasis. The mechanism for rescuing proteins from aggregation also involves the cooperation of another ATP-dependent chaperone system, the Hsp70. The Hsp100/Clp family solubilizes the aggregated protein and releases it in a state that can be refolded with the assistance of the Hsp70 system (Goloubinoff et al. 1999; Adam and Clarke 2002) Like many other Hsps/chaperones, Hsp100/Clp family chaperones are often constitutively expressed in plants, but their expression is developmentally regulated and is induced by different environmental assaults, such as heat, cold, dehydration and high salt or dark-induced etiolation. In addition to their normal cellular functions, these are now considered as a major group of stress related proteins, which function through cross-talk with other stress related proteins to decrease cellular damage.

In many studies, while analyzing global changes of gene expression analysis, the expression pattern of Hsps was found to be majorly altered under almost all type of abiotic stresses like salt, cold, drought and high light (Keeler et al. 2000; Adam et al. 2001; Queitsch et al. 2000). However, evidences for the direct involvement of these proteins under abiotic stresses except heat stress are very few. A study revealed that cisgenic *Arabidopsis* plants with altered AtHsp100 protein survived as high as 45 °C (1 h) temperature stress and also showed vigorous growth after the removal of stress (Katiyar-Agarwal et al. 2003). The transgenic rice lines over expressed with AtHsp101 showed re-growth in the post-high temperature stress recovery phase while the untransformed plants could not recover to the similar extents (Spiess et al. 2004). The updated list of the transgenic plants developed for Hsp100 is given in the Table 14.1.

14.7 Chaperonins

Molecular chaperonins are a part of cellular machinery that assists folding of newly synthesized proteins to their native state. Chaperonins are unique, high molecular weight cylindrical complexes which aid protein folding that is unmanageable by simpler chaperon systems (Hemmingsen et al. 1988). The term chaperonin was first suggested (Ranson et al. 1998) to describe proteins that are evolutionarily homologous to *E. coli* GroEL, a class of molecular chaperones found in prokaryotes and in the mitochondria and plastids of eukaryotes (Hartl 1996). Major examples of chaperonins include the prokaryotic GroEL and the eukaryotic equivalent Hsp60. Chaperonins are classified into two subfamilies, the GroE chaperonins (Group I) found in bacteria, mitochondria and chloroplasts (e.g. GroE and chCpn60) and the CCT chaperonins (Group II), found in Archaea and in the cytosol of eukaryotes (e.g. trigger factor 55, thermosomes and the TCP-1 ring complex) (Schroda 2004). Group I Cpn60 (also known as Hsp60), acts in the company of a co-chaperonin Cpn10 (Hsp10) in an ATP-dependent manner. While in bacteria, the Cpn10 is encoded by a single gene groES, in algae and plants, the plastid Cpn10 is encoded by multiple genes (Trosch et al. 2015). Although the bacterial Hsp10 is a ~10 kDa polypeptide, a ~20 kDa homologue comprising of two subunits is found in plastids. The two subunits are joined by a TDDVKD-linker sequence in head to tail fashion (Bukau and Horwich 1998). Hsp10 functions with Hsp60 as double-ring assemblies composed of back-to-back stacked rings of closely related rotationally symmetrical subunits (Kotak et al. 2007), assisting in folding, assembly and sorting of proteins.

There are Proteins with RNA chaperone activity that play important roles in cellular mechanisms (Semrad 2010). They prevent RNA from misfolding by loosening misfolded structures without ATP consumption. Oligonucleotide- or ribozyme-based assays were used to study RNA chaperone activity. Due to their functional as well as structural diversity, a common chaperoning mechanism or universal motif has not yet been identified. Although the exact mechanism is not yet understood, it is believed that disordered regions within proteins play an important role.

14.8 Heat Shock Transcription Factors (Hsfs)

Under heat stress, plant induces expression of Hsp's and other defensive genes. This happens due to the presence of conserved heat shock elements (HSEs) in the promoter region of gene, which triggers transcription of *Hsp* genes in response to heat. These *cis*-acting elements consist of the palindromic nucleotide sequence (5-AGAANNTTCT-3) that serve as recognizing as well as binding site for heat shock transcription factors or simply heat shock factors (HSFs) (Hasanuzzaman et al. 2013). As it is evident that Hsfs regulate *Hsp* genes, *Hsf* gene induction system has emerged as a powerful target for manipulating levels of Hsps through transgenic approach (Zhu et al. 2006; Zhu et al. 2009; Xin et al. 2010; Lee et al. 1995). Many researchers have opted for the transgenic approach to elucidate the function of *Hsp* and *Hsf* genes. The summary of these efforts is listed in Table 14.2. Over expression of *Arabidopsis HsfB4* resulted in altered root development and early duplication of endodermis cells, whereas impaired growth was observed in rice plants with suppressed HsfC1b. A group of researchers have successfully altered the expression of Hsps by making a change in the transcription factor (*AtHSF1*) responsible for activation of Hsps in *Arabidopsis* plants and able to produce heat stress tolerant

Table 14.2 Particulars on transgenic plants developed by using different classes of Hsf genes for high temperature tolerance

S. No	Gene	Source	Transgenic	Promoter	Reference
1	*Hsf1*	*A. thaliana*	*A. thaliana*	35s	Lee et al. (1995)
2	*Hsf3*	*A. thaliana*	*A. thaliana*	35s	Prandl et al. (1998)
3	*HsfA2*	*A. thaliana*	*A. thaliana*	35s	Li et al. (2005)
4	*HsfA2e*	*O. sativa*	*A. thaliana*	ZmUbi1	Yokotani et al. (2008)
5	*HsfA1*	*S. lycopersicon*	*S. lycopersicon*	35s	Mishra et al. (2002)
6	*Hsf3*	*A. thaliana*	*A. thaliana*	35s	Panchuk et al. (2002)
7	*HsfA1*	*G. max*	*G. max*	35s	Zhu et al. (2006)
8	*HsfA2*	*A. thaliana*	*A. thaliana*	35s	Ogawa et al. (2007)
9	*HsfA3*	*A. thaliana*	*A. thaliana*	35s	Yoshida et al. (2008)
10	*Hsf 7*	*O. sativa*	*A. thaliana*	35s	Liu et al. (2009)
11	*Hsf 1*	*B. hygrometrica*	*A. thaliana, N. tabacum*	35s	Zhu et al. (2009)
12	*HsfA2*	*L. longiflorum*	*A. thaliana*	35s	Xin et al. (2010)
13	*HsfC1b*	*O. sativa*	*O. Sativa*	ZmUbi1	Schmidt et al. (2012)
14	*Hsf A1a*	*A. thaliana*	*A. thaliana*	35s	Qian et al. (2014)
15	*Hsf A3*	*T. aestivum*	*A. thaliana*	35s	Zhang et al. (2013)
16	*Hsf A3*	*S. lycopersicon*	*A. thaliana*	35s	Li et al. (2013)
17	*HsfA1*	*L.longiflorum*	*A. thaliana*	35s	Gong et al. (2014)
18	*HsfA6f*	*T. aestivum*	*T. aestivum*	HVA1s	Xue et al. (2015)
19	*HsfA1d*	*T. salsuginea*	*A. thaliana*	35s	Higashi et al. (2013)
20	*HsfA2d*	*T. aestivum*	*A. thaliana*	35s	Chauhan et al. (2013)

Arabidopsis (Prandl et al. 1998). Over expression of *Athsf3* in *A. thaliana* using CaMV35 promoter showed a clearly enhanced thermotolerance in transgenic plants (Panchuk et al. 2002; Mishra et al. 2002). A study revealed that over expressed tomato *HsfA1* gene showed increased thermotolerance while transgenic lines in which transgene was silenced due to co-suppression were thermosensitive (Li et al. 2005). The *Glycine max* transgenics developed by over expressing *HsfA1* showed enhanced heat tolerance through activation of Hsp70 (Zhu et al. 2009). Constitutive expression of *HsfA2* in *A. thaliana* conferred enhanced basal and acquired thermotolerance (Yoshida et al. 2008; Zhu et al. 2006). The over expression of *AtHsfA3* in *A. thaliana* caused induction of a large number of heat stress associated genes that showed enhanced heat stress tolerance (Liu et al. 2009). Over expression of *OsHsp7* in *Arabidopsis* exhibited enhanced expression of certain Hsf target genes, concomitant to increased basal heat tolerance (Zhang et al. 2013). Hsf1 from resurrection plant *Boea hygrometrica* over expressed in *A. thaliana* and *N. tabacum* showed enhanced basal and acquired heat tolerance via regulation of genes involved in stress protection and mitotic cell cycle (Zhu et al. 2009). The over expression of *hsfA2* from *L. longiflorum* in *A. thaliana* activated *Hsp101*, *Hsp70*, *Hsp25.3* and *APX2* genes, resulting into heat tolerance of the transgenic plants (Lee et al. 1995). Transgenic *A. thaliana* over expressing wheat *HsfA3* showed increased thermotolerance (Li et al. 2013). Over expression of tomato *HsfA3* in *Arabidopsis* showed increased levels of several Hsp transcripts and increased heat tolerance (Higashi et al. 2013). Transgenic *A. thaliana* plants over expressing *HsfA1d* from *Thelluginella salsuginea* developed enhanced thermotolerance via induction of *AtHsfA1* regulon in the transgenic plants (Chauhan et al. 2013). Over expression of *TaHsfA2d*, which is expressed mainly in developing seeds, conferred higher tolerance to heat, salinity and drought stresses in *A. thaliana* in terms of higher survival rate, yield and biomass accumulation (Gong et al. 2014). Increased heat resistance was noted in transgenic *A. thaliana* plants over expressing a novel class of *AtHsfA1*, *LlHsfA1* from *L. longiflorum*, which was found to interact with *LlHsfA2* (Xue et al. 2014). Wheat plant over expressing *TaHsfA6f* showed tolerance to high temperature (Sakuma et al. 2006a). The updated list of the transgenic plants developed for Hsfs are summarized in the Table 14.2.

14.9 Heat Shock Promoters

During the last decade, several candidate genes, pathways and strategies have been identified by various groups across the globe and provided insight in plant heat stress adaptation. Nevertheless, we are still far from complete understanding of the molecular basis and regulatory mechanisms of abiotic stress adaptations, especially in crop plants. The regulated expression of transgenes in plants has attracted as one of the best approach in minimizing stress damage. Strong constitutive promoters are routinely used in plant transformation with a regulated expression of stress-responsive genes resulting in serious penalties on plant development with overall negative performance of transgenics. The use of stress inducible promoters may be more reliable for regulated expression of stress-responsive transgene for achieving the desired stress tolerance. Serious shortcomings on plant growth and development with overall negative performance of transgenics were observed when constitutive promoter was used for generation of transgenics (Sakuma et al. 2006b; Augustine et al. 2015b). Still, most of the researchers follow CaMV35S based expression for generation of stress tolerant transgenic plants (Table 14.3). Only few examples are available where investigators have examined alternative promoters like ubiquitin (Matsuura et al. 2013; Glover and Lindquist 1998). Since constitutive promoters are hampering the final productivity, it is important for us to identify and isolate heat-stress-inducible promoters and use them while developing transgenic crops. A typical *Hsp* gene is tightly regulated and rapidly and transiently activated upon stress. This happens as heat shock elements present in the promoter region of the *Hsp* genes, that makes Hsp promoter an ideal candidate for heat stress responsive promoter for generation of transgenic plants (Khurana et al. 2013). However, only few examples are available on the use of Hsp promoters for the transcriptional regulation of stress-related genes. The use of stress-related genes under transcriptional control of inducible promoters may minimize the adverse effect of the exogenous gene at phenotypic level. A prevailing approach for quantifying the activity of any

Table 14.3 Genetically modified plants advanced with diverse classes of Hsp promoters

S. no	Promoter	Source	Transgenic	Reference
1	Hsp18.2	*A. thaliana*	*A. thaliana*	Takahashi et al. (1992)
2	Hsp81	*A. thaliana*	*A. thaliana*	Yabe et al. (1994)
3	Hsp18.2	*A. thaliana*	*N. plumbaginifolia*	Moriwaki et al. (1999)
4	Hsp18.2	*A. thaliana*	*N. tabacum*	Lee et al. (2007)
5	HSP101	*O. sativa*	*O. sativa*	Proveniers and van Zanten (2013)
6	Hvhsp17	*H. vulgare*	*H. vulgare*	Freeman et al. (2011)
7	sHSP26	*T. aestivum*	*A. thaliana*	Khurana et al. (2013)

heat-shock promoter is by fusing the promoter of heat-shock gene to reporter genes such as GFP or GUS. This permits measuring the developmental and tissue-specific expression of genes with or without heat stress (Takahashi et al. 1992). There are few examples where Hsp promoters are fused with reporter or other gene. Hsp18.2 promoter fused to the *UidA* gene transgenic *Arabidopsis* plants showed that heat stress induced the *UidA* gene activity in almost all the organs of the plant (Lee et al. 2007). Similarly, *AtHsp18.2* promoter has been successfully used in *N. plumbaginifolia* (Moriwaki et al. 1999) and *N. tabacum* hairy roots (Yabe et al. 1994). Likewise, heat-shock-induced GUS activity was observed in transgenic *Arabidopsis* when the promoter of *Hsp81* gene was used (Crone et al. 2001). GmHsp17.5E promoter in all the organs and tissues of the flower is found to be differentially expressed in heat stress (Saidi et al. 2007). Moreover, the inducibility of GmHsp17.3B promoter was studied in the moss *Physcomitrella patens* (Proveniers and van Zanten 2013). This intricacy is now being divided into features like heat shock elements (HSEs), heat-shock factors (HSFs), and possible receptors of the heat-shock response, signaling components, and chromatin remodeling aspects (Wu et al. 2009). Transgenic rice seedlings expressing OsWRKY11 transcription factor under the rice HSP101 promoter were shown to survive longer and lose less water under a short, severe drought treatment, than wild type plants (Freeman et al. 2011). Transgenic wheat showed lower expression of *uidA* (beta-glucuronidase, GUS) reporter gene in older tissues, when *uidA* gene was fused with *HvHsp17* promoter but expression in other organs and tissues was normal. This observation was recorded upon induction of Hsp-GUS expressed transgenic plants (Nollen and Morimoto 2002). The deletion analysis of *Ta*Hsp26 promoter revealed the mechanism underlying *Ta*Hsp26 mediated regulation of heat tolerance. This study was done to characterize *Ta*Hsp26 promoter from wheat and *Arabidopsis* to generate transgenic plant (Takahashi et al. 1992). Although there are some reports on heat-stress inducible promoters, many gaps need to be filled to evaluate their role in crop plants. List of the transgenic plants developed for Hsp promoters is listed in the Table 14.3.

14.10 Signaling Molecules Involved in the Heat Stress Response

Acquired stress tolerance in plants is the result of various stress response mechanisms that act synergistically to bring favourable changes at physiological, biochemical and molecular level to prevent cellular damage during stress conditions. Substantial number of reports suggest that the Hsfs/Hsps interact with signalling molecules like growth hormones, protein kinases, cell cycle and cell death regulators and also with stress inducible proteins involved in redox regulation (glutathione and thioredoxin), antioxidants (ascorbate peroxidase) and osmolytes (trehalose, glycine-betaine and proline), and defense responses (Wang et al. 2014; Driedonks et al. 2015; Reddy et al. 2009; Baniwal et al. 2007). Interaction of Hsfs with other proteins determines their activity and function. For example, HsfA1 interact with HsfA2 to form super activator complex to induce expression of heat stress responsive genes. In contrary, interaction of HsfA5 with HsfA4 inhibits the activity of the HsfA4 through DNA binding (Lee et al. 2007; Fragkostefanakis et al. 2015). Members of class B Hsfs lack activation domain and therefore interaction with HsfA members is required for their function. In *Arabidopsis*, the activity of HsfA2 seems to be regulated by direct interaction of two co-chaperones, ROF1 and ROF2 with Hsp90 by either activating or repressing heat stress response respectively (Meiri et al. 2010). The regulation of Hsf activity is further complicated by interaction with non-chaperones like heat shock binding protein (HSBP). Hsfs exist as monomers and associate with Hsp70 and Hsp90 in the cytoplasm. The redox signalling moleculeH_2O_2 regulates Hsf activity through MAPK pathway during heat and oxidative stresses (Driedonks et al. 2015). Hsf interactions with ROS signalling molecules and scavenging enzymes have been well demonstrated (Jung et al. 2013). HsfA2 was found to be required for expression of H_2O_2 scavenging enzymes Apx1 and Apx2. In *Arabidopsis*, HsfA4a regulates expression of Apx1 through Zat12 transcription factor.

Our earlier work revealed the presence of Hsf binding *cis*-elements in the promoter region of PgApx, suggesting the interaction with ROS scavenging enzymes during heat stress. Apart from heat and oxidative stress, Hsfs involved in several stress responses including salinity and anoxia. The role of Hsf in calcium signalling is through interaction with both $Ca^{2+/}$calmodulin (CaM) and protein phosphatase (PP7). The mechanism by which CaM regulates Hsf is through interaction and phosphorylation of HsfA1a by CaM-binding protein kinase 3 (CBK3) that results in activation and binding of Hsf to HSE present in Hsp promoters (Liu et al. 2008). Wang et al. (2016) identified and validated 430 interactors of Hsp70 through co-localization and function based method in rice. Hsp90 associate with multichaperone complexes with Hsp70 and various co-chaperones such as HIP (Hsp70 interacting protein), HOP (Hsp70/Hsp90 organizing protein), Hsp40 and p23. The Hsp90 is regulated by different abiotic stresses and hormones indicating its role in stress tolerance networks. The plasma membrane H^+-ATPase (PM H^+-ATPase) plays an important role in signal transduction during cell expansion, intra cellular p^H

and stomata regulation during soil salinity. It has been shown that J3 chaperone (Hsp40-like) interact and repress the Salt Overly Sensitive2 (SOS2) like protein kinase5 that negatively regulates PM H^+-ATPase (Yang et al. 2010). Role of Hsps not only confined to countering abiotic stresses but also in biotic stress conditions. In an effector triggered immunity, precise regulation of R proteins is important for survival of plants. Studies support that Hsp90 plays crucial role along with RAR1 and suppressor of G2 allele of skp1 (SGT1) in regulation of R proteins (Seo et al. 2008). Hsp90-associated chaperonin activity is regarded to be an important factor for pathogen-triggered immunity. Defense against rice blast fungus requires chitin receptor (Cerk1) that transport from endoplasmic reticulum to the plasma membrane, which requires formation of Hsp90-HOP complex (Chen et al. 2010). In addition to Hsp90, Hsp70 is also important for defence response. From the available data, it is clear that plant immunity and heat response are connected through involvement of Hsfs and Hsps in defense response. The transition from vegetative to reproductive development in plants is controlled by multiple flowering pathways, which converge at the integrators, Flowering Locus T (FT) and Suppressor of over expression of Constans1 (SOC1). Expressions of these integrators are suppressed by flowering regulator Short Vegetative Phase (SVP). DNAJ HOMOLOG 3 (J3) of *Arabidopsis* expression is regulated by multiple flowering pathways and loss of function results in late flowering. It has been shown that J3 interacts directly with SVP and prevents binding of SVP to regulatory elements of SOC1 and FT there by promotes floral transition (Shen et al. 2011). During gametophyte development, abundant presence of Hsfs and Hsps supports the role of these proteins in floral development. Apart from this, sHsp's are also involved in early embryogenesis as evident in *Arabidopsis*, where double mutant for sHsps leads to seed abortion (Dafny-Yelin et al. 2008). Above evidence supports the role of Hsf/Hsp network in different plant developmental processes.

14.11 Genomic Approaches for Heat Stress Tolerance

DNA based molecular markers developed through contemporary technologies have become indispensable tools of plant breeding in enhancing genetic gains. Most of the studies on Hsps in relation to heat stress tolerance were either based on isolation and characterization of genes or *in vivo* expression analysis experiments but less attention has been paid towards marker assisted breeding compared to other abiotic traits like drought, salinity and cold. This could be due to the less availability of genetic resources and more complex nature of the trait. Linkage analysis based genetic mapping is the classical approach to identify QTLs related to quantitative traits. Mohammad et al. (2008) identified 3-heat stress tolerant QTLs in wheat RIL population based on stress susceptibility index (SSI) that explain 44.3 %, 27.3 % and 16.7 % phenotypic variance susceptibility. Apart from the markers associated with above QTLs. Yang et al. (2002) identified two more markers that could detect same QTLs but with additive effect for heat tolerance. In another independent study, five QTLs responsible for pollen stability at high temperature were identified in maize RIL population (Frova and Gorla 1993). But the recent revolutions in sequence technologies offered new genomic tools by which complex traits can be dissected and targeted more accurately and efficiently compared to SSR markers. In an independent study two QTLs related to heat tolerance were mapped in rice on chromosomes 3 and 4 using SSR markers (Lang et al. 2015). Using these markers, Lang et al. (2015) could successfully select homozygous plants through MABC program and this stood as successful example of molecular breeding. Hsp exhibit high genetic diversity that makes plants to behave differentially under heat stress. These allelic variations from natural populations can be captured using SNP markers and can be diploid in selection of superior genotypes in breeding programs. Identifying the naturally occurring allelic variations, that are functionally different from wild type and those that influence the target traits is really challenging. Using Eco-TILLING technology 11 SNP were identified in barley Hsp17.8 and their functional relevance to heat tolerance was evaluated. Garg et al. (2012) could identify a significant SNP that can change function of Hsp16.9 in wheat and successfully converted into breeder friendly marker. Ye et al. (2015) identified six-heat tolerance QTLs at flowering stage from two rice bi-parental populations using 6K SNP chip. Among these, two QTLs (q*HTS*1.2 & q*HTSF*6.1) contain Hsp genes, and this explains the role of Hsps in pollen fertility during heat stress in rice. This is supported by another independent study where Hsp101 was mapped on QTL region, identified for heat stress tolerance in *Arabidopsis* (Thudi et al. 2014).

Next Generation Sequencing (NGS) techniques can aid in the sequencing of condition, stage and tissue-specific transcriptome identification of heat, drought stress responsive genes, and helps in development of robust stress-associated molecular markers and construction of genetic and physical maps. This will help to elucidate key genes and metabolic pathways affected by heat and drought stresses, increase the adoptiveness and accuracy of breeding practices and accelerate crop improvement through genomics-assisted breeding. Thudi et al. (2014) identified significant SNPs associated with heat tolerance in chickpea using GBS based genome wide association studies and found few SNPs that fall in Hsp genes. Markers developed from these SNPs can be applied to select donors from germplasam for developing improved varieties through molecular breeding practices. But contrary results were obtained in GWAS for heat stress during flowering stage in *Arabidopsis* where no Hsps detected in genomic regions identified for heat stress tolerance. Only two Hsps were identified within 20 kb of moderately associated SNPs (threshold –log (P) = 4), suggesting that allelic variation in Hsps or Hsfs is not the main cause of natural variation in heat tolerance during flowering. Bulk segregation based sequence approach is another novel NGS method through which complex traits can be dissected in much simple way than map based studies. Epigenetic studies are required to detect genetic elements influenced by environmental factor (GXE) as heat stress response differs under different agro ecologies. The available whole genome sequence information and vast genetic data of crops like maize, rice can be exploited to use in less explored/orphan crops to identify the functional polymorphism in heat tolerant genes/QTLs. Studies of molecular genetic diversity among cultivars, wild accessions and ecotypes in crop species are useful for discovery of novel QTLs and alleles responsible for heat tolerance which can be further exploited in the programmes of thermotolerance improvement.

14.12 Conclusion

Understanding abiotic stress adaptations in plants is considered more challenging owing to polygenic nature of the trait and occurrence. Heat stress, being the major component of this complexity drags attention of researchers since long. Important molecules underlying heat stress tolerance identified are Hsps and Hsfs, showing chaperonin activity on various proteins of importance. Classification of different Hsps and the metabolic pathways involved are summarized to the best understanding. Role of Hsps and Hsfs as functional candidates in heat stress tolerance and other developmental pathways has been discussed with case studies. Though structural and functional characterization of Hsps/Hsfs established, their wide applicability in crop plants is still lagging due to unavailability of genetic and genomic resources. The recent revolutions in the field of genomics together with phenomics, offer exiting molecular tools which can be employed to breed heat tolerant crops. Further the cross talk molecules underlying heat stress tolerance during complex abiotic stress conditions need to be dissected.

Acknowledgements This work was supported partially from the Department of Biotechnology (Ministry of Science and Technology, Government of India) to MKR. PSR acknowledges the Department of Science and Technology, Govt. of India for the fellowship and research grant through the INSPIRE Faculty Award No. IFA11-LSPA-06 and Young Scientist Scheme SB/YS/LS-12/2013.

References

Adam Z, Clarke AK (2002) Cutting edge of chloroplast proteolysis. Trends Plant Sci 7:451–456

Adam Z, Adamska I, Nakabayashi K, Ostersetzer O, Haussuhl K, Manuell A, Zheng B, Vallon O, Rodermel SR, Shinozaki K, Clarke AK (2001) Chloroplast and mitochondrial proteases in Arabidopsis. A proposed nomenclature. Plant Physiol 125:1912–1918

Agarwal M, Katiyar-Agarwal S, Sahi C, Gallie DR, Grover A (2001) Arabidopsis thaliana Hsp100 proteins: kith and kin. Cell Stress Chaperones 6:219–224

Augustine SM, Cherian AV, Syamaladevi DP, Subramonian N (2015a) *Erianthus arundinaceus* HSP70 (EaHSP70) acts as a key regulator in the formation of Anisotropic Interdigitation in Sugarcane (*Saccharum* spp. hybrid) in response to drought stress. Plant Cell Physiol 56:2368–2380

Augustine SM, Narayan JA, Syamaladevi DP, Appunu C, Chakravarthi M, Ravichandran V, Subramonian N (2015b) *Erianthus arundinaceus* HSP70 (EaHSP70) overexpression increases drought and salinity tolerance in sugarcane (*Saccharum* spp. hybrid). Plant Sci 232:23–34

Baniwal SK, Chan KY, Scharf KD, Nover L (2007) Role of heat stress transcription factor HsfA5 as specific repressor of HsfA4. J Biol Chem 282:3605–3613

Bita CE, Gerats T (2013) Plant tolerance to high temperature in a changing environment: scientific fundamentals and production of heat stress-tolerant crops. Front Plant Sci 4:273

Boorstein WR, Ziegelhoffer T, Craig EA (1994) Molecular evolution of the Hsp70 multigene family. J Mol Evol 38:1–17

Buchberger A, Bukau B, Sommer T (2010) Protein quality control in the Cytosol and the endoplasmic reticulum: brothers in arms. Mol Cell 40:238–252

Bukau B, Horwich AL (1998) The Hsp70 and Hsp60 chaperone machines. Cell 92:351–366

Chang CC, Huang PS, Lin HR, Lu CH (2007a) Transactivation of protein expression by rice HSP101 in planta and using Hsp101 as a selection marker for transformation. Plant Cell Physiol 48:1098–1107

Chang HC, Tang YC, Hayer-Hartl M, Hartl FU (2007b) SnapShot: molecular chaperones, Part I. Cell 128:212

Chauhan H, Khurana N, Nijhavan A, Khurana JP, Khurana P (2012) The wheat chloroplastic small heat shock protein (sHSP26) is involved in seed maturation and germination and imparts tolerance to heat stress. Plant Cell Environ 35:1912–1931

Chauhan H, Khurana N, Agarwal P, Khurana JP, Khurana P (2013) A seed preferential heat shock transcription factor from wheat provides abiotic stress tolerance and yield enhancement in transgenic Arabidopsis under heat stress environment. PLoS One 8:e79577

Chen B, Zhong DB, Monteiro A (2006) Comparative genomics and evolution of the HSP90 family of genes across all kingdoms of organisms. BMC Genomics 7:156

Chen LT, Hamada S, Fujiwara M, Zhu TH, Thao NP, Wong HL, Krishna P, Ueda T, Kaku H, Shibuya N, Kawasaki T, Shimamoto K (2010) The Hop/Sti1-Hsp90 chaperone complex facilitates the maturation and transport of a PAMP receptor in rice innate immunity. Cell Host Microbe 7:185–196

Cho EK, Choi YJ (2009) A nuclear-localized HSP70 confers thermoprotective activity and drought-stress tolerance on plants. Biotechnol Lett 31:597–606

Chong LP, Wang Y, Gad N, Anderson N, Shah B, Zhao R (2015) A highly charged region in the middle domain of plant endoplasmic reticulum (ER) localized heat-shock protein 90 is required for resistance to tunicamycin or high calcium-induced ER stresses. J Exp Bot 66:113–124

Crone D, Rueda J, Martin KL, Hamilton DA, Mascarenhas JP (2001) The differential expression of a heat shock promoter in floral and reproductive tissues. Plant Cell Environ 24:869–874

Dafny-Yelin M, Tzfira T, Vainstein A, Adam Z (2008) Non-redundant functions of sHSP-CIs in acquired thermotolerance and their role in early seed development in Arabidopsis. Plant Mol Biol 67:363–373

Driedonks N, Xu JM, Peters JL, Park S, Rieu I (2015) Multi-level interactions between heat shock factors, heat shock proteins, and the redox system regulate acclimation to heat. Front Plant Sci 6:999

Ferradini N, Iannacone R, Capomaccio S, Metelli A, Armentano N, Semeraro L, Cellini F, Veronesi F, Rosellini D (2015) Assessment of heat shock protein 70 induction by heat in alfalfa varieties and constitutive overexpression in transgenic plants. PLoS One 10:e0126051

Fragkostefanakis S, Roth S, Schleiff E, Scharf KD (2015) Prospects of engineering thermotolerance in crops through modulation of heat stress transcription factor and heat shock protein networks. Plant Cell Environ 38:1881–1895

Freeman J, Sparks CA, West J, Shewry PR, Jones HD (2011) Temporal and spatial control of transgene expression using a heat-inducible promoter in transgenic wheat. Plant Biotechnol J 9:788–796

Frova C, Gorla MS (1993) Quantitative expression of maize HSPs: genetic dissection and association with thermotolerance. Theor Appl Genet 86:213–220

Garg D, Sareen S, Dalal S, Tiwari R, Singh R (2012) Heat shock protein based SNP marker for terminal heat stress in wheat (*Triticum aestivum* L.). Aust J Crop Sci 6:1516

Glover JR, Lindquist S (1998) Hsp104, Hsp70, and Hsp40: a novel chaperone system that rescues previously aggregated proteins. Cell 94:73–82

Goloubinoff P, Mogk A, Ben Zvi AP, Tomoyasu T, Bukau B (1999) Sequential mechanism of solubilization and refolding of stable protein aggregates by a bichaperone network. Proc Natl Acad Sci U S A 96:13732–13737

Gong BH, Yi J, Wu J, Sui JJ, Khan MA, Wu Z, Zhong XH, Seng SS, He JN, Yi MF (2014) LlHSFA1, a novel heat stress transcription factor in lily (*Lilium longiflorum*), can interact with LlHSFA2 and enhance the thermotolerance of transgenic *Arabidopsis thaliana*. Plant Cell Rep 33:1519–1533

Grover A, Mittal D, Negi M, Lavania D (2013) Generating high temperature tolerant transgenic plants: achievements and challenges. Plant Sci 205:38–47

Guy CL, Li QB (1998) The organization and evolution of the spinach stress 70 molecular chaperone gene family. Plant Cell 10:539–556

Hartl FU (1996) Molecular chaperones in cellular protein folding. Nature 381:571–580

Hasanuzzaman M, Nahar K, Alam MM, Roychowdhury R, Fujita M (2013) Physiological, biochemical, and molecular mechanisms of heat stress tolerance in plants. Int J Mol Sci 14:9643–9684

He ZS, Xie R, Wang YZ, Zou HS, Zhu JB, Yu GQ (2008) Cloning and characterization of a heat shock protein 70 gene, MsHSP70-1, in *Medicago sativa*. Acta Biochimica Et Biophysica Sinica 40:209–216

Hebert DN, Molinari M (2007) In and out of the ER: protein folding, quality control, degradation, and related human diseases. Physiol Rev 87:1377–1408

Hemmingsen SM, Woolford C, Vandervies SM, Tilly K, Dennis DT, Georgopoulos CP, Hendrix RW, Ellis RJ (1988) Homologous plant and bacterial proteins chaperone oligomeric protein assembly. Nature 333:330–334

Herrenkohl LR, Politch JA (1978) Effects of prenatal stress on the estrous cycle of female offspring as adults. Experientia 34:1240–1241

Higashi Y, Ohama N, Ishikawa T, Katori T, Shimura A, Kusakabe K, Yamaguchi-Shinozaki K, Ishida J, Tanaka M, Seki M, Shinozaki K, Sakata Y, Hayashi T, Taji T (2013) HsfA1d, a protein identified via FOX hunting using Thellungiella salsuginea cDNAs improves heat tolerance by regulating heat- stress- responsive gene expression. Mol Plant 6:411–422

Jiang CH, Xu JY, Zhang H, Zhang X, Shi JL, Li M, Ming F (2009) A cytosolic class I small heat shock protein, RcHSP17.8, of *Rosa chinensis* confers resistance to a variety of stresses to *Escherichia coli*, yeast and *Arabidopsis thaliana*. Plant Cell Environ 32:1046–1059

Johnson JL, Brown C (2009) Plasticity of the Hsp90 chaperone machine in divergent eukaryotic organisms. Cell Stress Chaperones 14:83–94

Jung HS, Crisp PA, Estavillo GM, Cole B, Hong F, Mockler TC, Pogson BJ, Chory J (2013) Subset of heat-shock transcription factors required for the early response of *Arabidopsis* to excess light. Proc Natl Acad Sci U S A 110:14474–14479

Katiyar-Agarwal S, Agarwal M, Grover A (2003) Heat-tolerant basmati rice engineered by overexpression of hsp101. Plant Mol Biol 51:677–686

Keeler SJ, Boettger CM, Haynes JG, Kuches KA, Johnson MM, Thureen DL, Keeler CL, Kitto SL (2000) Acquired thermotolerance and expression of the HSP100/ClpB genes of lima bean. Plant Physiol 123:1121–1132

Kerr RA (2007) Global warming is changing the world. Science 316:188–190

Khurana N, Chauhan H, Khurana P (2013) Wheat chloroplast targeted sHSP26 promoter confers heat and abiotic stress inducible expression in transgenic *Arabidopsis* plants. PLoS One 8:e54418

Kim KH, Alam I, Kim YG, Sharmin SA, Lee KW, Lee SH, Lee BH (2012) Overexpression of a chloroplast-localized small heat shock protein OsHSP26 confers enhanced tolerance against oxidative and heat stresses in tall fescue. Biotechnol Lett 34:371–377

Kim DH, Xu ZY, Hwang I (2013) AtHSP17.8 overexpression in transgenic lettuce gives rise to dehydration and salt stress resistance phenotypes through modulation of ABA-mediated signaling. Plant Cell Rep 32:1953–1963

Kong FY, Deng YS, Zhou B, Wang GD, Wang Y, Meng QW (2014) A chloroplast-targeted DnaJ protein contributes to maintenance of photosystem II under chilling stress. J Exp Bot 65:143–158

Kotak S, Vierling E, Baumlein H, von Koskull-Doring P (2007) A novel transcriptional cascade regulating expression of heat stress proteins during seed development of *Arabidopsis*. Plant Cell 19:182–195

Lang NT, Ha PTT, Tru PC, Toan TB, Buu BC, Cho YC (2015) Breeding for heat tolerance rice based on marker-assisted backcrosing in Vietnam. Plant Breed Biotechnol 3:274–281

Lavania D, Dhingra A, Siddiqui MH, Al-Whaibi MH, Grover A (2015) Current status of the production of high temperature tolerant transgenic crops for cultivation in warmer climates. Plant Physiol Biochem 86:100–108

Lee JH, Schoffl F (1996) An Hsp70 antisense gene affects the expression of HSP70/HSC70, the regulation of HSF, and the acquisition of thermotolerance in transgenic *Arabidopsis thaliana*. Mol Gen Genet 252:11–19

Lee GJ, Vierling E (2000) A small heat shock protein cooperates with heat shock protein 70 systems to reactivate a heat-denatured protein. Plant Physiol 122:189–197

Lee YRJ, Nagao RT, Key JL (1994) A Soybean 101-Kd heat-shock protein complements a yeast Hsp104 deletion mutant in acquiring thermotolerance. Plant Cell 6:1889–1897

Lee JH, Hubel A, Schoffl F (1995) Derepression of the activity of genetically engineered heat shock factor causes constitutive synthesis of heat shock proteins and increased thermotolerance in transgenic *Arabidopsis*. Plant J 8:603–612

Lee KP, Kim C, Landgraf F, Apel K (2007) EXECUTER1-and EXECUTER2-dependent transfer of stress-related signals from the plastid to the nucleus of *Arabidopsis thaliana*. Proc Natl Acad Sci U S A 104:10270–10275

Lee JH, Yun HS, Kwon C (2012) Molecular communications between plant heat shock responses and disease resistance. Mol Cells 34:109–116

Lee KW, Cha JY, Mun JY, Lee BH, Kim YG, Lee SH (2015) Heterologous expression of Mshsp23, a *Medicago Sativa* small heat shock protein, enhances heat stress tolerance in creeping bentgrass. J Anim Plant Sci 25:884–891

Li C, Chen Q, Gao X, Qi B, Chen N, Xu S, Chen J, Wang X (2005) AtHsfA2 modulates expression of stress responsive genes and enhances tolerance to heat and oxidative stress in *Arabidopsis*. Sci China C Life Sci 48:540–550

Li ZJ, Zhang LL, Wang AX, Xu XY, Li JF (2013) Ectopic overexpression of SlHsfA3, a heat stress transcription factor from Tomato, confers increased thermotolerance and salt hypersensitivity in Germination in transgenic *Arabidopsis*. PLoS One 8:e54880

Lindquist S, Craig EA (1988) The heat-shock proteins. Annu Rev Genet 22:631–677

Liu DL, Zhang XX, Cheng YX, Takano T, Liu SK (2006) rHsp90 gene expression in response to several environmental stresses in rice (*Oryza sativa* L.). Plant Physiol Biochem 44:380–386

Liu HT, Gao F, Li GL, Han JL, Liu DL, Sun DY, Zhou RG (2008) Thecalmodulin-binding protein kinase 3 is part of heat-shock signal transduction in *Arabidopsis thaliana*. Plant J 55:760–773

Liu JG, Qin QL, Zhang Z, Peng RH, Xiong AS, Chen JM, Yao QH (2009) OsHSF7 gene in rice, *Oryza sativa* L., encodes a transcription factor that functions as a high temperature receptive and responsive factor. BMB Rep 42:16–21

Mahesh U, Mamidala P, Rapolu S, Aragao FJL, Souza MT, Rao PJM, Kirti PB, Nanna RS (2013) Constitutive overexpression of small HSP24.4 gene in transgenic tomato conferring tolerance to high-temperature stress. Mol Breed 32:687–697

Malik MK, Slovin JP, Hwang CH, Zimmerman JL (1999) Modified expression of a carrot small heat shock protein gene, Hsp17.7, results in increased or decreased thermotolerance. Plant J 20:89–99

Masand S, Yadav SK (2016) Overexpression of MuHSP70 gene from *Macrotyloma uniflorum* confers multiple abiotic stress tolerance in transgenic *Arabidopsis thaliana*. Mol Biol Rep 43:53–64

Matsuura H, Takenami S, Kubo Y, Ueda K, Ueda A, Yamaguchi M, Hirata K, Demura T, Kanaya S, Kato K (2013) A computational and experimental approach reveals that the 5′-Proximal region of the 5′-UTR has a Cis-regulatory signature responsible for heat stress-regulated mRNA translation in *Arabidopsis*. Plant Cell Physiol 54:474–483

Meiri D, Tazat K, Cohen-Peer R, Farchi-Pisanty O, Aviezer-Hagai K, Avni A, Breiman A (2010) Involvement of Arabidopsis ROF2 (FKBP65) in thermotolerance. Plant Mol Biol 72:191–203

Mishra SK, Tripp J, Winkelhaus S, Tschiersch B, Theres K, Nover L, Scharf KD (2002) In the complex family of heat stress transcription factors, HsfA1 has a unique role as master regulator of thermotolerance in tomato. Genes Dev 16:1555–1567

Mittler R, Finka A, Goloubinoff P (2012) How do plants feel the heat? Trends Biochem Sci 37:118–125

Mohammad BA, Ibrahim AM, Hays D, Ristic Z (2008) Texas Plant Protection Association conferences (TPPA), Dec. 3–4, 2008, College Station, TX, USA

Montero-Barrientos M, Hermosa R, Cardoza RE, Gutierrez S, Nicolas C, Monte E (2010) Transgenic expression of the *Trichoderma harzianum* hsp70 gene increases *Arabidopsis* resistance to heat and other abiotic stresses. J Plant Physiol 167:659–665

Moriwaki M, Yamakawa T, Washino T, Kodama T, Igarashi Y (1999) Delayed recovery of beta-glucuronidase activity driven by an *Arabidopsis* heat shock promoter in heat-stressed transgenic *Nicotiana plumbaginifolia*. Plant Cell Rep 19:96–100

Mu CJ, Zhang SJ, Yu GZ, Chen N, Li XF (2013) Overexpression of small heat shock protein LimHSP16.45 in *Arabidopsis* enhances tolerance to abiotic stresses. PLoS One 8:e82264

Murakami T, Matsuba S, Funatsuki H, Kawaguchi K, Saruyama H, Tanida M, Sato Y (2004) Overexpression of a small heat shock protein, sHSP17.7, confers both heat tolerance and UV-B resistance to rice plants. Mol Breed 13:165–175

Napolitano EW, Pachter JS, Liem RKH (1987) Intracellular-distribution of mammalian stress proteins – effects of cytoskeletal-specific agents. J Biol Chem 262:1493–1504

Neta-Sharir I, Isaacson T, Lurie S, Weiss D (2005) Dual role for tomato heat shock protein 21: protecting photosystem II from oxidative stress and promoting color changes during fruit maturation. Plant Cell 17:1829–1838

Neuwald AF, Aravind L, Spouge JL, Koonin EV (1999) A class of chaperone-like ATPases associated with the assembly, operation, and disassembly of protein complexes. Genome Res 9:27–43

Nollen EAA, Morimoto RI (2002) Chaperoning signaling pathways: molecular chaperones as stress-sensing 'heat shock' proteins. J Cell Sci 115:2809–2816

Ogawa D, Yamaguchi K, Nishiuchi T (2007) High-level overexpression of the *Arabidopsis* HsfA2 gene confers not only increased themotolerance but also salt/osmotic stress tolerance and enhanced callus growth. J Exp Bot 58:3373–3383

Ono K, Hibino T, Kohinata T, Suzuki S, Tanaka Y, Nakamura T, Takabe T, Takabe T (2001) Overexpression of DnaK from a halotolerant cyanobacterium *Aphanothece halophytica* enhances the high-temperature tolerance of tobacco during germination and early growth. Plant Sci 160:455–461

Panchuk II, Volkov RA, Schoffl F (2002) Heat stress- and heat shock transcription factor-dependent expression and activity of ascorbate peroxidase in *Arabidopsis*. Plant Physiol 129:838–853

Park SM, Hong CB (2002) Class I small heat-shock protein gives thermotolerance in tobacco. J Plant Physiol 159:25–30

Pearl LH, Prodromou C (2006) Structure and mechanism of the Hsp90 molecular chaperone machinery. Annu Rev Biochem 75:271–294

Prandl R, Hinderhofer K, Eggers-Schumacher G, Schoffl F (1998) HSF3, a new heat shock factor from *Arabidopsis thaliana*, derepresses the heat shock response and confers thermotolerance when overexpressed in transgenic plants. Mol Gen Genet 258:269–278

Proveniers MCG, van Zanten M (2013) High temperature acclimation through PIF4 signaling. Trends Plant Sci 18:59–64

Qi YC, Wang HJ, Zou Y, Liu C, Liu YQ, Wang Y, Zhang W (2011) Over-expression of mitochondrial heat shock protein 70 suppresses programmed cell death in rice. FEBS Lett 585:231–239

Qian J, Chen J, Liu YF, Yang LL, Li WP, Zhang LM (2014) Overexpression of *Arabidopsis* HsfA1a enhances diverse stress tolerance by promoting stress-induced Hsp expression. Genet Mol Res 13:1233–1243

Queitsch C, Hong SW, Vierling E, Lindquist S (2000) Heat shock protein 101 plays a crucial role in thermotolerance in *Arabidopsis*. Plant Cell 12:479–492

Ranson NA, White HE, Saibil HR (1998) Chaperonins. Biochem J 333:233–242

Reddy GB, Das KP, Petrash JM, Surewicz WK (2000) Temperature-dependent chaperone activity and structural properties of human alpha A- and alpha B-crystallins. J Biol Chem 275:4565–4570

Reddy RA, Kumar B, Reddy PS, Mishra RN, Mahanty S, Kaul T, Nair S, Sopory SK, Reddy MK (2009) Molecular cloning and characterization of genes encoding *Pennisetum glaucum* ascorbate peroxidase and heat-shock factor: Interlinking oxidative and heat-stress responses. J Plant Physiol 166:1646–1659

Reddy PS, Mallikarjuna G, Kaul T, Chakradhar T, Mishra RN, Sopory SK, Reddy MK (2010) Molecular cloning and characterization of gene encoding for cytoplasmic Hsc70 from *Pennisetum glaucum* may play a protective role against abiotic stresses. Mol Genet Genomics 283:243–254

Reddy PS, Thirulogachandar V, Vaishnavi CS, Aakrati A, Sopory SK, Reddy MK (2011) Molecular characterization and expression of a gene encoding cytosolic Hsp90 from *Pennisetum glaucum* and its role in abiotic stress adaptation. Gene 474:29–38

Reddy PS, Kishor PBK, Seiler C, Kuhlmann M, Eschen-Lippold L, Lee J, Reddy MK, Sreenivasulu N (2014) Unraveling regulation of the small heat shock proteins by the heat shock factor HvHsfB2c in Barley: its implications in drought stress response and seed development. PLoS One 9:e89125

Reddy PS, Sharma KK, Vadez V, Reddy MK (2015) Molecular cloning and differential expression of cytosolic class I small Hsp gene family in *Pennisetum glaucum* (L.). Appl Biochem Biotechnol 176:598–612

Rhoads DM, White SJ, Zhou Y, Muralidharan M, Elthon TE (2005) Altered gene expression in plants with constitutive expression of a mitochondrial small heat shock protein suggests the involvement of retrograde regulation in the heat stress response. Physiol Plant 123:435–444

Saidi Y, Domini M, Choy F, Zryd JP, Schwitzguebel JP, Goloubinoff P (2007) Activation of the heat shock response in plants by chlorophenols: transgenic *Physcomitrella patens* as a sensitive biosensor for organic pollutants. Plant Cell Environ 30:753–763

Sakuma Y, Maruyama K, Osakabe Y, Qin F, Seki M, Shinozaki K, Yamaguchi-Shinozaki K (2006a) Functional analysis of an *Arabidopsis* transcription factor, DREB2A, involved in drought-responsive gene expression. Plant Cell 18:1292–1309

Sakuma Y, Maruyama K, Qin F, Osakabe Y, Shinozaki K, Yamaguchi-Shinozaki K (2006b) Dual function of an Arabidopsis transcription factor DREB2A in water-stress-responsive and heat-stress-responsive gene expression. Proc Natl Acad Sci U S A 103:18822–18827

Salas-Munoz S, Gomez-Anduro G, Delgado-Sanchez P, Rodriguez-Kessler M, Jimenez-Bremont JF (2012) The *Opuntia streptacantha OpsHSP18* gene confers salt and osmotic stress tolerance in *Arabidopsis thaliana*. Int J Mol Sci 13:10154–10175

Sangster TA, Queitsch C (2005) The HSP90 chaperone complex, an emerging force in plant development and phenotypic plasticity. Curr Opin Plant Biol 8:86–92

Sanmiya K, Suzuki K, Egawa Y, Shono M (2004) Mitochondrial small heat- shock protein enhances thermotolerance in tobacco plants. FEBS Lett 557:265–268

Santacruz H, Vriz S, Angelier N (1997) Molecular characterization of a heat shock cognate cDNA of zebrafish, hsc70, and developmental expression of the corresponding transcripts. Dev Genet 21:223–233

Sato Y, Yokoya S (2008) Enhanced tolerance to drought stress in transgenic rice plants overexpressing a small heat-shock protein, sHSP17.7. Plant Cell Rep 27:329–334

Schirmer EC, Lindquist S, Vierling E (1994) An *Arabidopsis* heat-shock protein complements a thermotolerance defect in yeast. Plant Cell 6:1899–1909

Schirmer EC, Glover JR, Singer MA, Lindquist S (1996) HSP100/Clp proteins: a common mechanism explains diverse functions. Trends Biochem Sci 21:289–296

Schmidt R, Schippers JHM, Welker A, Mieulet D, Guiderdoni E, Mueller-Roeber B (2012) Transcription factor OsHsfC1b regulates salt tolerance and development in *Oryza sativa* ssp japonica. AoB Plants 12:pls011

Schroda M (2004) The Chlamydomonas genome reveals its secrets: chaperone genes and the potential roles of their gene products in the chloroplast. Photosynth Res 82:221–240

Semrad K (2010) Proteins with RNA chaperone activity: a world of diverse proteins with a common task-impediment of RNA misfolding. Biochem Res Int 2011:532908

Seo NS, Lee SK, Song MY, Suh JP, Hahn TR, Ronald P, Jeon JS (2008) The HSP90-SGT1-RAR1 molecular chaperone complex: a core modulator in plant immunity. J Plant Biol 51:1–10

Shen LS, Kang YGG, Liu L, Yu H (2011) The J-domain protein J3 mediates the integration of flowering signals in *Arabidopsis*. Plant Cell 23:499–514

Song AP, Zhu XR, Chen FD, Gao HS, Jiang JF, Chen SM (2014) A Chrysanthemum heat shock protein confers tolerance to abiotic stress. Int J Mol Sci 15:5063–5078

Spiess C, Meyer AS, Reissmann S, Frydman J (2004) Mechanism of the eukaryotic chaperonin: protein folding in the chamber of secrets. Trends Cell Biol 14:598–604

Sun WN, Van Montagu M, Verbruggen N (2002) Small heat shock proteins and stress tolerance in plants. Biochim Biophys Acta Gene Struct Expr 1577:1–9

Sun LP, Liu Y, Kong XP, Zhang D, Pan JW, Zhou Y, Wang L, Li DQ, Yang XH (2012) ZmHSP16.9, a cytosolic class I small heat shock protein in maize (*Zea mays*), confers heat tolerance in transgenic tobacco. Plant Cell Rep 31:1473–1484

Takahashi T, Naito S, Komeda Y (1992) The *Arabidopsis* Hsp18.2 promoter/GUS gene fusion in transgenic *Arabidopsis* plants – a powerful tool for the isolation of regulatory mutants of the heat-shock response. Plant J 2:751–761

Thudi M, Upadhyaya HD, Rathore A, Gaur PM, Krishnamurthy L, Roorkiwal M, Nayak SN, Chaturvedi SK, Basu PS, Gangarao NVPR, Fikre A, Kimurto P, Sharma PC, Sheshashayee MS, Tobita S, Kashiwagi J, Ito O, Killian A, Varshney RK (2014) Genetic dissection of drought and heat tolerance in Chickpea through genome-wide and candidate gene-based association mapping approaches. PLoS One 9:e96758

Trosch R, Muhlhaus T, Schroda M, Willmund F (2015) ATP-dependent molecular chaperones in plastids – more complex than expected. BBA-Bioenerg 1847:872–888

Uchida A, Hibino T, Shimada T, Saigusa M, Takabe T, Araki E, Kajita H, Takabe T (2008) Overexpression of DnaK chaperone from a halotolerant cyanobacterium *Aphanothece halophytica* increases seed yield in rice and tobacco. Plant Biotechnol 25:141–150

Usman MG, Rafii MY, Ismail MR, Malek MA, Latif MA, Oladosu Y (2014) Heat shock proteins: functions and response against heat stress in plants. Int J Sci Technol Res 3:204–218

Veinger L, Diamant S, Buchner J, Goloubinoff P (1998) The small heat- shock protein IbpB from *Escherichia coli* stabilizes stress-denatured proteins for subsequent refolding by a multichaperone network. J Biol Chem 273:11032–11037

Wang WX, Vinocur B, Shoseyov O, Altman A (2004) Role of plant heat-shock proteins and molecular chaperones in the abiotic stress response. Trends Plant Sci 9:244–252

Wang Y, Lin S, Song Q, Li K, Tao H, Huang J, Chen X, Que S, He H (2014) Genome-wide identification of heat shock proteins (Hsps) and Hsp interactors in rice: Hsp70s as a case study. BMC Genomics 15:344

Wang X, Yan B, Shi M, Zhou W, Zekria D, Wang H, Kai G (2015) Overexpression of a *Brassica campestris* HSP70 in tobacco confers enhanced tolerance to heat stress. Protoplasma 1–9

Wang XL, Yang J, Li XB, Zhou Q, Guo C, Bao MZ, Zhang JW (2016) Over expression of PmHSP17.9 in transgenic *Arabidopsis thaliana* confers thermotolerance. Plant Mol Biol Report (in press)

Waters ER (2003) Molecular adaptation and the origin of land plants. Mol Phylogenet Evol 29:456–463

Waters ER, Lee GJ, Vierling E (1996) Evolution, structure and function of the small heat shock proteins in plants. J Exp Bot 47:325–338

Wu X, Shiroto Y, Kishitani S, Ito Y, Toriyama K (2009) Enhanced heat and drought tolerance in transgenic rice seedlings overexpressing OsWRKY11 under the control of HSP101 promoter. Plant Cell Rep 28:21–30

Xin H, Zhang H, Chen L, Li X, Lian Q, Yuan X, Hu X, Cao L, He X, Yi M (2010) Cloning and characterization of HsfA2 from Lily (*Lilium longiflorum*). Plant Cell Rep 29:875–885

Xu J, Xue C, Xue D, Zhao J, Gai J, Guo N, Xing H (2013) Overexpression of GmHsp90s, a heat shock protein 90 (Hsp90) gene family cloning from soybean, decrease damage of abiotic stresses in *Arabidopsis thaliana*. PLoS One 8:e69810

Xue Y, Peng R, Xiong A, Li X, Zha D, Yao Q (2010) Over-expression of heat shock protein gene hsp26 in *Arabidopsis thaliana* enhances heat tolerance. Biol Plant 54:105–111

Xue GP, Sadat S, Drenth J, McIntyre CL (2014) The heat shock factor family from *Triticum aestivum* in response to heat and other major abiotic stresses and their role in regulation of heat shock protein genes. J Exp Bot 65:539–557

Xue GP, Drenth J, McIntyre CL (2015) TaHsfA6f is a transcriptional activator that regulates a suite of heat stress protection genes in wheat (*Triticum aestivum* L.) including previously unknown Hsf targets. J Exp Bot 66:1025–1039

Yabe N, Takahashi T, Komeda Y (1994) Analysis of tissue-specific expression of *Arabidopsis thaliana* Hsp90-family gene Hsp81. Plant Cell Physiol 35:1207–1219

Yang J, Sears RG, Gill BS, Paulsen GM (2002) Quantitative and molecular characterization of heat tolerance in hexaploid wheat. Euphytica 126:275–282

Yang YQ, Qin YX, Xie CG, Zhao FY, Zhao JF, Liu DF, Chen SY, Fuglsang AT, Palmgren MG, Schumaker KS, Deng XW, Guo Y (2010) The Arabidopsis chaperone J3 regulates the plasma membrane H + −ATPase through interaction with the PKS5 kinase. Plant Cell 22:1313–1332

Ye CR, Tenorio FA, Argayoso MA, Laza MA, Koh HJ, Redona ED, Jagadish KSV, Gregorio GB (2015) Identifying and confirming quantitative trait loci associated with heat tolerance at flowering stage in different rice populations. BMC Genet 16:1–10

Yokotani N, Ichikawa T, Kondou Y, Matsui M, Hirochika H, Iwabuchi M, Oda K (2008) Expression of rice heat stress transcription factor OsHsfA2e enhances tolerance to environmental stresses in transgenic *Arabidopsis*. Planta 227:957–967

Yoshida T, Sakuma Y, Todaka D, Maruyama K, Qin F, Mizoi J, Kidokoro S, Fujita Y, Shinozaki Y, Yamaguchi-Shinozaki K (2008) Functional analysis of an *Arabidopsis* heat-shock transcription factor HsfA3 in the transcriptional cascade downstream of the DREB2A stress-regulatory system. Biochem Biophys Res Commun 368:515–521

Zhang SX, Xu ZS, Li PS, Yang L, Wei YQ, Chen M, Li LC, Zhang GS, Ma YZ (2013) Overexpression of TaHSF3 in transgenic *Arabidopsis* enhances tolerance to extreme temperatures. Plant Mol Biol Report 31:688–697

Zhou Y, Chen H, Chu P, Li Y, Tan B, Ding Y, Tsang EW, Jiang L, Wu K, Huang S (2012) NnHSP17.5, a cytosolic class II small heat shock protein gene from *Nelumbonucifera*, contributes to seed germination vigor and seedling thermotolerance in transgenic *Arabidopsis*. Plant Cell Rep 31:379–389

Zhu B, Ye C, Lu H, Chen X, Chai G, Chen J, Wang C (2006) Identification and characterization of a novel heat shock transcription factor gene GmHsfA1, in soybeans (*Glycine max*). J Plant Res 119:247–256

Zhu Y, Wang Z, Jing YJ, Wang LL, Liu X, Liu YX, Deng X (2009) Ectopic over-expression of BhHsf1, a heat shock factor from the resurrection plant *Boeahygrometrica*, leads to increased thermotolerance and retarded growth in transgenic *Arabidopsis* and tobacco. Plant Mol Biol 71:451–467

Zou J, Liu CF, Liu AL, Zou D, Chen XB (2012) Overexpression of OsHsp17.0 and OsHsp23.7 enhances drought and salt tolerancc in rice. J Plant Physiol 169:628–635

Chapter 15
NGS-Based Expression Profiling of HSP Genes During Cold and Freeze Stress in Seabuckthorn (*Hippophae rhamnoides* L.)

P.C. Sharma and Saurabh Chaudhary

Abstract Expression of many Heat Shock Protein (HSP) genes triggers in plant cells when exposed to abiotic stresses. An important exercise towards understanding the mechanism of plants response to cold and freeze stress is the identification of genes responsible for cold and freeze tolerance in plants adapted to low temperature climates. Seabuckthorn (*Hippophae rhamnoides* L.) is one such plant species that sustains growth in extreme temperature environments. We performed complete transcriptome profiling of leaf and root tissues of seabuckthorn using Illumina Next Generation Sequencing. The seabuckthorn transcriptome data revealed the presence of 205 heat shock protein and their co-protein coding genes. Later, DeepSAGE, a tag based approach, was used to identify differentially expressed genes coding for HSPs and their co-proteins under cold and freeze stress. The DeepSAGE data revealed differential expression of 13 genes under cold stress (CS) and 9 genes under freeze stress (FS) with respect to control (CON). Similarly, 14 genes were differentially expressed under FS with respect to CS. Expression of the most abundant *hsp70* gene was validated using qRT-PCR under different stress treatments. Our results may assist future efforts aiming towards understanding the role of HSP genes in mediating cold and freeze stress in crop plants.

Keywords DeepSAGE • Gene expression analysis • Heat shock proteins • Next generation sequencing • Seabuckthorn • Transcriptome profiling

---

P.C. Sharma (✉) • S. Chaudhary
University School of Biotechnology, Guru Gobind Singh Indraprastha University, New Delhi, India
e-mail: prof.pcsharma@gmail.com

A.A.A. Asea et al. (eds.), *Heat Shock Proteins and Plants*,
Heat Shock Proteins 10, DOI 10.1007/978-3-319-46340-7_15

Abbreviations

ABA	abscisic acid
CA	cold acclimation
CTAB	cetyl trimethylammonium bromide
DeepSAGE	deep serial analysis of gene expression
DEGs	differentially expressed genes
ER	endoplasmic reticulum
EST	expressed sequence tag
FDR	false discovery rate
GEO	gene expression omnibus
GO	gene ontology
GPAT	glycerol-3-phosphate acyltransferase
HSF	heat shock factor
HSP	heat shock protein
hsp	heat shock protein gene
KEGG	Kyoto encyclopedia of genes and genomes
LT	low temperature
MPSS	massive parallel signature sequencing
NGS	next generation sequencing
RIN	RNA integrity number
SAGE	serial analysis of gene expression
SRA	short read archive
SSR	simple sequence repeats

15.1 Introduction

Plants being sessile are exposed to a variety of abiotic stresses such as drought, salinity, extreme temperatures (low and high), osmotic stress, and many more. To adapt these stress conditions, plants not only respond at cellular and molecular level but also undergo various physiological and biochemical changes. Among all stresses, cold and freeze stresses are the major environmental constraints that significantly limit growth and productivity of crop plants worldwide (Chinnusamy et al. 2007). At chilling (<20°C) and freezing (<0°C) temperatures, plant cells show significant changes in the expression pattern of many genes and their protein products to rehabilitate the internal structure of cells to mitigate the cold and freeze induced injuries (Chinnusamy et al. 2006). Plants inhabiting low temperature regions show an increase in freezing tolerance on exposure to low non-freezing temperature with an adaptive phenomenon called cold acclimation (CA) (Levitt 1980). CA or low temperature resistance in plants is a complex trait involving many genes, metabolic pathways and cellular organelles (Hannah et al. 2006). Considering the importance of cold and freeze tolerance, a series of attempts have been made to develop stress tolerant crops using conventional breeding without much success. In contrast, advent of molecular biology tools in the recent past has made it possible to identify, isolate and characterize large number of cold and freeze responsive genes in crop plants successfully. Nevertheless, to understand the mechanism of cold and freeze tolerance in depth and to develop cold and freeze tolerant crop plants using transgenic approach, identification and selection of genes for cold and freeze tolerance from the plant native to cold climate remains a prime and potent approach.

Seabuckthorn (*Hippophae rhamnoides* L.), a member of the family Elaegnaceae, is one such plant native to cold regions that has the ability to adapt to extreme environmental conditions with temperature ranging from 40 to −40°C (Lu 1992). Moreover, seabuckthorn is known for its immense medicinal and nutritional value (Teng et al. 2006; Zeb 2006; Suryakumar and Gupta 2011). The common seabuckthorn, *Hippophae rhamnoides* L., is the most widely spread species, distributed in the cold regions of Indian Himalayas, China, Russia, Mongolia and many other countries in Europe and North America (Yang et al. 2005). Seabuckthorn is also considered to be an ecosystem restorer by virtue of its ability to prevent soil erosion, land reclamation and nitrogen fixation (Trajkovsk and Jeppsson 1999). Seabuckthorn genome is expected to harbour genes imparting tolerance to various abiotic stresses as the plant is highly adaptable to varying and extreme environmental conditions specifically cold and freeze stress. In the recent past, extensive studies have been made at transcriptome and proteome level, suggesting role of many genes and proteins under cold and freeze stress in seabuckthorn (Gupta et al. 2009; Ghangal et al. 2012; Gupta and Deswal 2012; Chaudhary and Sharma 2015).

Heat Shock Proteins (HSPs) having molecular chaperons activity are considered to be the key components in maintaining cellular homeostasis under stress as well as normal conditions. HSPs are responsible for protein folding, assembly, translocation, stabilization, and degradation during normal cellular processes and in response to various stresses (Wang et al. 2004; Vinocur and Altman 2005). Initially, HSPs were considered to be induced in response to heat stress only and hence studied extensively in relation to heat shock. However, now they are well known to respond under a variety of stresses such as cold, freeze, drought, salinity, osmotic stress, etc. (Lindquist and Craig 1988; Boston et al. 1996; Wang et al. 2004). Many studies have suggested the accumulation of HSPs in response to cold and freeze stress in a number of plant species (Bae et al. 2003; Kawamura and Uemura 2003; Lopez-Matas et al. 2004; Renaut et al. 2004; Taylor et al. 2005). The induced expression of HSPs during cold and freeze stress is the result of change in expression of HSP coding genes (*hsp*). It has been estimated that about 4 % of the total *Arabidopsis* genome might alter its expression during low temperature stress (Fowler and Thomashow 2002). Large number of genes including genes coding for chaperons of the HSP subfamily are involved in protein stabilization and show increased expression in response to freeze stress. Now, it is a well established fact that cold and freeze tolerance is a multigenic trait and among various genes involved, genes coding for HSPs play crucial role in such stress tolerance in plants. Thus, identification and expression of these genes in different plant tissues at different times and variable temperature needs to be explored in non model but important plant species such as seabuckthorn.

Transcriptome is the set of all RNAs including mRNA, tRNA, rRNA and other non coding RNAs transcribed in a cell/tissue in a particular environment. Transcriptome analysis has been the major research thrust for gene discovery and gene function annotation during the last three decades, specifically for non model but important plant species. The research in the field of transcriptomics and genomics has progressed from single gene based detections using blotting techniques to high throughput whole transcriptome profiling using Next Generation Sequencing (NGS) technologies. The advent of NGS technologies has revolutionized the transcriptomics and genomics research and outperformed the standard Sanger's Sequencing technology in terms of massive data generation at a much reduced cost and labour (Morozova et al. 2009; Thakur and Varshney 2010). The transcriptome data generated using NGS technologies may be useful for a variety of analysis including genome annotation, discovery of alternative isoforms, and gene expression profiling. Moreover, the technology has also helped researchers to gain deep insight into the mechanisms underlying processes of gene expression and secondary metabolism, and development of genomic resources for diversity characterization, evolutionary analysis and marker assisted crop breeding (Unamba et al. 2015). Among NGS technologies, the commercially available major sequencing platforms for transcriptome sequencing of non-model organisms include Roche-454, Illumina-Solexa, ABI-SOLiD and HeliScope (Morozova et al. 2009; Thakur and Varshney 2010; Unamba et al. 2015). We exploited paired end sequencing using Illumina HiSeq 2000 platform for complete transcriptome profiling of seabuckthorn (Ghangal et al. 2013).

The advancement in the NGS technology further provides a better alternative for global gene expression profiling as compared to hybridization based microarrays and has revolutionized the sequencing based approaches for gene expression profiling (Morozova et al. 2009; Jain 2011; Ozsolak and Milos 2011). Gene expression profiling using sequencing based approaches such as Serial Analysis of Gene Expression (SAGE) (Velculescu et al. 1995), Massive Parallel Signature Sequencing (MPSS) (Brenner et al. 2000) and RNA-Seq (Ozsolak and Milos 2011) present a global picture of expression of thousand of genes simultaneously, without prior knowledge of the genome by comparing the abundance of transcripts in two or more experimental conditions. We employed Deep Serial Analysis of Gene Expression (DeepSAGE) (Nielsen et al. 2006), a combination of classical SAGE and NGS, for identification of differentially expressed genes (DEGs) during cold and freeze stress in seabuckthorn (Chaudhary and Sharma 2015). DeepSAGE provides high sensitivity, advancement in robustness and resolution, and simplifies experimental steps over the gene expression profiling techniques including classical SAGE and its modifications (Nielsen et al. 2006; t'Hoen et al. 2008). Our DeepSAGE data (Chaudhary and Sharma 2015) provide a valuable resource for future studies on functional genomics of abiotic stresses in plants, specifically seabuckthorn.

The aim of the present chapter is to focus on the expression of HSPs and co-protein encoding genes during cold and freeze stress in seabuckthorn, a plant known for its ability to adapt to extreme environmental conditions. We reanalyzed the seabuckthorn transcriptome and DeepSAGE data to identify and predict the role of HSP genes under cold and freeze stress. Moreover, the article also focuses on the recent advancement in the NGS and gene expression technologies, and their applications in the transcriptome studies of unexplored important non model plant species.

15.2 Heat Shock Proteins: Role Under Cold/Freeze Stress in Plants

The molecular chaperon activity of heat shock proteins (HSPs) is well known to maintain the homeostasis in cell under both normal as well as stress conditions. Molecular chaperons in normal cellular processes help in folding, assembly, translocation and degradation of other proteins, whereas in stress conditions, they play a significant role in the stabilization of membrane and cellular proteins and also in refolding of the proteins (Li and Srivastava 2004; Wang et al. 2004). In general, molecular chaperons are stress proteins and most of them are HSPs (Lindquist and Craig 1988). The expression of heat shock protein genes (*hsp*) and co-protein genes vary in pattern depending upon the nature and level of stress and species under study. Moreover, the genes coding for HSPs are highly conserved and occur in the genome of almost every species ranging from simplest bacteria to complex eukaryotes. On the basis of sequence homology and molecular weight of their encoded product, the *hsp* genes are classified in different families including *hsp110, hsp100, hsp90, hsp70, hsp60, hsp10* and small *hsp* family (Schlesinger 1990). These families are further subdivided into subfamilies and have multiple members depending upon the intercellular localization and functions. Among these, HSP90, HSP 70, and HSP60 family proteins and their coding genes are most widely studied (Li and Srivastava 2004). Wang et al. (2004) have reviewed the HSPs and molecular chaperons and their role in abiotic stress response. They classified HSPs in plants into five major families on the basis of intercellular localization and functions including HSP70 (subfamily DnaK), HSP60 (chapronins), HSP90, HSP100 (Clp), and small HSP family. In a particular cell, the HSPs may be widely distributed in the cytoplasm and cellular organelles such as mitochondria, chloroplast, endoplasmic reticulum (ER) and nucleus (Boston et al. 1996).

The high production of heat shock proteins triggers in plants when they are exposed to various environmental stresses such as extreme temperature (high or low), drought, salinity, osmotic and oxidative stress (Wang et al. 2003, 2004). As a consequence, the heat shock proteins are also referred to as stress responsive proteins and their elevated expression is generally referred to as stress response (Santoro 2000). Among all, the expression patterns of *hsp90, hsp70* and *hsp60* genes have been extensively studied and functionally characterized in many model plant species under cold stress. Plants when exposed to stress (cold/freeze), HSP70 starts playing a significant role in the transport of native proteins across membranes into organelles, the folding of newly translated proteins, the repair of misfolded proteins, and targeting damaged proteins for degradation (Mayer and Bukau 2005). On the other hand, HSP60 has been shown to function as a chaperonin in the assembly of mitochondrial enzyme complexes composed of proteins encoded by nuclear genes and imported from the cytosol (Cheng et al. 1989). Moreover, HSP90 has been found to be highly conserved and essential molecular chaperone involved in maturation and activation of signalling proteins under stress conditions in plants (Kadota and Shirasu 2012).

15.2.1 Seabuckthorn: Cold and Freeze Stress Adapted Plant

Seabuckthorn (*Hippophae rhamnoides* L.) is a spiny, hardy, deciduous, wind pollinated shrub, with $2n = 24$ chromosomes (Elena et al. 2011). Seven species have been included in the genus *Hippophae*, two of them probably of hybrid origin (Bartish et al. 2002). Among seven species, *Hippophae rhamnoides* is the most common and abundant in nature. Natural habitat of seabuckthorn extends widely across the colder regions of Europe, North America and Asia including countries like China, Mongolia, India, Russia, Sweden, Finland and Norway (Lu 1997; Lian and Chen 2000). In India, seabuckthorn is widely distributed on hilly tracks of Himalayan belt covering Ladakh in Jammu and Kashmir, Lahaul-Spiti in Himachal Pradesh and some parts of Uttarakhand, Arunachal Pradesh and Sikkim (Stobdan et al. 2008). Seabuckthorn is known from ancient times for its immense medicinal and nutritional value. The medicinal value of different parts of the seabuckthorn plant has been exploited in traditional medicinal systems in China, Turkey, Mongolia, Tibet, Uzbekistan, Pakistan, Russia and India to treat many diseases such as bowel irregularities, gastric ulcers, skin infections/wounds, influenza, cough and cold. The pharmacological benefits of different seabuckthorn preparations include anti-tumor, anti-carcinogenic, anti-atherogenic and antimicrobial activities, immuno-modulatory and radio-protective properties (Teng et al. 2006; Zeb 2006; Suryakumar and Gupta 2011). Besides medicinal and nutritional value, seabuckthorn has great ecological importance as an ideal plant for prevention of soil erosion due to its ability to stabilize mobile sand dunes and to develop an extensive root system (Ledwood and Shimwell 1971; Trajkovsk and Jeppsson 1999). Moreover, the plant is known for land reclamation as it harbors *Frankia*, a nitrogen fixing actinomycetes, in its roots (Akkermans et al. 1983; Trajkovsk and Jeppsson 1999).

The high nutritional, medicinal and ecological value of seabuckthorn has attracted the attention of many researchers worldwide. Many programs have been completed and many others are in pipeline to promote domestication and breeding of this multipurpose plant. The main focus of research has been limited to the study of biochemical characterization and documentation of medicinal uses of seabuckthorn extracts. Nevertheless, molecular biology research has been initiated in the recent past in different directions and important data have been now made available in the public domain. The major efforts on molecular biology research included cloning of glycerol-3-phosphate acyltransferase (GPAT) gene and validation of its increased expression in cold-stressed leaves of seabuckthorn (Gupta et al. 2009), development of EST (EST) based simple sequence repeat (SSR) markers (Jain et al. 2010), generation of ESTs (Ghangal et al. 2012), transcriptome study of mature seeds and fatty acid composition of seabuckthorn berries (Fatima et al. 2012), secretome and antifreeze protein analysis (Gupta and Deswal 2012), *de novo* assembly of short reads of seabuckthorn transcriptome (Ghangal et al. 2013), mining of microsatellites from next generation sequencing derived seabuckthorn transcriptome (Jain et al. 2014), differential gene expression analysis under cold and freeze stress in seabuckthorn (Chaudhary and Sharma 2015), and study of sex-biased temporal gene expression (Chawla et al. 2015).

The climatic conditions of the native habitat of seabuckthorn are very harsh and extreme such as low temperature, arid soil, high altitude, UV-exposure, low precipitation, and low oxygen. The exposure to such harsh conditions makes seabuckthorn a good model plant for the study of various abiotic stress tolerance properties especially for extreme temperature as it has the ability to survive in the temperature ranging from −40 to 40°C (Lu 1992). Earlier, Li et al. (2005) correlated the cold and freeze tolerance with sex of the plant as well as with the two ecotypes (northern and southern hemisphere region) of *Hippophae rhamnoides*. The study concluded that the northern region male plants have more freezing tolerance than southern hemisphere female plants because of early cold acclimation in male plants of northern region which results from the change in abscisic acid (ABA) levels. In another study, Ghangal et al. (2012) identified some cold inducible elements in seabuckthorn using ESTs. On the basis of similarity search and Gene Ontology (GO) annotation, they identified 43 genes responsive to various biotic and abiotic stresses in seabuckthorn. The expression of 16 genes was further validated under cold and freeze stress using qRT-PCR. Among 16 genes, three genes were reported to be seabuckthorn specific which differentially expressed in seabuckthorn under cold and freeze stress. Later, a proteomics study carried out by Gupta and Deswal (2012) reported 61 low temperatures (LT) responsive extracellular proteins from LT treated secretome in seabuckthorn. We have recently reported a large number of differentially expressed genes under cold and freeze stress using transcriptome profiling, DeepSAGE and qRT-PCR (Chaudhary and Sharma 2015). These important findings in the past at transcriptome and proteome level support the notion to consider seabuckthorn as a model plant for the study of cold and freeze tolerance. Although large number of genes and proteins were found to be differentially expressed in response to cold and freeze stress in seabuckthorn, here we discuss only those with reference to HSPs/chaperons and their co-proteins/co-chaperons. In addition, we examine the expression of HSP genes (*hsp*) under cold and freeze stress in seabuckthorn.

15.3 Transcriptome Analysis: Abundance of Heat Shock Protein Genes in Seabuckthorn Transcriptome

The advent of Next Generation Sequencing technology has revolutionized the sequencing scenario. The technology shift from Sanger's sequencing method to high throughput next generation sequencing technology has offered an opportunity to obtain enormous sequence information in a time and cost effective manner for any organism. The potential of NGS technologies facilitate many areas of advanced research such as large scale re-sequencing of sequenced genome, transcriptome and micro RNA sequencing, DNA methylation studies and so on. The emergence of NGS has significantly improved the efficiency and speed of gene discovery (Schuster 2008; Ansorge 2009). Moreover, NGS technologies have expanded the field of transcriptomics dramatically by generating opportunities for multidimensional examination of transcriptome thereby opening an important area of biological research (Morozova et al. 2009). The various platforms developed for next generation sequencing including Illumina/Solexa, 454/Roche, ABI-SOLiD, HeliScope, etc. have potential to elucidate the complete transcriptome profiling of any organism. However, after generation of millions of short read tags during NGS data generation as compared to Sanger's capillary sequencing method, the *de novo* assembly and annotation of short reads into well representing form is still a major challenging task. For seabuckthorn transcriptome profiling, we optimized a *de novo* method for assembly of short reads generated by using Illumina sequencing (Ghangal et al. 2013). After reanalysis of seabuckthorn transcriptome data, we identified many heat shock proteins (*hsp*)/chaperons and their co-proteins/co-chaperons genes.

The seabuckthorn seedlings were grown hydroponically in plant growth chamber. Total RNA was isolated from the root and leaf tissues, harvested from seabuckthorn plantlets using a modified CTAB method (Ghangal et al. 2009). The quality and quantity of RNA isolated was checked by using agarose gel electrophoresis as well as bioanalyzer. RNA samples having more than 8.0 RIN (RNA Integrity Number) value were further processed for tag preparation and Illumina sequencing. In brief, mRNA was purified from total RNA using oligo(dT) beads and fragmented to generate short mRNAs. The short fragments were used as template for synthesis of first strand cDNA using random hexamer primers. Purified double stranded cDNAs containing sequencing adaptors were sequenced using Illumina HiSeq 2000. To gain a comprehensive view of the seabuckthorn transcriptome, 86,253,874 high quality sequences were generated using next generation massive parallel sequencing. The short reads generated were then processed for assembly, annotation and analysis using various bioinformatics tools. The systematic diagram for pipeline of bioinformatics analysis is summarized and shown in the flow chart (Fig. 15.1). The short reads were subjected to quality check by using NGS QC Tool Kit developed by Patel and Jain (2012). For *de novo* short read assembly, we studied and compared the performance of six commonly used *de novo* short read assemblers including Velvet (Zerbino and Birney 2008), ABySS (Simpson et al. 2009), SOAPdenovo (Li et al. 2009), Trinity (Grabherr et al. 2011), Oases (Schulz et al. 2012),

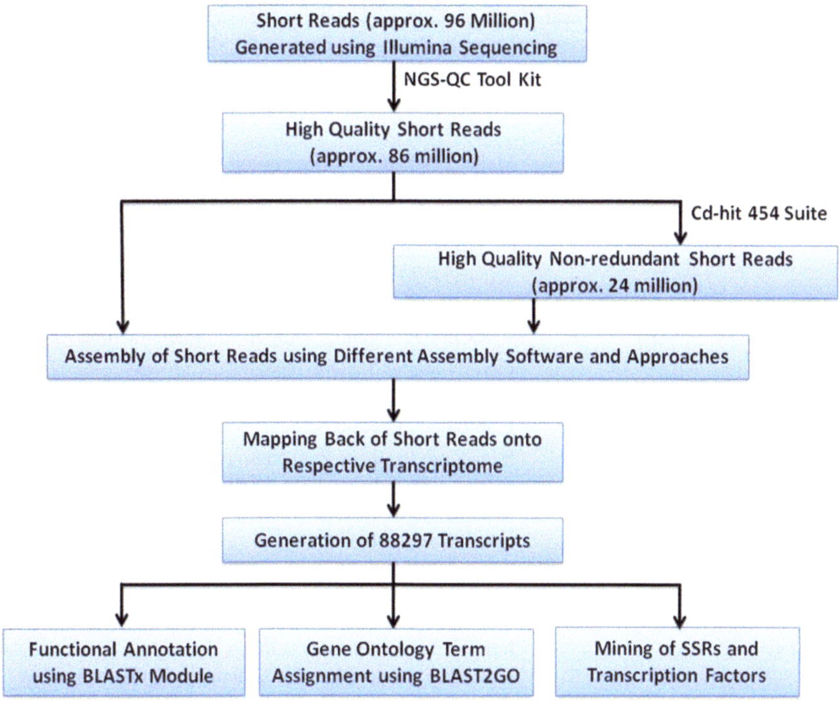

Fig. 15.1 The pipeline for bioinformatics analysis for short reads generated using Illumina Hiseq 2000

and CLC Genomics Workbench (commercially available) with different approaches. The use of different approaches and six assembly tools also proved to be beneficial for the assembly of short reads. Among all, the *de novo* assembly using ABySS tool with additive k-mer followed by TGCIL suite was found to be most suitable approach, which yielded 88,297 transcripts (>100 bp). This assembly represented about 53 Mb of seabuckthorn transcriptome with an average transcript length of 610 bp, N50 length 1198 bp with 91 % of the short reads uniquely mapping back to seabuckthorn transcriptome (Ghangal et al. 2013). The data generated in the study can be accessed from Short Read Archive (SRA) of NCBI under study accession number "SRP011938".

For similarity search and functional annotation, BLASTX algorithm was used to annotate 88,297 seabuckthorn assembled transcripts against non-redundant (nr) protein database of NCBI. A total of 41,340 (46.8 %) transcripts showed significant similarity with sequences present in nr protein databases of NCBI (E-value < 1E−06). The significant sequence homology showed top-hits at NCBI database with *Vitis vinifera*, followed by *Ricinus communis, Populus trichocarpa* and *Glycine max*. BLAST2GO tool (Conesa et al. 2005) was used to assign Gene Ontology (GO) terms associated with top 20 BLAST hits for each transcript. The seabuckthorn transcripts represented different categories viz. biological process, molecular

Table 15.1 Distribution of heat shock proteins and sub families with major gene ontology description in seabuckthorn transcriptome. P: Biological Process, F: Molecular Function and C: Cellular Component

Family/co-chaperons of HSPs	Total number of genes	Major gene ontology description
HSP-70	34	P:auxin biosynthetic process; P:protein folding; P:response to cadmium ion; P:oxidation reduction; P:response to heat; C:plasma membrane; C:nucleus; F:2-alkenal reductase activity; F:ATP binding
DnaJ	51	P:protein folding; P:response to stress; C:nucleus; F:heat shock protein binding; F:unfolded protein binding
DnaK	3	F:ATP binding; P:auxin biosynthetic process
HSP-90	17	P:protein folding; P:response to stress C:cytoplasm; F:unfolded protein binding; F:ATP binding
Small Heat Shock Proteins (sHSPs)	11	P:response to stress; P:response to cadmium ion; P:response to heat; P:response to high light intensity; P:response to hydrogen peroxide; C:chloroplast stroma; C:cytoplasm; C:mitochondrion; C:plastid
Heat shock transcription factors/ heat shock factors (HSFs)	18	P:regulation of transcription; P:response to heat; P:regulation of transcription, DNA-dependent; P:response to heat; C:nucleus; F:transcription factor activity; F:transcription repressor activity; F:sequence-specific DNA binding; F:transcription factor activity; F:protein binding

function and cellular component. In biological process category, "primary metabolic process", "cellular metabolic process" and "biosynthetic process" processes were predominant. However, presence of groups like "cellular response to stimulus", "response to external stimulus", "response to biotic stimulus", "response to endogenous stimulus", "response to abiotic stimulus" and "response to stress" in our dataset indicated that a large number of transcripts are expressed in response to environmental stresses, which further supported the hypothesis to consider seabuckthorn as a model plant for abiotic stress tolerance. In molecular function category, unigenes with "catalytic activity", "protein binding" and "nucleotide binding" formed the largest groups.

The seabuckthorn transcriptome data having 88,297 putative unigenes were further reanalyzed for the presence of genes coding for HSPs and their co-proteins (Table 15.1). On the basis of similarity search and functional annotation, the unigenes were assigned sequence description and putative function. In total, 205 heat shock protein binding and their co-protein activity genes were identified on the basis of functions assigned to putative unigenes. As expected, gene description suggested HSP-70 family as the most abundant with 34 representatives and their subfamilies DnaJ and DnaK were represented by 51 and three, members, respectively. Moreover, the HSP-90 and small heat shock proteins (sHSPs) families were identified to be 17 and 11 in number, respectively. The transcription factors such as heat shock factors (HSFs) also play significant role in the regulation of heat shock protein (*hsp*) genes. In total, 18 such transcription factors were identified in seabuckthorn transcriptome. The involvement of heat shock protein (*hsp*) genes under abiotic stress and presence of their transcripts in seabuckthorn transcriptome, therefore, further support the seabuckthorn ability to sustain in harsh environmental conditions.

15.4 Differential Expression of HSP Genes Under Cold and Freeze Stress in Seabuckthorn

For DeepSAGE analysis, 30 days old seabuckthorn plantlets grown in laboratory conditions were subjected to cold stress treatment at 4°C and freeze stress at −10°C for 6 h. The seedlings grown at 28°C were taken as control. For qRT-PCR analysis, the plantlets were subjected to cold (4°C) and freeze stress (−10°C) for three time courses i.e. 2 h, 4 h and 6 h with biological duplicates. Total RNA was isolated from the leaf tissues of all the treated samples using modified CTAB method and quantified by gel electrophoresis and nanodrop readings (A260/A280). The RNA samples with A260/A280 ratio from 1.9 to 2.1, A260/A230 ratio from 2.0 to 2.5 and RIN (RNA Integrity Number) value of more than or equal to 8.0 were processed for further analysis. The 49 base pair tags were prepared and sequenced using Illumina Gene Expression Sample Prep Kit and Solexa Sequencing Chip (Flowcell). The main instruments used for sequencing included Illumina Cluster Station and Illumina HiSeqTM2000 System. Various bioinformatics tools and parameters were used to process raw sequences generated by Illumina sequencing. The systemic representation of generation of tags, sequencing and bioinformatics analysis is given in Fig. 15.2. In brief, the raw tags were cleaned and analysed. Tools such as BLAST and BLAST2GO were used for alignment and Gene Ontology term (GO-term), assignment, respectively. All clean tags were mapped on to the reference sequences. The number of unambiguous clean tags for each gene was calculated and normalized to TPM (number of transcripts per million clean tags) (t'Hoen et al. 2008; Morrissy et al. 2009).

Next, the differentially expressed genes (DEGs) were identified on the basis of variation in the counts of their related sequence tags using an algorithm developed by Audic and Claverie (1997). The FDR (false discovery rate)_0.01, and absolute value of log2Ratio_1 were used as threshold to judge the significance of gene expression differences (Benjamini and Yekutieli 2001). These stringent criteria and a relative threshold of twofold change in the sequence were used to identify DEGs in different libraries. Pathway enrichment analysis was done for differentially expressed genes by using KEGG (Kyoto Encyclopedia of Genes and Genomes) (Kanehisa et al. 2008).

DeepSAGE approach used to identify differentially expressed genes under cold and freeze stress in seabuckthorn was found to be promising as large amount of data was genrated. In total, 36.2 million raw tags including 13.9 million distinct tags were generated using Illumina sequencing platform for three leaf tissue libraries

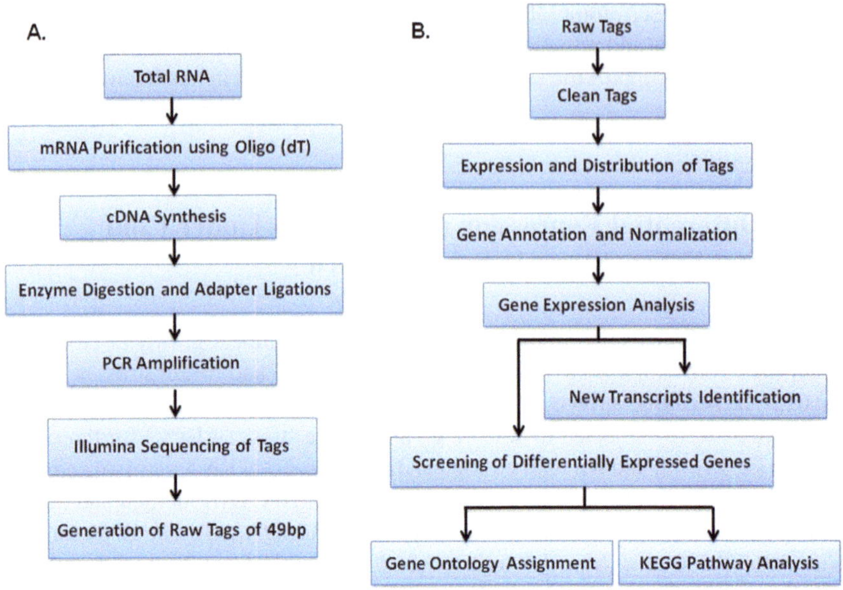

Fig. 15.2 The pipeline of generation and sequencing of tags (**a**) and bioinformatics analysis of generated raw tags (**b**)

including control, cold stress, and freeze stress. A total of 11,922 differentially expressed genes (DEGs) including 6539 up regulated and 5383 down regulated genes were identified. The Gene Ontology (GO) functional analysis of the differentially expressed genes suggested that metabolic, cellular, primary metabolic, cellular metabolic and macromolecule metabolic processes genes were found to be abundant. Moreover, approximately 28 % and 15 % of the differentially expressed genes were categorized as genes responsive to stimulus and stress, respectively. The relevant raw and processed data generated for differential expression analysis can be accessed from NCBI-Gene Expression Omnibus (GEO) with study accession number "GSE62489".

On the basis of GO terms assigned to differentially expressed genes in various experimental setups i.e. control vs cold stress, control vs freeze stress and cold stress vs freeze stress, a number of heat shock protein (*hsp*) and their related protien genes were identified (Table 15.2). In control vs cold stress setup, 11 heat shock protein (*hsp*) genes, one heat shock factor (*htf*) gene and one chaperone protein (*dna*J) gene, were found to be upregulated as shown in Fig. 15.3. Among all, *hsp-70* gene is highly expressed around seven fold under cold stress with respect to control (Table 15.2a). In the other comparison of control vs freeze stress, total nine heat shock protein and related genes were identified including two heat shock factor protein (*hsf*) genes as differentially expressed (Table 15.2b). Among all, six genes were found to be upregulated and three were down regulated as shown in (Fig. 15.3). Moreover, we also compared cold stress vs freeze stress, to identify genes

Table 15.2 Differential expression of heat shock protein (*hsp*) and their related protein genes identified in various experimental setups. (a) Control vs cold stress, (b) Control vs freeze stress, and (c) Cold stress vs freeze stress. *TPM*- Transcripts per million

(a) Control vs cold stress

TPM-control	TPM-cold stress	log2 ratio/fold change	P-value	FDR	Homology on the basis of BLASTnr
0.01	1.25	6.965784	4.11E-05	0.00031	Heat shock protein 70 (HSP70)-interacting protein, putative (*Ricinus communis*)
0.17	3.08	4.179324	0.000111	0.000761	Heat shock protein 81–1 (*Arabidopsis lyrata*)
0.26	3.25	3.643856	6.95E-05	0.000499	Heat shock protein, putative (*Ricinus communis*)
0.35	3.58	3.354533	4.26E-05	0.000318	Heat shock protein, putative (*Ricinus communis*)
0.78	5.99	2.94101	2.46E-06	2.32E-05	Heat shock factor protein, putative (*Ricinus communis*)
1.21	9.08	2.907685	1.08E-07	1.22E-06	Small heat shock protein 17.3 kDa (*Vitis vinifera*)
23.6	100.91	2.09621	3.37E-13	7.57E-12	Heat shock protein binding protein, putative (*Ricinus communis*)
3.98	14.49	1.864217	1.02E-13	2.68E-12	Heat shock protein 70 (HSP70)-interacting protein, putative (*Ricinus communis*)
2.68	9.49	1.824175	8.99E-12	1.54E-10	Chaperone protein dnaJ 8, chloroplast precursor, putative (*Ricinus communis*)
53.26	152.2	1.514844	8.12E-13	1.61E-11	Heat shock protein (*Glycine max*)
5.79	16.32	1.495006	0	0	Heat shock protein (*Glycine max*)
46.69	106.91	1.195211	1.09E-12	2.09E-11	Mitochondrial small heat shock protein (*Capsicum annuum*)
9.94	20.9	1.072185	8.74E-12	1.50E-10	Heat shock protein, putative (*Ricinus communis*)

(b) Control vs freeze stress

TPM-control	TPM-freeze stress	log2 ratio/fold change	P-value	FDR	Homology on the basis of BLASTnr
0.01	1.6	7.321928	2.55E-06	2.48E-05	Heat shock factor protein, putative (*Ricinus communis*)
0.26	2.6	3.321928	7.10E-05	0.00052	Heat shock cognate protein 70 (*Thellungiella halophila*)

(continued)

Table 15.2 (continued)

(b) Control vs freeze stress

TPM-control	TPM-freeze stress	log2 ratio/ fold change	P-value	FDR	Homology on the basis of BLASTnr
0.26	2.02	2.957772	0.000108	0.00076	Heat shock protein, putative (*Ricinus communis*)
4.15	16.79	2.016419	9.13E-14	2.27E-12	Heat shock 70 kDa protein
1.21	3.69	1.608614	0.000103	0.000727	Heat shock protein 70 (*Gossypium hirsutum*)
3.63	10.16	1.484859	9.77E-10	1.44E-08	Heat shock protein binding protein, putative (*Ricinus communis*)
3.89	1.09	−1.83544	1.17E-05	9.97E-05	Low molecular weight heat-shock protein (*Corylus avellana*)
108.68	31.07	−1.80649	9.67E-119	8.19E-117	Heat-shock protein, putative (*Ricinus communis*)
7.61	3.44	−1.14549	1.48E-05	0.000125	Heat shock factor protein HSF30, putative (*Ricinus communis*)

(c) Cold stress vs freeze stress

TPM-cold stress	TPM-freeze stress	log2 ratio/ fold change	P-value	FDR	Homology on the basis of BLASTnr
0.01	1.09	6.768184	0.000115	0.000811	Similar to DNAJ heat shock N-terminal domain-containing protein (*Vitis vinifera*)
3.16	10.16	1.684904	1.27E-11	2.24E-10	Heat shock protein binding protein, putative (*Ricinus communis*)
1.25	0.01	−6.96578	3.27E-05	0.000261	Heat shock protein 70 (HSP70)-interacting protein, putative (*Ricinus communis*)
548.79	31.07	−4.14266	0	0	Heat-shock protein, putative (*Ricinus communis*)
13.57	1.09	−3.63802	3.68E-34	1.78E-32	Low molecular weight heat-shock protein (*Corylus avellana*)
3.16	0.34	−3.21632	3.61E-08	4.52E-07	Heat shock protein, putative (*Ricinus communis*)
24.06	2.77	−3.11868	5.29E-52	3.25E-50	Heat-shock protein, putative (*Ricinus communis*)
100.91	21.5	−2.23066	3.24E-147	2.76E-145	Heat shock protein binding protein, putative (*Ricinus communis*)

(continued)

Table 15.2 (continued)

(c) Cold stress vs freeze stress					
TPM-cold stress	TPM-freeze stress	log2 ratio/fold change	P-value	FDR	Homology on the basis of BLASTnr
10.07	3.44	−1.54958	2.18E-10	3.44E-09	Heat shock factor protein HSF30, putative (*Ricinus communis*)
44.05	15.12	−1.54268	3.22E-40	1.71E-38	Heat-shock protein, putative (*Ricinus communis*)
106.91	37.03	−1.52963	6.21E-94	4.72E-92	Mitochondrial small heat shock protein (*Capsicum annuum*)
9.08	3.36	−1.43423	1.29E-08	1.70E-07	Small heat shock protein 17.3 kDa (*Vitis vinifera*)
14.49	5.46	−1.40808	1.25E-12	2.49E-11	Heat shock protein 70 (HSP70)-interacting protein, putative (*Ricinus communis*)
5.99	2.52	−1.24913	3.28E-05	0.000262	Heat shock factor protein, putative (*Ricinus communis*)

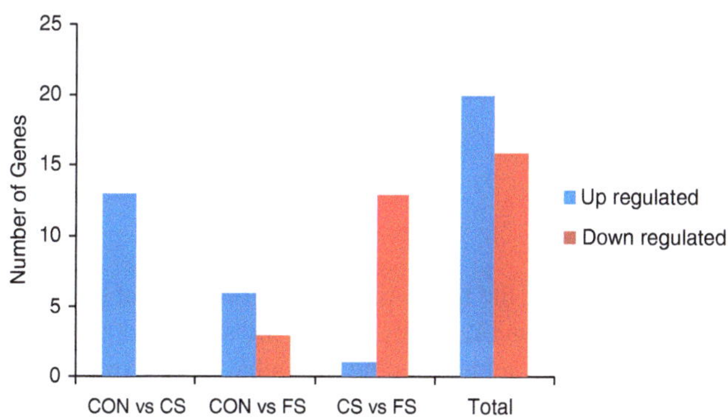

Fig. 15.3 Differentially expressed heat shock protein (*hsp*) genes between control (*CON*, treatment 28°C, 6 h), cold stress (*CS*, treatment 4°C, 6 h) and freeze stress (*FS*, treatment −10°C, 6 h) libraries

which are differentially expressed under freeze stress as compare to cold stress. A total of 14 genes were identified, however, the expression of 13 genes was down regulated in freeze stress as compared to cold stress (Fig. 15.3). The only gene which is up regulated was *Dna*J gene for heat shock N-terminal domain containing protein (Table 15.2c).

To validate the results of DeepSAGE, we selected 22 genes including one *hsp70* gene for further expression analysis using qRT-PCR (Chaudhary and Sharma 2015). The expression analysis of *hsp-70* gene is reproduced in the present chapter. The expression was analysed at different treatment i.e. cold stress (4°C) and freeze stress

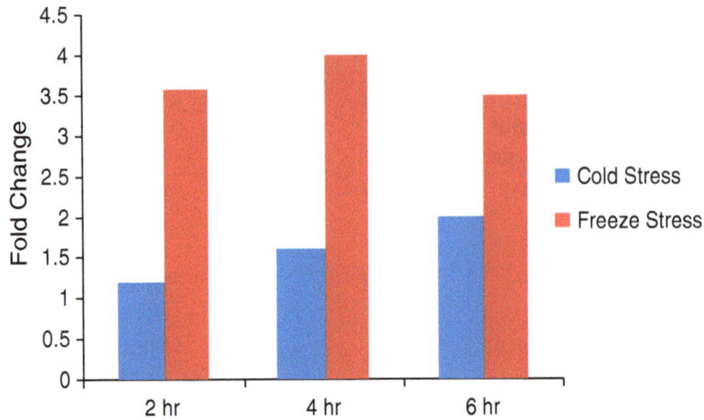

Fig. 15.4 qRT-PCR based differential expression pattern of heat shock protein-70 (*hsp-70)* gene during cold stress (4°C), and freeze stress (−10°C) in seabuckthorn

(−10°C) and different time intervals i.e. 2 h, 4 h and 6 h as shown in Fig. 15.4. The qRT-PCR gene expression analysis suggested that the *hsp-70* gene is highly expressed under freeze stress condition. This gene showed 3–4 fold upregulation as compared to control whereas in case of cold stress the upregulation was a marginal 1.2–2.5 fold. These results suggest a significant role of *hsp*-70 gene in mediating freezing tolerance in seabuckthorn possibly by stabilizing proteins against freeze induced denaturation or injury.

15.5 Conclusion

Cold and freeze stresses are the major limiting factors in plant growth and productivity in many areas of the world. Being multigenic trait, cold and freeze tolerance phenomenon requires deep understanding of global gene expression profile in a plant native to cold regions. The advancement in sequencing and gene expression technologies has facilitated study of abundance and expression of large number of genes simultaneously that too in a affordable, accessible and robust manner. The seabuckthorn transcriptome data generated using Illumina Next Generation Sequencing technology was reanalyzed to identify the expression of heat shock protein (*hsp*) genes. The ability of seabuckthorn to sustain growth in adverse environment and the presence of large number of genes encoding HSPs and their related submembers in seabuckthorn transcriptome, suggested the implication of HSPs in abiotic stress tolerance. Further, DeepSAGE, a tag based profiling approach proved to be an efficient and powerful tool for the study of global gene expression in seabuckthorn providing a comprehensive view of genes differentially expressed during cold and freeze stress. Isolation and further characterization of *hsp* genes will help researchers in understanding their role in cold and freeze tolerance in seabuckthorn and may provide important gene resources to be exploited for the development of stress tolerant crop plants in future.

Acknowledgements The authors acknowledge the financial support provided by the Department of Biotechnology, Government of India.

References

Akkermans ADL, Roelofsen W, Blom J, Hussdanell K, Harkink R (1983) Utilization of carbon and nitrogen compounds by *Frankia* in synthetic media and in root nodules of *Alnus glutinosa*, *Hippophae rhamnoides*, and *Datisca cannabina*. Can J Bot 61:2793–2800

Ansorge WJ (2009) Next-generation DNA sequencing techniques. New Biotechnol 25:195–203

Audic S, Claverie JM (1997) The significance of digital gene expression profiles. Genome Res 7:986–995

Bae MS, Cho EJ, Choi EY, Park OK (2003) Analysis of the Arabidopsis nuclear proteome and its response to cold stress. Plant J 36:652–663

Bartish IV, Jeppsson N, Nybom H, Swenson U (2002) Phylogeny of *Hippophae* (Elaeagnaceae) inferred from parsimony analysis of chloroplast DNA and morphology. Syst Bot 27:41–54

Benjamini Y, Yekutieli D (2001) The control of the false discovery rate in multiple testing under dependency. Ann Stat 29:1165–1188

Boston RS, Viitanen PV, Vierling E (1996) Molecular chaperones and protein folding in plants. In: Post-transcriptional control of gene expression in plants. Springer, Dordrecht, pp 191–222

Brenner S, Johnson M, Bridgham J, Golda G, Lloyd DH, Johnson D, Luo S, McCurdy S, Foy M, Ewan M, Roth R (2000) Gene expression analysis by massively parallel signature sequencing (MPSS) on microbead arrays. Nat Biotechnol 18:630–634

Chaudhary S, Sharma PC (2015) DeepSAGE based differential gene expression analysis under cold and freeze stress in seabuckthorn (*Hippophae rhamnoides* L.). PLoS One 10:e0121982

Chawla A, Stobdan T, Srivastava RB, Jaiswal V, Chauhan RS, Kant A (2015) Sex-based temporal gene expression in male and female floral buds of seabuckthorn (*Hippophae rhamnoides*). PLoS One 10:e0124890

Cheng MY, Hartl FU, Martin J, Pollock RA, Kalousek F, Neupert W, Hallberg EM, Hallberg RL, Horwich A (1989) Mitochondrial heat-shock protein hsp60 is essential for assembly of proteins imported into yeast mitochondria. Nature 6208:585–674

Chinnusamy V, Zhu J, Zhu JK (2006) Gene regulation during cold acclimation in plants. Physiol Plant 126:52–61

Chinnusamy V, Zhu J, Zhu JK (2007) Cold stress regulation of gene expression in plants. Trends Plant Sci 12:444–451

Conesa A, Gotz S, Garcia-Gomez JM, Terol J, Talon M, Robles M (2005) Blast2GO: a universal tool for annotation, visualization and analysis in functional genomics research. Bioinformatics 21:3674–3676

Elena T, Capraru G, Rosu CM, Zamfirache MM, Olteanu Z, Manzu C (2011) Morphometric pattern of somatic chromosomes in three Romanian seabuckthorn genotypes. Caryologia 64:189–196

Fatima T, Snydrez CL, Schoeder WR, Cram D, Datla R, Wishat D, Krishna P (2012) Fatty acid composition of developing Seabuckthorn (*Hippophae rhamnoides* L.) berry and transcriptome of the mature seeds. PLoS One 7:e34099

Fowler S, Thomashow MF (2002) Arabidopsis transcriptome profiling indicates that multiple regulatory pathways are activated during cold acclimation in addition to the CBF cold response pathway. Plant Cell 14:1675–1690

Ghangal R, Raghuvanshi S, Sharma PC (2009) Isolation of good quality RNA from a medicinal plant seabuckthorn, rich in secondary metabolites. Plant Physiol Biochem 47:1113–1115

Ghangal R, Raghuvanshi S, Sharma PC (2012) Expressed sequence tag based identification and expression analysis of some cold inducible elements in seabuckthorn (*Hippophae rhamnoides* L.). Plant Physiol Biochem 51:123–128

Ghangal R, Chaudhary S, Jain M, Purty RS, Sharma PC (2013) Optimization of *de novo* short read assembly of seabuckthorn (*Hippophae rhamnoides* L.) transcriptome. PLoS One 8:e72516
Grabherr MG, Haas BJ, Yassour M, Levin JZ, Thompson DA, Amit I, Adiconis X, Fan L, Raychowdhury R, Zeng Q, Chen Z (2011) Full-length transcriptome assembly from RNA-Seq data without a reference genome. Nat Biotechnol 26:644–652
Gupta R, Deswal R (2012) Low temperature stress modulated secretome analysis and purification of antifreeze protein from *Hippophae rhamnoides*, a Himalayan wonder plant. J Proteome Res 11:2684–2696
Gupta SM, Ahmed Z, Kumar N (2009) Isolation of cDNA fragment of glycerol-3-phosphate acyl-transferase gene from seabuckthorn. Def Sci J 59:147–151
Hannah MA, Wiese D, Freund S, Fiehn O, Heyer AG, Hincha DK (2006) Natural genetic variation of freezing tolerance in Arabidopsis. Plant Physiol 142:98–112
Jain M (2011) Next-generation sequencing technologies for gene expression profiling in plants. Brief Funct Genomics 11:63–70
Jain A, Ghangal R, Grover A, Raghuvanshi S, Sharma PC (2010) Development of EST-based new SSR markers in seabuckthorn. Physiol Mol Biol Plants 16:375–378
Jain A, Chaudhary S, Sharma PC (2014) Mining of microsatellites using next generation sequencing of seabuckthorn (*Hippophae rhamnoides* L.) transcriptome. Physiol Mol Biol Plants 20:115–123
Kadota Y, Shirasu K (2012) The HSP90 complex of plants. Biochim Biophys Acta 1823:689–697
Kanehisa M, Araki M, Goto S, Hattori M, Hirakawa M, Itoh M, Katayama T, Kawashima S, Okuda S, Tokimatsu T, Yamanishi Y (2008) KEGG for linking genomes to life and the environment. Nucleic Acids Res 36:D480–D484
Kawamura Y, Uemura M (2003) Mass spectrometric approach for identifying putative plasma membrane proteins of Arabidopsis leaves associated with cold acclimation. Plant J 36:141–154
Ledwood JS, Shimwell DW (1971) Growth rates of *Hippophae rhamnoides* L. Ann Bot 35:1053–1058
Levitt J (1980) Responses of plants to environmental stress, vol 1: chilling, freezing, and high temperature stress. Academic, New York
Li Z, Srivastava P (2004) Heat-shock proteins. Curr Protoc Immunol A-1T
Li C, Yang Y, Junttila O, Palva ET (2005) Sexual differences in cold acclimiation and freezing tolerance development in seabuckthorn (*Hippophae rhamnoides* L.) ecotypes. Plant Sci 168:1365–1370
Li R, Zhu H, Ruan J, Qian W, Fang X, Shi Z, Li Y, Li S, Shan G, Kristiansen K, Li S (2009) *De novo* assembly of human genomes with massively parallel short read sequencing. Genome Res 20:265–272
Lian YS, Chen XL (2000) The regular patterns of distribution on the natural components in plants of the genus *Hippophae* L. J Northwest Normal Univ (Nat Sci Ed) 36:113–128
Lindquist S, Craig EA (1988) The heat-shock proteins. Annu Rev Genet 22:631–677
Lopez-Matas MA, Nuñez P, Soto A, Allona I, Casado R, Collada C, Guevara MA, Aragoncillo C, Gomez L (2004) Protein cryoprotective activity of a cytosolic small heat shock protein that accumulates constitutively in chestnut stems and is up-regulated by low and high temperatures. Plant Physiol 134:1708–1717
Lu R (1992) Seabuckthorn – a multipurpose plant for fragile mountains, International Centre for Integrated Mountain Development (ICIMOD) occasional paper no. 20. International Centre for Integrated Mountain Development, Kathmandu
Lu R (1997) Eco-geographical distribution of seabuckthorn and prospects of international cooperation. In: Worldwide research & development of seabuckthorn. Science Publishing House & Technology Press, Beijing, pp 11–22
Mayer M, Bukau B (2005) Hsp70 chaperones: cellular functions and molecular mechanism. Cell Mol Life Sci 62:670–684

Morozova O, Hirst M, Marra MA (2009) Application of new sequencing technologies for transcriptome analysis. Annu Rev Genomics Hum Genet 10:135–151
Morrissy S, Morin RD, Delaney A, Zeng T, McDonald H, Jones S, Zhao Y, Hirst M, Marra MA (2009) Next-generation tag sequencing for cancer gene expression profiling. Genome Res 19:1825–1835
Nielsen KL, Hogh AL, Emmersen J (2006) DeepSAGE—digital transcriptomics with high sensitivity, simple experimental protocol and multiplexing of samples. Nucleic Acids Res 34:e133
Ozsolak F, Milos PM (2011) RNA sequencing: advances, challenges and opportunities. Nat Rev Genet 12:87–98
Patel RK, Jain M (2012) NGS QC toolkit: a toolkit for quality control of next generation sequencing data. PLoS One 7:e30619
Renaut J, Lutts S, Hoffmann L, Hausman JF (2004) Responses of poplar to chilling temperatures: proteomic and physiological aspects. Plant Biol 6:81–90
Santoro MG (2000) Heat shock factors and the control of the stress response. Biochem Pharmacol 59:55–63
Schlesinger MJ (1990) Heat shock proteins. J Biol Chem 265:12111–12114
Schulz MH, Zerbino DR, Vingron M, Birney E (2012) Oases: robust *de novo* RNA-seq assembly across the dynamic range of expression levels. Bioinformatics 28:1086–1092
Schuster SC (2008) Next-generation sequencing transforms today's biology. Nat Methods 5:16–18
Simpson JT, Wong K, Jackman SD, Schein JE, Jones SJ, Birol I (2009) ABySS: a parallel assembler for short read sequence data. Genome Res 19:1117–1123
Stobdan T, Angchuk D, Singh SB (2008) Seabuckthorn: an emerging storehouse for researchers in India. Curr Sci 94:1236–1237
Suryakumar G, Gupta A (2011) Medicinal and therapeutic potential of Seabuckthorn (*Hippophae rhamnoides* L.). J Ethnopharmacol 138:268–278
t'Hoen PA, Ariyurek Y, Thygesen HH, Vreugdenhil E, Vossen RH, deMenezes RX, Boer JM, van Ommen GJB, den Dunnen JT (2008) Deep sequencing based expression analysis shows major advances in robustness, resolution and inter-lab portability over five microarray platforms. Nucleic Acids Res 36:e141
Taylor NL, Heazlewood JL, Day DA, Millar AH (2005) Differential impact of environmental stresses on the pea mitochondrial proteome. Mol Cell Proteomics 4:1122–1133
Teng BS, Lu YH, Wang ZT, Tao XY, Wei DZ (2006) *In vitro* anti-tumor activity of isorhamnetin isolated from *Hippophae rhamnoides* L. against BEL-7402 cells. Pharmacol Res 54:186–194
Thakur V, Varshney R (2010) Challenges and strategies for next generation sequencing (NGS) data analysis. J Comput Sci Syst Biol 3:040–042
Trajkovsk V, Jeppsson N (1999) Domestication of seabuckthorn. Bot Lith 2:37–46
Unamba CIN, Nag A, Sharma RK (2015) Next generation sequencing technologies: the doorway to the unexplored genomics of non-model plants. Front Plant Sci 6:1074
Velculescu VE, Zhang L, Vogelstein B, Kinzler KW (1995) Serial analysis of gene expression. Science 270:484–487
Vinocur B, Altman A (2005) Recent advances in engineering plant tolerance to abiotic stress: achievements and limitations. Curr Opin Biotechnol 16:123–132
Wang W, Vinocur B, Altman A (2003) Plant responses to drought, salinity and extreme temperatures: towards genetic engineering for stress tolerance. Planta 218:1–14
Wang W, Vinocur B, Shoseyov O, Altman A (2004) Role of plant heat-shock proteins and molecular chaperones in the abiotic stress response. Trends Plant Sci 9:244–252
Yang Y, Yao Y, Xu G, Li C (2005) Growth and physiological responses to drought and elevated ultraviolet-B in two contrasting populations of *Hippophae rhamnoides*. Physiol Plant 124:431–440
Zeb A (2006) Anticarcinogenic potential of lipids from *Hippophae*; Evidence from the recent literature. Asian Pac J Cancer Prev 7:32–35
Zerbino DR, Birney E (2008) Velvet: algorithms for *de novo* short read assembly using de Bruijn graphs. Genome Res 18:821–829

Chapter 16
Epigenetic Regulation of Plant Heat Shock Protein (HSP) Gene Expression

Yuanyuan Ren and Yan Zhu

Abstract Heat shock proteins (HSPs) are the key components of plants' adaptive mechanism to maintain protein homeostasis. HSPs play critical roles not only in response to unfavorable environmental conditions, but also during normal plant growth and development. Compared with yeast and animals, plants' HSP-encoding genes have expanded into a large gene family, likely because of their sessile lifestyle. These genes have divergent expression patterns in various plant organs, and respond at various levels to heat treatment. While the genetic control of plant HSP genes by upstream heat shock factors has been well studied, the epigenetic regulation of these stress-responsive genes has emerged as a critical pathway implicated in the plant response to environmental changes. In this chapter, we summarize the progress regarding the epigenetic regulation of plant HSP gene expression. The common and disparate epigenetic pathways and factors in the heat response in plants and other organisms are also discussed.

Keywords Epigenetic regulation • Histone chaperone • Histone modification • Histone variant

Abbreviations

ARP6	ACTIN-related protein 6
ASF1	anti-silencing function 1
FACT	facilitates chromatin transcription
HSF	heat shock factor
HSP	heat shock protein

Y. Ren • Y. Zhu (✉)
State Key Laboratory of Genetic Engineering, Collaborative Innovation Center of Genetics and Development, Department of Biochemistry, Institute of Plant Biology, School of Life Sciences, Fudan University, Shanghai 20043, China
e-mail: zhu_yan@fudan.edu.cn

A.A.A. Asea et al. (eds.), *Heat Shock Proteins and Plants*,
Heat Shock Proteins 10, DOI 10.1007/978-3-319-46340-7_16

NRP1	NAP1-related protein 1
PARP	poly (ADP-ribose) polymerase
Pol II	Polymerase II
PP2A	protein phosphatase type 2A
SPT16	suppressor of TY16
SWR1	SWI2/SNF2-related 1
TSS	transcription start site

16.1 Introduction

Ongoing global warming has led to rising ambient temperatures, which has become a major worldwide concern, especially for agriculture. Heat stress is expected to occur more frequently and to challenge the demography and crop yield of ecological plant species (Grover et al. 2013). High temperature destroys the homeostasis of cellular proteins, resulting in non-functional and potentially harmful mis-folded proteins that are prone to aggregation, greatly inhibiting each stage of plant growth and development, especially plant sexual reproduction (Zinn et al. 2010). To cope with heat stress, plants have evolved various cellular and physiological mechanisms. At the molecular level, a protein quality control system is necessary to maintain protein homeostasis to detect mis-folded proteins and promote their re-folding by chaperones and degradation by ATP-dependent proteases. Considerable evidence shows that various HSPs and their upstream transcription factors, heat shock factors (HSFs), are the critical components of the adaptive mechanism by which cells respond to suboptimal or even stressfully high temperatures. HSPs and HSFs play decisive roles in the heat stress response in both the vegetative and reproductive stages of plant growth (Frank et al. 2009; Sarkar et al. 2014; Liu et al. 2015).

HSPs are molecular chaperones that regulate protein localization, accumulation, and degradation, assist in protein re-folding, and prevent irreversible aggregation of denaturing proteins, conferring the property of thermotolerance on cells. Based on their molecular sizes, the HSP proteins are classified into a number of conserved protein families, including HSP20, HSP70, HSP90, and HSP100, which are evolutionarily unrelated to each other, despite their intimate functional interplay. The protein structure of each HSP family is largely conserved in eukaryotes. The characteristics of the protein architectures of these chaperones have been well reviewed (Kadota and Shirasu 2012; Waters 2013; Mogk et al. 2015) and thus are not discussed in this chapter.

16.2 Large Gene Family of Plant Heat Shock Proteins (HSPs) and Their Divergent Expression Pattern in Arabidopsis

HSP20s range from 15 to 42 kDa, and are thus also named as small HSPs. In many higher plants, the HSP20 gene family is the most prevalent and abundant family induced by heat stress, playing a critical role in the acquisition of plant thermotolerance and is thus regarded as the first line of stress defense in cells (Haslbeck and Vierling 2015). Compared with yeast and animals, plant HSP20 genes have expanded into a large gene family, and the gene number varies among plant species: the model dicot plant Arabidopsis thaliana encodes 19 HSP20s (Scharf et al. 2001), and the model monocot rice (Oryza sativa) possesses 23 (Sarkar et al. 2009). The large families of higher plant HSP20 genes are thought to have arisen by duplication and divergence. Plant HSP20s are divided into various sub-families based on their function, sequence homology, and more importantly, cellular location, including cytoplasm-nucleus (CI/CII), plastid (P), endoplasmic reticulum (ER), and mitochondria (M) proteins (Fig. 16.1).

To reverse harmful protein aggregation, plant cells have evolved a powerful bi-chaperone system comprising two cooperating ATP-driven machines, the HSP70 chaperone system and the hexameric chaperone HSP100s (Mogk et al. 2015). Apart from their protective role in thermotolerance, HSP70s also exert a regulatory effect on HSF activity, leading to auto-regulation of the heat shock response in plants. They make up the second largest protein family, with 18 members in Arabidopsis (Lin et al. 2001) and 23 members in rice (Sarkar et al. 2013). By contrast, fewer HSP100 ($n=7$ in Arabidopsis and $n=5$ in rice) chaperones have been identified. HSP101, a representative member of the Arabidopsis HSP100 family, genetically interacts with the HSP20 chaperone system to re-solubilize protein aggregates after heat stress (Lee et al. 2005). Moreover, pollen from AtHSP101 transgenic plants exhibited significantly higher germination rates and much better pollen tube vigor under elevated temperatures or after a heat exposure (Burke and Chen 2015).

HSP90 is a highly conserved protein involved in the assembly, maturation, stabilization, and activation of key signaling proteins in eukaryotic cells. Arabidopsis and rice genomes encode seven and nine HSP90s located in different sub-cellular compartments, respectively (Xu et al. 2012), and many HSP90 members play important roles in plant development and growth (Kadota and Shirasu 2012). In tomato (Solanum lycopersicum), HSP90 protein modulates HSFA2 transcript degradation, and interacts with HSFB1 to target it for proteasomal degradation (Hahn et al. 2011).

We surveyed the developmental expression pattern of most HSP genes in major organs of Arabidopsis using a public database (Fig. 16.1). Compared with other HSP gene families, most HSP20 members are generally expressed at a low level. Notably, several HSP20 genes, such as AtHSP17.4-CI, AtHSP17.7-CII, and AtHsp26.5-P(r), are expressed at a relatively higher level in siliques, which may represent a desiccating biological process. Many members of the AtHSP70,

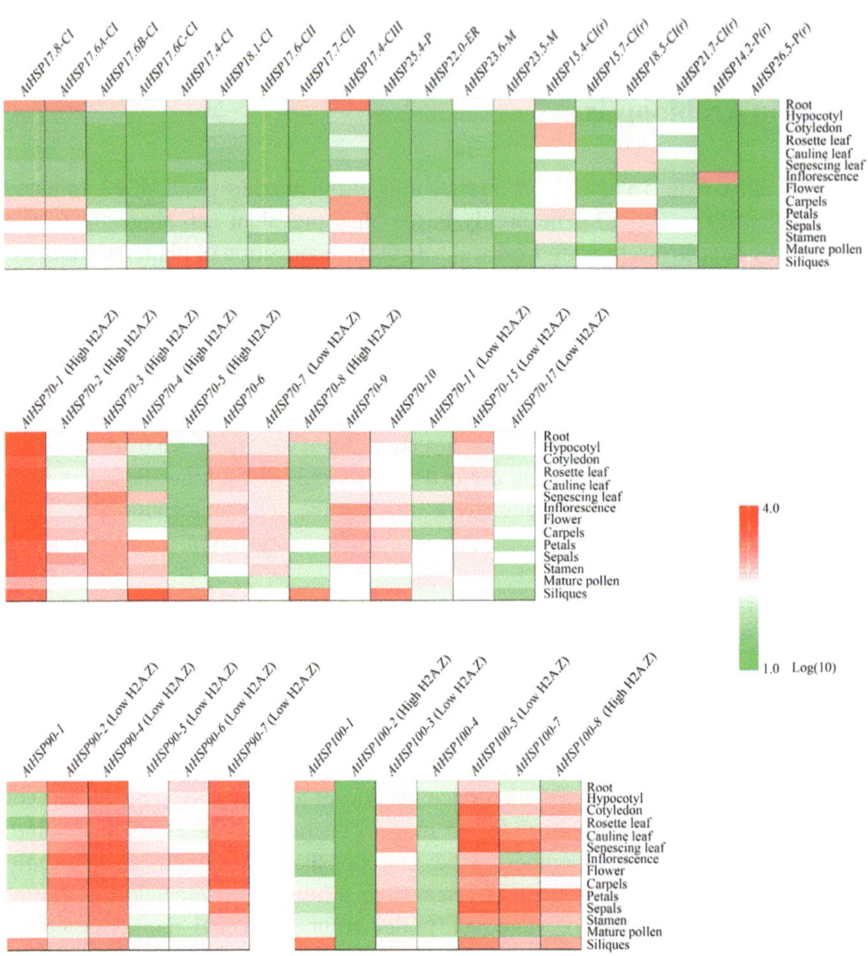

Fig. 16.1 Gene expression of selected *Arabidopsis HSP* genes in different plant organs. Gene expression data were extracted from the Arabidopsis eFP Browser (http://bar.utoronto.ca/efp/cgi-bin/efpWeb.cgi). The data are normalized by the robust multi-array average (RMA) method and shown as log10-transformed levels. The color gradient represents the normalized data in a log (10) space; i.e., the red tile marked as 4.0 represents 1000 times level of the green tile marked as 1.0. The *HSP* genes identified with high or low levels of H2A.Z enriched within their gene bodies in a previous study (Coleman-Derr and Zilberman 2012) are labeled as 'High H2A.Z' and 'Low H2A.Z', respectively

AtHSP90, and AtHSP100 families show a ubiquitous high-level expression pattern during the whole plant lifespan, implying their wide participation in plant organogenesis. In addition, certain other members display a silique-preferential expression pattern, such as AtHSP70-5, AtHSP70-8, and AtHSP100-1. In this survey, we failed to identify a simple correlation between a certain expression pattern of an HSP gene and the sub-cellular compartment of its protein product.

Transcriptional profiling of Arabidopsis HSP genes in response to different abiotic and biotic stress treatments have been reported, revealing extensive overlap between heat and non-heat stress response pathways. High temperature is associated with a rapid activation for all gene families, except HSP100, in both roots and shoots. Nevertheless, the heat-induced activation levels of individual HSP genes are largely non-uniform (Swindell et al. 2007). The homogeneous and simultaneous induction in response to heat stress of many HSP genes suggest that are ideal model genes for the study of chromatin changes during gene activation and subsequent gene repression. Although the regulation of most HSP genes by their upstream multiple HSF transcription factors is well studied (Kotak et al. 2007), the multiple epigenetic mechanisms underlying the activation of plant HSP genes appear sophisticated and are discussed below.

16.3 H2A.Z, the Thermosensor for Increased Ambient Temperature

The nucleosome, the basic repeating unit of chromatin, consists of approximately 147 bp of DNA wrapped around a globular histone octamer containing two molecules each of histones H2A, H2B, H3, and H4. The octamer is structured by the (H3-H4)2 tetramer at the center and the two H2A-H2B dimers attached symmetrically on either side. H2A.Z is one highly conserved histone variant that diverged early from the canonical H2A histones during eukaryotic evolution. H2A.Z variants from different organisms show a higher level of sequence similarity than between H2A.Z and H2A within the same organism. At the level of primary protein sequence, H2A.Z variants differ from the canonical H2A at many positions, mainly in the docking domain, a structure involved in the interaction of the H2A-H2B dimer with the centrally located (H3-H4)2 tetramer in the nucleosome (March-Diaz and Reyes 2009). Likely based on these differences, or other unknown mechanisms, in vivo H2A.Z confers distinct DNA-unwrapping properties on the assembled nucleosomes. Compared with the H2A-containing nucleosome, the H2A.Z-containing nucleosome is more inaccessible for RNA Pol II, which needs to invade into the nucleosomal DNA for transcription (Kumar and Wigge 2010).

AtHSP70-4 (At3g12580) is expressed at a level proportionate to the increase of ambient temperature in the non-stress range (i.e., 12–27 °C); therefore, Kumar and Wigge (2010) used this gene as an output measure of the ambient temperature-sensing pathway to unravel the molecular mechanism of the ambient temperature response. In the survey of factors regulating AtHSP70-4 expression during ambient temperature increase, ACTIN-RELATED PROTEIN 6 (ARP6) was screened out as the decisive factor that regulates AtHSP70-4 expression in a temperature-responsive manner. By contrast, the mutant defective in ARP6 function displayed constitutively high AtHSP70-4 expression. ARP6 is a conserved and critical subunit of the SWI2/SNF2 chromatin remodeling complex SWI2/SNF2-RELATED 1 (SWR1), which shows conserved activity of specifically incorporating H2A.Z into nucleosomes (March-Diaz and Reyes 2009). Moreover, ARP6-mediated incorporation of H2A.Z is highly related to the whole transcriptome response to temperature increase. Notably, when the ambient temperature increases, H2A.Z occupancy within the AtHSP70-4 promoter is decreased. Similar chromatin remodeling was detected widely in many genes responsive to high temperature, independent of their transcriptional change. It is now clear that the presence of H2A.Z-containing nucleosomes is rate limiting for the expression of the majority of the genes in the ambient temperature transcriptome. H2A.Z-containing nucleosomes located at the transcription start site (TSS) may maintain RNA Pol II in a poised state, or else, occupancy of H2A.Z-nucleosomes at the TSS restricts access of the transcriptional machinery to the gene body. When the temperature rises, H2A.Z is evicted from the first nucleosomes after the TSS and the occupancy of RNA Pol II is shifted into place and contributes to the upregulation of HSP70 expression (Kumar and Wigge 2010). As a result, H2A.Z-associated regulation of HSP gene expression represents a functional chromatin response to ambient temperature change.

Brachypodium distachyon is a model plant for monocot crops. Distinct from the dicot Arabidopsis, the effects of temperature increase (17–27 °C) on the phenology of Brachypodium are more pronounced in the developing grain compared with vegetative seedlings, implying a different adaptive mechanism for these plants to ambient temperature during evolution. Despite the absence of large-scale phenological changes in vegetative plants at 27 °C, a considerable acceleration in grain development is observed at this higher temperature. Consistently, H2A.Z-nucleosome occupancy is more responsive to increases in ambient temperature in the reproductive tissue of developing grains compared with vegetative seedlings (Boden et al. 2013), suggesting that the H2A.Z-associated molecular mechanism of thermally responsive transcription could be tissue specific or developmental stage specific in different plants.

16.4 H2A.Z and Gene Responsiveness to Heat

A systematic analysis of H2A.Z global chromatin distribution in Arabidopsis showed that H2A.Z enrichment across gene bodies is inversely correlated with the transcription levels. Intriguingly, loss of H2A.Z causes mis-regulation of many genes that are disproportionately associated with response to environmental and developmental stimuli, including heat. In contrast to H2A.Z deposition near the TSS, H2A.Z deposition in gene bodies promotes variability in the levels and patterns of gene expression, and likely facilitates rapid activation or inactivation of genes, which represents higher gene responsiveness (Coleman-Derr and Zilberman 2012).

Based on the analysis of this study, several HSP genes have been identified as having high or low levels of H2A.Z enrichment within their gene bodies. These genes include members of the HSP70, HSP90, and HSP100, but not the HSP20,

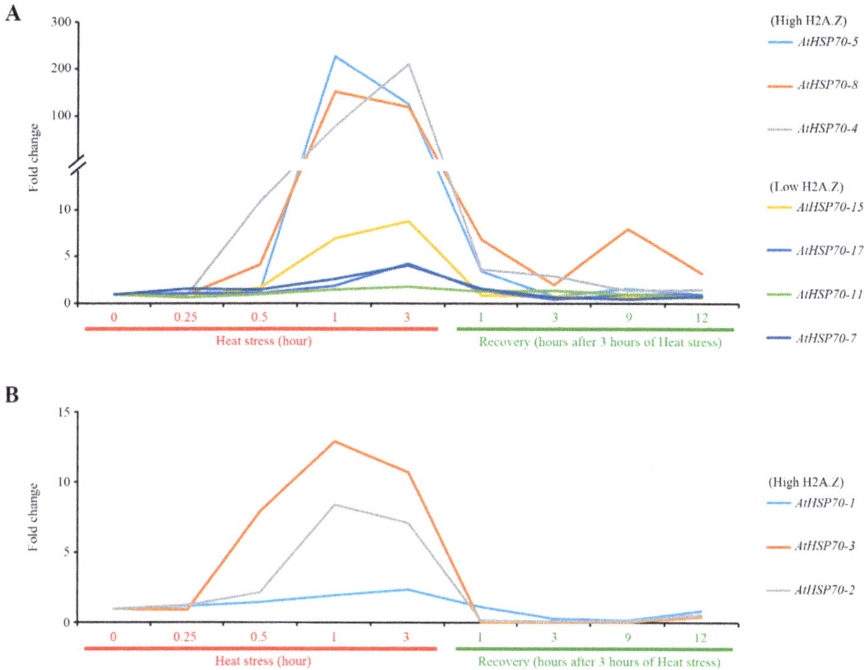

Fig. 16.2 Comparison of gene expressions of distinct *AtHSP70* genes with 'Low H2A.Z' or 'High H2A.Z' during 3 h of heat stress and subsequent 12 h of recovery. Gene expression data were extracted from the AtGenExpress Visualization Tool (AVT, http://jsp.weigelworld.org/expviz/expviz.jsp). The values were referenced to the expression level in untreated plants, which was set as 1

gene families (Fig. 16.1), suggesting that epigenetic regulation via gene body-enriched H2A.Z may not have been integrated during the evolutionary expansion of the huge Arabidopsis HSP20 gene family. Intriguingly, all the H2A.Z body-depleted HSP genes, including four HSP70 members (Fig. 16.2), five HSP90 members, and two HSP100 members, display a relatively low fold-change in gene activation in response to heat. This is consistent with the proposed role of H2A.Z in promoting gene responsiveness. However, not all the H2A.Z body-enriched HSP genes show high activation. For instance, AtHSP70-4, -5, and -8 are potently induced by heat by more than 100-fold (the data were extracted from the database, and the transcriptional analysis in our study using quantitative RT-PCR showed an even higher level), whereas the activations of AtHSP70-1, -2, and -3 are relatively mild (Fig. 16.2). These observations suggested that histone variant H2A.Z is unlikely to be the only epigenetic component determining the responsiveness of target genes, and crosstalk with other factors or components may interfere with the outcome of H2A.Z regulation.

16.5 H3K56 Acetylation in HSP Gene Activation

H2A.Z incorporation into the nucleosome is mediated by a conserved chromatin remodeling complex, with SWR1 as its catalytic subunit. A breakthrough study showed that acetylation of histone H3 on lysine 56 (H3K56ac) alters the substrate specificity of the SWR1 complex, leading to promiscuous dimer exchange, in which either H2A.Z or H2A could be exchanged from nucleosomes (Watanabe et al. 2013). The N-terminal tails of the histones protrude from the spherical nucleosomes and can be subjected to various covalent modifications, such as acetylation, phosphorylation, and others. Conserved K56 is located within the N-terminal α-helix of H3, and its acetylation interrupts histone-DNA contact and increases the rate of nucleosomal DNA ends unwrapping from the histone octamer (Neumann et al. 2009), thereby promoting nucleosome disassembly during transcriptional activation.

In Arabidopsis, H3K56ac levels drastically increased at several heat-induced genes including HSFA2 and Hsa32 gene loci in heat-treated compared with untreated wild-type plants. Similar increases were observed for several examined HSP gene loci (personal communication). In yeast and animal cells, H3K56 acetylation is catalyzed by histone acetyltransferase Rtt109 and CBP/p300 proteins. Rtt109 is a yeast-specific histone acetyltransferase, while CBP/p300 has several plant homologs in Arabidopsis and rice. Notably, such acetylation also requires the presence of histone chaperone ANTI-SILENCING FUNCTION 1 (ASF1). ASF1 is a conserved chaperone with affinity for the core histone H3-H4 in all eukaryotes. While yeast and Drosophila possess a single ASF1 copy, higher plant species, exemplified by Arabidopsis thaliana, encode two homologs with high protein identity and redundant function, AtASF1A and AtASF1B. The histone-binding activity of ASF1 is thought to help the enzymes to exert their acetyltransferase activity. Consequently, increases in H3K56 acetylation in heat-induced gene activation were significantly impaired in the two AtASF1 genes double mutant. Moreover, although the exact contribution by which plant CBP/p300 homologs in H3K56 acetylation is still obscure. AtASF1 could aid the acetyltransferase activity of yeast Rtt109 in vitro (Weng et al. 2014). Our study highlighted the crucial role of AtASF1 in H3K56ac stimulation and the consequent nucleosome removal during heat stress gene activation. Nevertheless, when the ambient temperature increases to a harmful state as heat stress, the molecular interplay between the dynamic H2A.Z variant maintained by the SWR1 complex and the increase of H3K56ac modification is still not clear and remains as an interesting issue for future research.

16.6 Roles of Other Histone Acetylation and Poly (ADP-Ribose) Modifications

Besides H3K56, H3K9 and H3K14 are also acetylation modification sites in the H3 tail. The unicellular green alga Chlamydomonas reinhardtii contains a single canonical HSF (HSF1), representing the key regulator of the stress response. Detailed analysis revealed that HSF1 binds at the promoter of its target gene HSP22F, which precedes histone acetylation, at least at H3K9 and H3K14, and consequent histone eviction and transcriptional activation (Strenkert et al. 2011). This finding suggested that HSF1, as a conserved activating transcription factor, is capable of triggering chromatin re-organization for subsequent epigenetic change mediated by other factors. Consistently, human HSF1 can partially open the chromatin structure of its target gene promoter, which is likely required for the subsequent recruitment of histone modification machineries (Inouye et al. 2007).

Although histone acetylation is advantageous for transcriptional activation, the components and the underlying molecular mechanism involved in their establishment and removal during certain HSP gene activations and attenuations are only beginning to be uncovered. In Saccharomyces cerevisiae, the Gcn5-containing SAGA complex responsible for histone acetylation and Rpd3 complexes acting in histone deacetylation, are rapidly and synchronously recruited to the HSP82 genes activated by the sole HSF1. Their opposing activities on acetylation status can modulate the local HSP gene chromatin structure, and are likely critical for the fine tuning of transcriptional output (Kremer and Gross 2009). During Drosophila HSP70 gene activation, histone H2A lysine 5 is acetylated by Tip60 and deacetylated by HDAC3 (Petesch and Lis 2012). The above-mentioned histone acetylation-related proteins (or complexes) are largely conserved in eukaryotes, including plants. Their functional conservation in plant HSP transcriptional modulation needs to be verified by further genetic and molecular analyses. Nevertheless, the complexity of the plant HSF-HSP gene families likely adds an additional layer of demand for specificity in individual HSP gene modulations.

In Drosophila, the acetylation of H2A at K5 (H2AK5ac) also plays a critical role in the activation of POLY(ADP-RIBOSE) POLYMERASE (PARP) upon heat shock, which implies a functional interplay of two different epigenetic modifications: acetylation and poly(ADP-ribose) (PAR), in heat-induced HSP gene activation (Petesch and Lis 2012). PARP binds to the HSP70 gene promoter before heat stress, and rapidly spreads across the gene loci upon heat treatment. The catalytic activity of PARP is not required for the initial binding status, but is critical for its heat-induced spread, which is consistent with the absence of PAR modification in HSP70 loci before heat stress, as well as the rapid accumulation of PAR following heat treatment (Petesch and Lis 2012). More importantly, HSF is critical for the dynamic changes in the spread of the PARP protein during heat stress, consistent with the vital role of HSF1 in Chlamydomonas reinhardtii for triggering chromatin re-organization for subsequent epigenetic change (Strenkert et al. 2011). It would be interesting to verify whether the involvement of PAR modification is conserved in plant species during heat treatment, as well as the functional interplay of different epigenetic modifications.

16.7 H3 Phosphorylation in HSP Gene Activation

In eukaryotic cells, the heat stress response is not always related to histone acetylation. For instance, in the polytene chromosomes of Drosophila salivary glands, the acetylation of histones H3 and H4 does not change during heat shock. Instead, the global level of phosphorylated H3 at serine 10 (H3S10p) is remarkably sensitive to heat shock, as observed by its dramatic decrease. However, a local increase in H3S10p is instead detected at the heat-induced loci (Nowak and Corces 2000). H3S10p is involved in regulation of chromatin organization (Johansen and Johansen 2006). Therefore, transcriptional activation of the heat-induced loci upon heat shock, as well as repression of non-heat-shock genes in Drosophila, correlates with the dynamic modulation of histone H3 phosphorylation. PROTEIN PHOSPHATASE TYPE 2A (PP2A) is thought to be largely responsible for the global de-phosphorylation of H3 in Drosophila because its mutation leads to reduced genome-wide H3 de-phosphorylation, resulting in the sites of H3 phosphorylation in the non-heat-shock genes remaining transcriptionally active during heat shock in PP2A mutants (Nowak et al. 2003).

In animals, oncoprotein SET was identified as an effective and highly specific inhibitor of PP2A activity, which inhibits H3S10 de-phosphorylation. Drosophila SET was located in transcriptionally active regions of polytene chromosomes, and as a result, might be implicated in the activation and repression of gene expression during heat shock by controlling PP2A activity (Nowak et al. 2003). In support of this hypothesis, the Drosophila SET protein dramatically accumulates at HSP70 loci, accompanied by enhanced H3S10p upon heat shock.

NAP1-RELATED PROTEIN 1 and 2 (NRP1/2) are the redundant plant homologs of the SET protein and also act as histone chaperones. They are predominantly located in nuclei, and display significantly reduced mobility when the temperature rises. This molecular feature is thought to be achieved by the crosslinking of NRP proteins to other nuclear proteins (Ayaydin et al. 2015). Notably, overexpression of Arabidopsis NRP1 increased the heat tolerance of plants. Arabidopsis NRP1/2 interacted with animal PP2A and histone H3 modified as phosphorylated Ser10 in a co-immunoprecipitation assay (Biro et al. 2012). Upon heat treatment, promoters of Arabidopsis HSP genes, such as HSP18.2, HSP70.1, and HSP100, are associated with H3S10p-marked chromatin (Biro et al. 2012). However, H3S10p enrichment was not impaired in the NRP1/2 double mutant. Besides, NRP1/2 are dispensable for the heat-induced expression of HSP genes in Arabidopsis, suggesting a divergent role of plant NRPs as H3S10p regulators in HSP gene regulation in response to heat. The factor or mechanism responsible for the specific phosphorylation of H3 at plant heat shock loci is intriguing because it represents a conserved local histone modification coordinated with HSP gene activation.

Kinase JIL-1 phosphorylates H3S10, localizes to transcriptionally active regions, and is required for activation of the HSP70 gene in Drosophila (Ivaldi et al. 2007). Nevertheless, JIL-1 is a protein kinase that is likely unique to insects. Currently, the kinase(s) responsible for these modifications in plants and their biological function in plants' heat response remain obscure and need to be identified in future studies.

16.8 Histone Chaperones Implicated in Histone Turnover During HSP Expression

A related study in yeast showed that, after heat treatment, half of the nucleosome eviction at the promoter of the HSP82 gene occurred in the first 30 s, and peaked at approximately 95 % after just 8 min (Erkina and Erkine 2006). Rapid loss of nucleosomes was also observed over a large chromatin domain at Drosophila HSP70 loci (Petesch and Lis 2008). Interestingly, the nucleosome loss that occurs within the first minute of heat stress is dependent on HSF and is uncoupled from activation of HSP70. PARP proteins are then required as necessary factors for the changes in nucleosome structure observed by 2 min of heat stress. Actually, in a very early study, the 5′ ends of Drosophila HSP genes in chromatin were found to be hypersensitive to DNase I (Wu 1980). DNase I specifically hydrolyze non-nucleosomal DNA, suggesting that the regulatory region of HSP genes is less organized in the form of nucleosomal structure or is subjected to highly dynamic regulation. Nucleosomes are verified to be sufficient to block transcription mediated by RNA Polymerase II (Pol II) in vitro. To efficiently transcribe genes, RNA Polymerase II (Pol II) must overcome the nucleosome barrier to gain access to the DNA template. For this purpose, eukaryotic cells adapt multiple molecular mechanisms to remove, transfer, or slide a nucleosome along a DNA template (Li et al. 2007).

Our previous study in Arabidopsis indicated that histone occupancy in examined heat-responsive genes quickly declined upon heat stress, which correlated with the recruitment of the ASF1 chaperone and Pol II onto these chromatin regions. As mentioned above, in the AtASF1 double mutant, besides the decrease of H3K56ac modification and histone occupancy, the accumulation of Pol II in response to heat stress was also impaired. Consistently, the basal transcription levels, as well as their potent activation, of various HSF and HSP genes were obviously reduced in the mutant. Notably, not all the examined HSF and HSP genes are associated with ASF1 function, suggesting that the chaperone role of ASF1 is selective for a certain gene set in response to heat stress. The molecular mechanism conferring such specificity will form an intriguing subject of future study to unravel the differential expression and activation pattern of various plant heat-responsive genes.

The role of ASF1 in nucleosome displacement during heat response is also conserved in yeast (Erkina and Erkine 2006). In this study, chaperone ASF1 was found to act synergistically with the ATP-dependent, chromatin remodeling complex SWI/SNF. Chromatin remodeling factors are ATP-using enzymes, usually in the form of protein complexes, which, through interaction with other subunits, fine tune its activity and specificity. Because plant HSP genes have evolved into a large gene family with largely independent expression patterns, whether such coordination of chromatin-remodeling factors in the large-scale modulation of a gene set is conserved in plants remains an open question.

In yeast, the depletion of the SUPPRESSOR OF TY16 (SPT16)-encoding gene, which is a major subunit of the FACILITATES CHROMATIN TRANSCRIPTION (FACT) histone chaperone complex, leads to a severe growth defect phenotype associated with unusual thermotolerance. The acquired thermotolerance in the SPT16-depleted strain is associated with a defect in the reassembly of nucleosomes at the promoters of HSP genes during sustained heat stress, leading to increased recruitment of HSF and RNA polymerase II (Erkina and Erkine 2006).

16.9 Conclusion

For the sustainability of food production, one highly anticipated agronomic trait is to produce modern crops with enhanced heat stress tolerance. More and more evidence supports the view that the accumulation of HSPs and the acquisition of heat tolerance are correlated processes (Sarkar et al. 2014; Lavania et al. 2015). Epigenetic regulation by multiple pathways has a profound impact on gene transcription levels and patterns. The study of epigenetic pathways that control plant responses to abiotic stress, including heat stress, is useful to improve plant breeding and biotechnology towards more stress-resistant cultivars, and is also helpful to understand plant ecology and evolution under unfavorable conditions.

Acknowledgements This work was supported by the National Basic Research Program of China (973 Program; Grant 2012CB910500), and by the National Natural Science Foundation of China (Grant NSFC31271374). We thank many members of our laboratory and colleagues for productive collaborations in our studies of the plant heat response over many years.

References

Ayaydin F, Bíró J, Domoki M, Ferenc G, Fehér A (2015) *Arabidopsis* NAP-related proteins (NRPs) are soluble nuclear proteins immobilized by heat. Acta Physiol Plant 37:1–8

Biro J, Farkas I, Domoki M, Otvos K, Bottka S, Dombradi V, Feher A (2012) The histone phosphatase inhibitory property of plant nucleosome assembly protein-related proteins (NRPs). Plant Physiol Biochem 52:162–168

Boden SA, Kavanova M, Finnegan EJ, Wigge PA (2013) Thermal stress effects on grain yield in *Brachypodium distachyon* occur via H2A.Z-nucleosomes. Genome Biol 14:R65

Burke JJ, Chen J (2015) Enhancement of reproductive heat tolerance in plants. PLoS One 10:e0122933

Coleman-Derr D, Zilberman D (2012) Deposition of histone variant H2A.Z within gene bodies regulates responsive genes. PLoS Genet 8:e1002988

Erkina TY, Erkine AM (2006) Displacement of histones at promoters of *Saccharomyces cerevisiae* heat shock genes is differentially associated with histone H3 acetylation. Mol Cell Biol 26:7587–7600
Frank G, Pressman E, Ophir R, Althan L, Shaked R, Freedman M, Shen S, Firon N (2009) Transcriptional profiling of maturing tomato (*Solanum lycopersicum L.*) microspores reveals the involvement of heat shock proteins, ROS scavengers, hormones, and sugars in the heat stress response. J Exp Bot 60:3891–3908
Grover A, Mittal D, Negi M, Lavania D (2013) Generating high temperature tolerant transgenic plants: achievements and challenges. Plant Sci 205–206:38–47
Hahn A, Bublak D, Schleiff E, Scharf KD (2011) Crosstalk between Hsp90 and Hsp70 chaperones and heat stress transcription factors in tomato. Plant Cell 23:741–755
Haslbeck M, Vierling E (2015) A first line of stress defense: small heat shock proteins and their function in protein homeostasis. J Mol Biol 427:1537–1548
Inouye S, Fujimoto M, Nakamura T, Takaki E, Hayashida N, Hai T, Nakai A (2007) Heat shock transcription factor 1 opens chromatin structure of interleukin-6 promoter to facilitate binding of an activator or a repressor. J Biol Chem 282:33210–33217
Ivaldi MS, Karam CS, Corces VG (2007) Phosphorylation of histone H3 at Ser10 facilitates RNA polymerase II release from promoter-proximal pausing in *Drosophila*. Genes Dev 21:2818–2831
Johansen KM, Johansen J (2006) Regulation of chromatin structure by histone H3S10 phosphorylation. Chromosome Res 14:393–404
Kadota Y, Shirasu K (2012) The HSP90 complex of plants. Biochim Biophys Acta 1823:689–697
Kotak S, Larkindale J, Lee U, von Koskull-Doring P, Vierling E, Scharf KD (2007) Complexity of the heat stress response in plants. Curr Opin Plant Biol 10:310–316
Kremer SB, Gross DS (2009) SAGA and Rpd3 chromatin modification complexes dynamically regulate heat shock gene structure and expression. J Biol Chem 284:32914–32931
Kumar SV, Wigge PA (2010) H2A.Z-containing nucleosomes mediate the thermosensory response in *Arabidopsis*. Cell 140:136–147
Lavania D, Dhingra A, Siddiqui MH, Al-Whaibi MH, Grover A (2015) Current status of the production of high temperature tolerant transgenic crops for cultivation in warmer climates. Plant Physiol Biochem 86:100–108
Lee U, Wie C, Escobar M, Williams B, Hong SW, Vierling E (2005) Genetic analysis reveals domain interactions of *Arabidopsis* Hsp100/ClpB and cooperation with the small heat shock protein chaperone system. Plant Cell 17:559–571
Li B, Carey M, Workman JL (2007) The role of chromatin during transcription. Cell 128:707–719
Lin BL, Wang JS, Liu HC, Chen RW, Meyer Y, Barakat A, Delseny M (2001) Genomic analysis of the Hsp70 superfamily in *Arabidopsis thaliana*. Cell Stress Chaperones 6:201–208
Liu J, Feng L, Li J, He Z (2015) Genetic and epigenetic control of plant heat responses. Front Plant Sci 6:267
March-Diaz R, Reyes JC (2009) The beauty of being a variant: H2A.Z and the SWR1 complex in plants. Mol Plant 2:565–577
Mogk A, Kummer E, Bukau B (2015) Cooperation of Hsp70 and Hsp100 chaperone machines in protein disaggregation. Front Mol Biosci 2:22
Neumann H, Hancock SM, Buning R, Routh A, Chapman L, Somers J, Owen-Hughes T, van Noort J, Rhodes D, Chin JW (2009) A method for genetically installing site-specific acetylation in recombinant histones defines the effects of H3 K56 acetylation. Mol Cell 36:153–163
Nowak SJ, Corces VG (2000) Phosphorylation of histone H3 correlates with transcriptionally active loci. Genes Dev 14:3003–3013
Nowak SJ, Pai CY, Corces VG (2003) Protein phosphatase 2A activity affects histone H3 phosphorylation and transcription in *Drosophila melanogaster*. Mol Cell Biol 23:6129–6138

Petesch SJ, Lis JT (2008) Rapid, transcription-independent loss of nucleosomes over a large chromatin domain at *Hsp70* loci. Cell 134:74–84
Petesch SJ, Lis JT (2012) Activator-induced spread of poly(ADP-ribose) polymerase promotes nucleosome loss at *Hsp70*. Mol Cell 45:64–74
Sarkar NK, Kim YK, Grover A (2009) Rice sHsp genes: genomic organization and expression profiling under stress and development. BMC Genomics 10:393
Sarkar NK, Kundnani P, Grover A (2013) Functional analysis of Hsp70 superfamily proteins of rice (*Oryza sativa*). Cell Stress Chaperones 18:427–437
Sarkar NK, Kim YK, Grover A (2014) Coexpression network analysis associated with call of rice seedlings for encountering heat stress. Plant Mol Biol 84:125–143
Scharf KD, Siddique M, Vierling E (2001) The expanding family of *Arabidopsis thaliana* small heat stress proteins and a new family of proteins containing alpha-crystallin domains (Acd proteins). Cell Stress Chaperones 6:225–237
Strenkert D, Schmollinger S, Sommer F, Schulz-Raffelt M, Schroda M (2011) Transcription factor-dependent chromatin remodeling at heat shock and copper-responsive promoters in *Chlamydomonas reinhardtii*. Plant Cell 23:2285–2301
Swindell WR, Huebner M, Weber AP (2007) Transcriptional profiling of *Arabidopsis* heat shock proteins and transcription factors reveals extensive overlap between heat and non-heat stress response pathways. BMC Genomics 8:125
Watanabe S, Radman-Livaja M, Rando OJ, Peterson CL (2013) A histone acetylation switch regulates H2A.Z deposition by the SWR-C remodeling enzyme. Science 340:195–199
Waters ER (2013) The evolution, function, structure, and expression of the plant sHSPs. J Exp Bot 64:391–403
Weng M, Yang Y, Feng H, Pan Z, Shen WH, Zhu Y, Dong A (2014) Histone chaperone ASF1 is involved in gene transcription activation in response to heat stress in *Arabidopsis thaliana*. Plant Cell Environ 37:2128–2138
Wu C (1980) The 5′ ends of *Drosophila* heat shock genes in chromatin are hypersensitive to DNase I. Nature 286:854–860
Xu ZS, Li ZY, Chen Y, Chen M, Li LC, Ma YZ (2012) Heat shock protein 90 in plants: molecular mechanisms and roles in stress responses. Int J Mol Sci 13:15706–15723
Zinn KE, Tunc-Ozdemir M, Harper JF (2010) Temperature stress and plant sexual reproduction: uncovering the weakest links. J Exp Bot 61:1959–1968

Index

A.A.A. Asea et al. (eds.), *Heat Shock Proteins and Plants*,
Heat Shock Proteins 10, DOI 10.1007/978-3-319-46340-7

MIX
Papier aus verantwortungsvollen Quellen
Paper from responsible sources
FSC® C105338

If you have any concerns about our products,
you can contact us on
ProductSafety@springernature.com

In case Publisher is established outside the EU,
the EU authorized representative is:
Springer Nature Customer Service Center GmbH
Europaplatz 3, 69115 Heidelberg, Germany

Printed by Libri Plureos GmbH
in Hamburg, Germany